Richard Eddy

Alcohol in History

Richard Eddy

Alcohol in History

ISBN/EAN: 9783743376885

Manufactured in Europe, USA, Canada, Australia, Japa

Cover: Foto ©berggeist007 / pixelio.de

Manufactured and distributed by brebook publishing software (www.brebook.com)

Richard Eddy

Alcohol in History

ALCOHOL IN HISTORY,

AN

AC OUNT OF INTEMPERANCE IN ALL AGES;

TOGETHER WITH A

HISTORY OF THE VARIOUS METHODS EMPLOYED FOR ITS REMOVAL.

BY

RICHARD EDDY, D. D.

———◇◇———

NEW YORK:
The National Temperance Society and Publication House,
No. 58 READE STREET,
1887.

PRIZE ESSAYS.

JNO. N. STEARNS, ESQ.
    Cor. Sec'y. and Pub. Agent,
        Nat. Temp. Soc'y and Pub. House,
            58 Reade Street, New York.

DEAR SIR—The undersigned members of the Committee appointed by the National Temperance Convention, held at Saratoga, N.Y., August 26 and 27, 1873, to aid Mr. Job H. Jackson of West Grove, Pa., in securing an "American Standard work on Temperance," having completed the duty assigned them, desire herewith to present through you to the National Temperance Society and Publication House, Parts II. and III. of such work, with their recommendation for publication.

Part I. "The Scientific: embracing the Chemical, Physiological and Medical Relations of the Temperance Question," written by Dr. William Hargreaves of Philadelphia, was published by the National Society under the title of "Alcohol and Science."

Part II. "The Historical, Statistical, Economical and Political," and,

Part III. "The Social, Educational and Religious Aspects," written by Rev. Richard Eddy, of Massachusetts, are herewith presented in MS.

The MSS. presented under the prize competitive plan not proving satisfactory to the Committee, that method was abandoned, and Dr. Eddy was engaged by Mr. Jackson to prepare Part II.; and that work receiving the approval of the Committee, he was also engaged to write Part III., which after pains-taking care in reading and examination by each member, received the unanimous favora   judg-ment of the Committee.

In literary execution, clearness of statement, comprehen siveness in research, fact and reasoning, Parts II. and III. will we think commend themselves as "eminently satisfactory;" and "a matter for congratulation that this protracted enterprise comes to so acceptable a close." These three general parts, will, we believe, constitute the most complete work on Temperance produced in the United States; and will prove a valuable addition to our literature on that subject, furnishing an arsenal from which the educator, legislator and philanthropist, may draw lessons from history, philosophy, experience and statistics, to be used in the warfare against the "drink system."

Ten years have passed since the Committee, consisting of A. M. Powell, Gen. Neal Dow, Judge R. C. Pitman, Rev. A. A. Miner and James Black were appointed. The entire Committee passed upon and approved Part I. Subsequently Gen. Dow and Judge Pitman finding the labor required too severe a tax upon time and strength, fully required for other duties, declined to further serve. Since their retirement, the undersigned have endeavored to meet the labor and responsibilities, and have borne the personal expenses incident to their appointment, and with the passing of these MSS. to the National Temperance Society they judge their duties will end.

We deem it proper to say that beside the initial conception of this work (mapped out and broadened by another,) the whole financial burden in the payment of prizes offered, and compensation to writers, has been paid by Mr. Jackson, to whose courage, devotedness, and perseverance in overcoming great obstacles, *the honor* of this work is greatly due.

We trust the silent influences of this work may enlighten and preserve many, and redeem some from the drink curse; and that it may prove under God an efficient agency in directing the moral and political power of the people for the banishment of the alcoholic drink trade from the recognition and protection of our national and state governments.

Very respectfully,

A. M. POWELL.
A. A. MINER.
JAMES BLACK.

PREFACE.

—◇◇◇—

THE following pages are the result of an effort to furnish what will supply a want long felt by those who labor in the temperance cause, as well as by the general reader on the subject, viz.:—a book that shall contain in orderly array, the many facts in the history of Intemperance which are scattered in numerous volumes, many of them not accessible to the general reader, and some not to be found except in the large libraries of Colleges and other Public Institutions.

In arranging these pages the writer has desired to set forth facts, rather than to minister to pride of authorship; and, disclaiming originality, is satisfied to be known only as a Compiler of the various chapters in the story of the world's great curse, as it has been told in so many climes, and through the most distant ages.

While it is not claimed that this historical field is exhausted—since no one can know better than he who has attempted the exploration of any portion of it, what vast regions are yet unexamined—it is believed that there are brought together in these pages a more full statement of reliable facts in regard to the extent and uniform consequences of intemperance, than can be found elsewhere.

In preparing the portion devoted to the history of efforts to suppress intemperance, an examination has been made of the great mass of conjecture which has accumulated on the subject. Unsupported traditions have been discarded, and only well-attested facts have been recorded. What is given, is, therefore, believed to be worthy of credit; and copious references to the sources of information on any portion of the subject of the volume will be found in the marginal notes.

The writer has also aimed at candor and impartiality in analyzing the causes which have led to radical changes of policy in temperance work, as also in setting forth and considering Objections to special methods of operation; and he humbly trusts that these pages, which cost him many researches, and much and long continued labor, may be of service to the workers in the field of temperance, as well as to the general reader, helping each and all to hopeful and persistent effort in battling against intemperance. To this end he invokes the blessing of God on his work.   R. E.

CONTENTS.

(ix)

CHAPTER III.

CHAPTER IV.

CHAPTER V,

ALCOHOL IN HISTORY.

CHAPTER I.

Temperance and Intemperance Defined—The Temperate and Moderation Plea Examined—List of the Chief Substances Employed in Producing Intoxicants—Fermented and Unfermented Wines—Malt Liquors—Distilled Liquors—Adulterations of Liquors.

XENOPHON, in his Memorabilia of Socrates, represents that great philosopher as making no distinction between wisdom and temperance, but as teaching that "He who knows what is good and obeys it, and what is bad and avoids it, is both wise and temperate."*

The statement is significant in the most general application of these terms to the largest fields where wisdom and temperance are demanded, but it is especially pertinent in its application to the particular use of the word temperance, to denote abstinence from intoxicating drinks as a beverage; for if temperance is determined by the avoidance of what is bad, any use, even the most infrequent, of the bad, must be a degree of intemperance; a moderate temperance in regard to a bad thing, being an absurdity, a folly that never can be called wise.

There is, no doubt, a safe and wise, and even necessary use of alcohol, the chief intoxicating agent of our day; but that use is confined to the arts and sciences, possibly

---

* Book III. c. ix., v. 4.

including medicine in the last named;* but certainly it is never wise, useful nor safe as a beverage.

The so-called moderate, *i. e.*, occasional drinkers of intoxicants, and also a few who do not drink at all, claim that intemperance can be charged only to those who become sots, unable to take care of themselves, or unwilling to restrain their appetites; while those who take but little at a time, or who are infrequent in their indulgences, may claim to be, if not advocates, at least examples of temperance. More than this is declared to be unwarranted by any just use of these terms; and even some professed Christian men and women say, is contrary to the demand of the Bible; temperance, not total abstinence, being the requirement of the Gospel. Such statements, however honestly made, have no foundation save in thoughtlessness and ignorance.

I.   The word temperance occurs but three times in the New Testament; the original Greek word in its various forms, but seven times; and in no instance does it conflict with Socrates' idea of it as denoting the avoidance of what is bad.   If in any instance it possibly suggests moderation, or a moderate use of, the connection is such as to exclude the idea of its allowing any indulgence whatever in that which is tainted with evil. †

When Paul "reasoned" with Felix ‡ "of temperance," he meant, as all scholars concede, chastity, intending to rebuke the adulterous lives of Felix and Drusilla.   To interpret the apostle as meaning less than total abstinence

---

* "Its application as an agent that shall enter the living organization is properly limited by the learning and skill possessed by the physician—a learning that itself admits of being recast and revised in many important details, and perhaps in principles."—Dr. B. W. Richardson's *Cantor Lectures on Alcohol*. American edition. p. 178.   See also, numerous facts and authorities cited in Dr. Hargreaves' Essay, "*Alcohol, What it Is and What it Does.*"

† See Galatians v. 23.   Titus i. 8.   2 Peter i. 6.

‡ Acts xxiv. 25.

from this immorality, would be to make him the encoura-
ger of moderate criminal indulgence, a corrupter and not a
purifier of men.

When the same apostle institutes a comparison between
the Grecian and the Christian race, and says of the compe-
titor in the former: "Every man that striveth for the mastery is temperate in all things,"* it must be conceded—
because the rules enforced by the trainer of the athlete are
well known—that modify the word temperate as we may,
to denote moderation in eating, exercise or repose, it stands
for absolute abstinence from intoxicating drinks. That
was not only the ancient rule,† it is as imperatively demanded by all modern trainers of men for contests of physical
agility, strength and endurance.

The only remaining passage where the word rendered
temperate, occurs in the New Testament, is in Paul's advice on the subject of celibacy and marriage, the word
translated " contain." ‡ The same reasons which necessitate the idea of entire abstinence in the case of Felix and
Drusilla, necessitate it here; and no further argument is
necessary, since less than this involves the Scriptures in
allowing some degree of indulgence in immorality,—a sup-
position which is as impious as it is absurd.

II. Nor is it any less unwise to say that temperance
consists in, or properly allows moderation in the use of intoxicants, on the ground that it is the immoderate use
alone which is injurious to ourselves or to the community.
It would hardly be possible to come in more direct conflict
with well-established facts. Let us look at a few of these
facts as attested by the most competent and unimpeachable
witnesses.

---

* 1 Cor. ix. 25.
† Epictetus, Enchiridion, chap. xxxv.
‡ 1 Cor. vii. 9.   Macknight renders the verse, " If they cannot live continently." Dean Alford: " If they have not con-
tinency."

Dr. Trotter says : " It is not drinking spirituous liquors to the length of intoxication, that alone constitutes intemperance. A man may drink a great deal—pass a large portion of his time at the bottle, and yet be able to fill most of the avocations of life. There are certainly many men of this description, who have never been so transformed with liquor as to be unknown to their own house dog, or so foolish in their appearance, as to be hooted by school-boys, that are yet to be considered as intemperate livers. These 'sober drunkards,' if I may be allowed the expression, deceive themselves as well as others; and though they pace slowly along the road to ruin, their journey terminates at the goal—bad health."

Says Dr. Gordon, "When I was studying at Edinburgh, I had occasion to open a great many bodies of persons who had died of various diseases, in a population much more renowned for sobriety and temperance than that of London ; but the remarkable fact was, that in all these cases there was more or less some affection of the liver. I account for it, from the fact, that these moral and religious people were in the habit of drinking a small quantity of spirits every day, some one or two glasses. They were not in any shape or form intemperate, and would have been shocked at the imputation." "Leaving drunkenness out of the question, the frequent consumption of a small quantity of spirits, gradually increased, is as surely destructive of life as is more habitual intoxication."

Dr R. G. Dods, bore this testimony before Parliament: "No one is safe from the approach of countless maladies, who is in the daily habit of using even the smallest portion of ardent spirit. The practice cannot possibly do any good, and it has often done much harm."

Dr. Copland, in his Dictionary of Practical Medicine, says: "There can be no doubt, that, as expressed by the late Dr. Gregory, an occasional excess is, upon the whole, less injurious to the constitution than the practice of daily taking a moderate quantity of any fermented liquor or spirit."

Says Dr. Benjamin Rush : " I have known many persons destroyed by ardent spirits who were never completely intoxicated during the whole course of their lives."

So Dr. Harris, of the United States Navy: " The moderate use of spirituous liquors has destroyed many who were never drunk." And Dr. Ramsay, of Charleston: " Health is much injured by those who are frequently sipping strong liquors, though they are never intoxicated."

Prof. Henry Munroe, M. D., says: "Alcohol, whether taken in large or small doses, immediately disturbs the functions of the body and the mind."

Dr. Macnish, in his Anatomy of Drunkenness, says: "Men indulge habitually, day by day, not perhaps to the extent of producing any evident effect, either upon the body or mind at the time, and fancy themselves all the while strictly temperate, while they are, in reality, undermining their constitution by slow degrees—killing themselves by inches, and shortening their existence several years."

Of the following statement of Rev. Dr. Lyman Beecher, Dr. Macnish says: " I fully concur with him."

" It is a matter of undoubted certainty that habitual tippling is worse than periodical drunkenness. The poor Indian, who once a month drinks himself dead, all but simple breathing, will outlive for years the man who drinks little and often, and is not perhaps suspected of intemperance." *

To the following, the signatures of eighty eminent English physicians and surgeons were appended: " An opinion, handed down from rude and ignorant times, and imbibed by Englishmen from their youth, has become very general, that the habitual use of some portion of alcoholic drink, as of wine, beer, or spirit, is beneficial to health, and even necessary for those subjected to habitual labor.

" Anatomy, physiology, and the experience of all ages and countries, when properly examined, must satisfy every mind well informed in medical science, that the above opinion is altogether erroneous. Man, in ordinary health, like other animals, requires not any such stimulants, and cannot be benefitted by the habitual employment of any quantity of them, large or small; nor will their use during his life-time increase the aggregate amount of his labor. In whatever quantity they are employed, they will rather tend to diminish it.

" When he is in a state of temporary debility from illness or other causes, a temporary use of them, as of other stimulant medicines, may be desirable; but as soon as he is

---

* Beecher's Six Sermons on Intemperance, 1827, p. 4.

2

raised to his natural standard of health, a continuance of their use can do no good to him, even in the most moderate quantities; while larger quantities, (yet such as by many persons are thought moderate,) do sooner or later prove injurious to the human constitution, without any exceptions."

The highest medical authorities of Great Britain, on being examined in large numbers before a committee appointed by the British Parliament to inquire into the causes of drunkenness, unanimously testified : " Ardent spirits are absolutely poisonous to the human constitution; in no case whatever are they necessary or even useful to persons in health ; but are always, in every case, and to the smallest extent, deleterious, pernicious, or destructive, according to the proportions in which they may be taken into the system." *

Prof. James Miller says : " Alcohol is a luxury in one sense, no doubt. Its first effects are pleasurable ; and to some frames intensely so. But its *tendencies*, even in truly 'moderate' allowance, are *always* evil." †

The famous authority on Physiology, Dr. Wm. B. Carpenter, says : " My position is, that in the discharge of the ordinary duties of life, Alcohol is not necessary, but injurious, in so far as it acts at all. Even in small quantities, habitually taken, it perverts the ordinary functions by which the body is sustained in health." And again : " We maintain that the action of the excessive or of the moderate use of Alcohol upon the healthy body is a question of degree alone ; its immediate effect being essentially the same in the one case as in the other." ‡

Dr. Chas. Wilson, in his " Pathology of Drunkenness," says that "no circumstances of ordinary life can render even the moderate use of ardent spirits or other intoxicating fluids either beneficial or necessary, or even innocuous. The disordered functions of nutrition caused indirectly by its action on the stomach, and directly by its own absorption and diffusion

---

* Quoted by Rev Marcus E. Cross, Mirror of Intemperance, p. 30.

† Alcohol: Its Place and Power, p. 208.

‡ Essay on the Use and Abuse of Alcoholic Liquors.

throughout the system, contribute to the production of an ill-assimilated blood, and tend to attach new forms of danger to every description of disease or accident."

Prof. Youmans says of Alcohol : "It is an inveterate foe of the intellectual and moral principle of man. In all its numberless forms, and in every quantity, it is the potent adversary of the mind."

Dr. James Johnson, physician to King William IV., said : "A very considerable proportion of the middling and higher classes of life, as well as the lower, commit serious depredations on their constitutions, when they believe themselves to be sober citizens, and really abhor debauch. This is by drinking ale or other malt liquor to a degree far short of intoxication indeed, yet from long habit producing a train of effects that embitter the ulterior periods of existence."

Dr. Macroie of the Liverpool Hospital, says : Having treated more than 300,000 patients, I give it as my decided opinion that the constant moderate use of stimulating drinks is more injurious physically than the now-and-then excessive indulgence in them."

Dr. Maudsley bears this testimony : "If men took careful thought of the best use which they could make of their bodies, they would probably never take alcohol, except as they would take a dose of medicine, in order to serve some special purpose. It is idle to say that there is any real necessity for persons who are in good health to indulge in any kind of alcoholic liquor. At the least it is an indulgence which is unnecessary : at the worst, it is a vice which occasions infinite misery, sin, crime, madness, and disease. Short of the patent and undeniable ills which it is admitted on all hands to produce, it is at the bottom of manifold mischiefs that are never brought directly home to it. How much ill-work would not be done, how much good work would be better done, but for its baneful inspiration! Each act of crime, each suicide, each outbreak of madness, each disease, occasioned by it, means an infinite amount of suffering endured and inflicted before matters have reached that climax." *

Chas. Buxton, Esq., M. P., a well-known London brewer, says in his essay on " How to Stop Drunkenness : " " Dr. Carpenter gives a fearful list of the diseases that are generated by alcohol,—delirium tremens, insanity, oinomania, idiocy, apo-

* Responsibility in Mental Disease, p. 285.

plexy, paralysis, epilepsy, moral perversion, irritation of the mucous membrane of the stomach, gastric dyspepsia, congestion of
the liver and a multitude more.   And he shows that even moderate doses of the poison, regularly taken, tend to produce the same
result; and also to elicit all kinds of diseases that might else
have lain dormant, and slowly to sap the faculties of body and
mind.   There is no doubt that a large amount of suffering is
caused by drinking, even when it does not by any means bulge
out into drunkenness." (pp. 11, 12.)

Dr. Chambers says: "The action of frequent small divided
drams, is to produce the *greatest amount of harm* of which alcohol is capable, with the least amount of good."

Says Dr. Andrew Combe: "I regard even the temperate
use of wine, when not required by the state of the constitution,
as always more or less injurious."

An article in the Second Report of the Board of State Charities of Massachusetts, on "Alcohol as a cause of Vitiation of Human Stock," after showing how rapidly alcohol as compared
with other poisons is eliminated from the system, suggests:
Whether this peculiarity of alcohol does not make its constant
use in small doses worse for posterity than its occasional use in
large quantities; that is, whether tippling is not worse than
drunkenness, as far as it affects the number and the condition
of the offspring."

More recently, Sir Henry Thompson, in a letter to the Archbishop of Canterbury, says: "I have long had the conviction
that there is no greater cause of evil, moral and physical, in
this country, than the use of alcoholic beverages.   I do not
mean by this that extreme indulgence which produces drunkenness.   The habitual use of fermented liquors to an extent far
short of what is necessary to produce that condition, and such
as is quite common in all ranks of society, injures the body and
diminishes the mental power to an extent which, I think, few
people are aware of.   Such, at all events, is the result of observation during more than twenty years of professional life, devoted to hospital practice, and to private practice in every rank
above it.   Thus I have no hesitation in attributing a very large
proportion of some of the most painful and dangerous maladies
which have come under my notice, as well as those which
every medical man has to treat, to the ordinary and daily use
of fermented drinks taken in the quantity which is conventionally deemed moderate.   Whatever may be said in regard to
its evil influences on the mental and moral faculties, as to the
fact above stated, I feel that I have a right to speak with author-

ity; and I do so solely because it appears to me a duty, especially at this moment, not to be silent on a matter of such extreme importance. I know full well how unpalatable is such a truth, and how such a declaration brings me into painful conflict, I had almost said with the national sentiments and the time-honored and prescriptive usages of our race. * * * My main object is to express my opinion as a professional man in relation to the employment of fermented liquor as a beverage. But, if I venture one step further, it would be to express a belief that there is no single habit in this country which so much tends to deteriorate the qualities of the race, and so much disqualifies it for endurance in that competition which in the nature of things must exist, and in which struggle the prize of superiority must fall to the best and to the strongest." *

And the last words which have been spoken on this subject from the physiological standpoint, are the utterance of the highest authority of his day, Dr. Benjamin W. Richardson: "If it be really a luxury for the heart to be lifted up by alcohol; for the blood to course more swiftly through the brain; for the thoughts to flow more vehemently; for words to come more fluently; for emotions to rise ecstatically, and for life to rush on beyond the pace set by nature; then those who enjoy the luxury must enjoy it, with the consequences." "If this agent do really for the moment cheer the weary and impart a flush of transient pleasure to the unwearied who crave for mirth, its influence (doubtful even in these modest and moderate degrees) is an infinitesimal advantage, by the side of an infinity of evil for which there is no compensation, and no human cure." †

"The evils, in the slighter stages of alcoholic diseases, are often connected with others, which are perhaps passing, but which give rise to very unpleasant phenomena. There is what is called a dyspepsia or indigestion, to relieve which the sufferer too frequently resorts to the actual cause of it as a cure for it. There is thirst, there is uneasiness of the stomach, flatulency, and a set of so-called nervous phenomena, which keep the mind irritable, and make trifling cares and anxieties assume an exaggerated and unnatural character. From the earliest period in the history of the drinking of alcohol these phenomena have been observed. 'Who,' says Solomon, referring to this action, 'who hath woe? who hath contentions,

* Open Letter to the Abp. of Canterbury.
† Cantor Lectures, pp. 121, 179.

who hath babbling, who hath wounds without cause? who hath redness of eyes ?' What modern physiologist could define better the steady and progressive effect of alcohol upon those who, even under the guise of temperate men, trust to it as a support? And yet these evils are minor, compared with certain I have to bring forward.*

"Listen carefully to the whole argument of science as she tells you her mind fairly and faithfully. She tells you nothing whatsoever about the devil and his devices, but that there is, as claimed, a certain degree of moderation which does not seem to be attended with much evil, if it be closely followed. She grants that the moderate of the moderates may have a *rule nisi.* She says to a man of sound health: if you are in a first-rate condition of body, if you can throw off freely a cause of oppression and depression, if you are actively engaged in the open air, if you have nothing to do that requires great exactitude or precision of work, if you are not subjected to any worry of mind or mental strain, if you sleep well, if you are properly clothed and are not exposed to excesses of heat or cold, if your appetite is good and you can get plenty of wholesome food; if you are favored with all these advantages, then you may indulge in Dr. Parker's moderate potation of wine, or beer, or spirit. But these favorable conditions are all necessary. If you are limited in respect to exercise, if you are of sedentary habits, if you are much worn or reduced in mind, body, or estate, then that small amount of alcohol is adding to all your troubles, and you will leave it off if you are wise.

"I can imagine with what pleasure some of the world of pleasure may receive such tidings as these. The salt of the earth, and the salt is good, can then enjoy its luxury, just as it can keep a carriage, a livery servant, a horse, or any other unnecessary, but pleasant extravagance. It can take wine in moderation. What more is required? Science, in her most puritanical utterances, gives, so far, her consent.

"It is quite true, but take her consent with her provisions, equally true and very solemn.

"Science says, you who can afford the luxury may use it with the perfect understanding that it is a luxury. Positively, solemnly, it is never a necessity, and if the expression of truth be absolutely rendered, you are better and safer without even the moderate indulgence." †

"To conclude. From my readings of Science, she gives no countenance to the use of strong drink in any sense, except med-

---

* Ibid. p. 148.   † Lecture on Moderate Drinking, pp. 31, 32.

ically and under scientific direction. She faithfully records its evils; she honestly exposes its dangers; she exposes the gross and vain fallacies by which it is supported: and if, in her absolute fairness, she admits it under certain arbitrary restrictions as a luxury, she condemns it as a traitorous evil." *

"The physician can find no place for alcohol as a necessity of life. In whatever direction he turns his attention to determine the value of alcohol to man, beyond the sphere of its value as a drug, which he may at times prescribe, he sees nothing but a void ; in whatever way he turns his attention to determine the persistent effects of alcohol, he sees nothing but disease and death ; mental disease, mental death ; physical disease, physical death." †

III. The stubborn facts brought to light in the experiences of Life Insurance Companies, facts elicited and published, not in the interests of philanthropy, but as the basis of economic business transactions, confirm the foregoing statements of physicians.

One of the oldest and most successful life insurance societies in the old world, is largely indebted for its success to its requirement of eleven per cent. extra on the annual premiums of beer drinkers. When this demand was first made, it so excited the hostility of the publicans and their customers, that they formed a new company, exclusively for themselves. So great and so rapid, however, was the mortality, that the company failed in five years.

In 1840, "The United Kingdom Temperance and General Provident Institution," was organized in London. For the first ten years of its existence policies were issued to total abstainers only; since 1850, moderate drinkers have been allowed to insure, their accounts being kept separate and distinct from the total 'abstainers' accounts. The Actuary of the Company, into whose hands the books are placed once in five years, is not a total abstainer ; his figures, therefore, are not open to the suspicion of being made in the interests of total abstinence. The following is one of his reports :

---

* Ibid, p. 46. † The Diseases of Modern Life, pp. 209, 210.

MORTALITY, 1871-75.

YEAR.	TEMPERANCE.				GENERAL.			
	EXPECTED.		ACTUAL.		EXPECTED.		ACTUAL.	
	No.	Amount	No.	Amount.	No.	Amount	No.	Amount.
1871...	127	£24,051	72	£13,065	233	£46,105	217	£40,158
1872...	137	26,058	90	13,005	244	48,883	282	50,575
1873...	144	28,052	118	22,860	253	51,463	246	49,840
1874...	153	29,648	110	24,683	263	54,092	288	57,006
1875...	162	32,010	121	24,160	273	56,907	297	57,483
5 Years	723	£139,819	511	£97,773	1,266	£257,450	1,330	£255,062

The result is an unmistakable argument for total abstinence, and a plain warning against moderate drinking : the
deaths in the Temperance Section being 212 less than was
expected, while in the Moderates Section they were 64
more than were expected ! * The Company has accumulated
a surplus of £348,458, which is distributed as a bonus on
the policies in force, at the rate of from 35 to 114 per cent.
to the abstainers, and from 20 to 64 per cent. to the non-
abstainers, both classes being governed by age and the
amount paid by them in premiums. For five years the
bonus additions on ordinary whole-life policies for £1,000,
have been according to the following examples :

Date of Policy.	Age at Entrance.	Premiums Paid, 1871-1875.			Bonus added to each £1,000 in Temperance Section			Bonus added to each £1 000 in General Section.		
		£	s.	d.	£	s.	d.	£	s.	d.
1871	15	83	2	6	76	14	0	43	10	0
"	20	93	6	8	80	16	0	46	1	0
"	25	106	9	2	85	16	0	48	17	0
"	30	122	1	8	90	8	0	51	9	0
"	35	138	19	2	94	18	0	54	1	0
"	40	162	5	10	100	12	0	57	3	0
"	45	188	10	10	107	10	0	61	0	0
"	50	226	5	0	118	3	0	67	6	0
"	55	284	3	4	136	10	0	77	10	0

* From 1876 to 1881 the same uniformity of difference was
manifest, 68 per cent. of mortality among the total abstainers,
97 per cent. among the moderate drinkers.

These facts require no comment. They are demonstrations of the folly of all mere theorizing on the absence of harm in moderate drinking. For it must not be forgotten that all in the above tables who are not total abstainers, are classed as moderate drinkers; those who use intoxicants beyond the bounds of so-called moderation being barred out from the possibility of being insured.

In comparing the number of deaths which occurred in several of the most eminent Life Assurance Companies of England during the first five years of their existence, with the number occurring during the same period in the membership of the Temperance Provident Institution, a period when the latter insured total abstainers only, we have the following showing of the dangers and folly of moderation, and the wisdom and safety of total abstinence :

First Office issued 838 Policies, and had 11 deaths, being 13 per thousand.
Second " " 1901 " " " 27 " " 14 " "
Third " " 944 " " " 14 " " 15 " "
Fourth " " 2470 " " " 65 " " 26 " "
Tem.Prov.Ins. " 1596 " " " 12 " " 7½ " "

Dr. Richardson was not out of the way, then, when he said :

"I do not over-estimate the facts when I say that if such a miracle could be performed in England as a general conversion to temperance, the vitality of the nation would rise one-third in value; and this without reference to the indirect advantages which would of necessity follow." *

In our country we are not yet able to present such classified tables, the only company that makes total abstinence a condition of insurance having but recently followed the English example of insuring moderate drinkers also. †   But we are not wholly dependent on foreign sources for the principal fact which we are now seeking to state. "The Mutual Life Insurance Company of New York," the largest

* Lecture on Vitality in Men and Races, 1875.
† See however, for further facts on this head, "Alcohol, What it is and What it Does," by Dr. Hargreaves.

of our life insurance organizations, has compelled its managers to cancel a large number of policies, and to make
more stringent regulations for the future. It has found
that to the interrogatory in the application for insurance :
" Do you use intoxicating liquors ? " only one applicant in
ten answers " no," the others replying, " occasionally," or
" moderately." Not one admits that he is an habitual
drinker, yet the death losses show that six-tenths are
traceable directly or indirectly to the use of intoxicants ;
and about the same proportion of the contested cases are
from the same cause. In view of these facts, a circular was
issued to the policy holders on the 17th of January, 1878,
announcing the intention of the company to cancel all
policies held by those who " practice habits which obviously
tend to the shortening of life." In this circular they say :

"This company contemplates no invasion of the sanctity of
private life, and no interference with the legitimate rights of
the individual; but it cannot be blind to the fact that large
numbers of deaths occur every year among those it has insured
which are the direct results of intemperance ; that still larger
numbers of deaths attributed to accidents, fevers, pneumonia,
liver complaints, and disorders of the brain, stomach, and kidneys, are the sequences of intemperate habits, and that it is
under no legal liability to pay claims by deaths which are de-
monstrably due to these causes.

"At a meeting of the board of trustees, held in the month of
December, 1877, these subjects were referred for consideration to
a special committee, who, after due deliberation, unanimously
adopted the following preamble and resolution :

" ' *Whereas*, The mortuary statistics of this company unmis-
takably point to an alarming and steadily increasing mortality
from the use of intoxicating drinks, thereby prejudicing the interests of the policy holders of the company ; it is therefore

" ' *Resolved*, That the executive officers be, and they are hereby
instructed to strictly enforce the conditions contained in the
application and policies; and with that end in view, that they
be instructed to prepare a circular letter, setting forth the duties
and obligations of the assured.'

" The board of trustees have since unanimously adopted the
action of the committee, and the executive officers were ordered
to send such circular to every policy holder.

"In taking this action the board of trustees are not to be understood as casting any imputation upon the integrity or the habits of the great body of the insured. It is believed that the membership of this company, as a class, is superior in intelligence, sobriety, and thrift to that of any similar organization in this country, and any intention to enter the arena of debatable questions in religion, morals, or political economy is expressly disavowed; this is purely a matter of business, in which the company relies for its protection on a proper administration of the law of contracts.

IV. It is further obvious that so-called moderate drinking differs in nothing from, and is therefore included in, any just definition of Intemperance, from the fact that it is impossible to state what the moderate use of intoxicants is, even its advocates failing to agree among themselves with regard to it.

The late Dr. Anstie placed the moderate use of alcoholic drinks at "three-quarters of an ounce for an adult female, and an ounce and a-half for an adult male; beyond this, is excess and intemperance." He adds: "For youths, say under twenty-five, whose bodily frame is as yet not fully consolidated, the proper rule is, either no alcohol, or very little indeed."

The "Lancet," a high medical authority, attempts to lay down the rule, "That for young and active men a glass of beer, or one or two of claret, at dinner, is, we believe, an ample supply; while men of middle age may, with advantage, stop at the third glass of claret, sherry or port, and feel no ill result." Yet the writer is forced to admit that no accurate definition of moderation can be given; that "the ultimate test in every case, must be experience; and until men have enough moral control and discretion to limit their drinking to that which they absolutely require, all direction and rebuke will be thrown away."

Dr. Sewell says that "the taking of a glass of mint-sling in the morning, of toddy at night, or two or three glasses of Madeira at dinner, is in common parlance termed 'temperate drinking.'" The London Standard "affirms that the taking of half a dozen glasses of wine, a glass of brandy and water, or two glasses of ale daily, is temperate drinking." Dr. Hun, still more liberal in his allowances, says, that "the drinking just so much as promotes the comfort and well-being of an individual, at any particular time, of which each person must be his own judge, is temperate drinking."

Pliny tells us that Democritus wrote a volume to show that "no person ought to exceed four or six glasses of wine." Epictetus declares, "That man is a drunkard who takes more than three glasses; and though he be not drunk, he hath exceeded moderation." A Temperance Society of the sixteenth century, of which more anon, allowed its members to drink fourteen glasses of wine daily. The Moderation Society started in New York, a few years ago, from which such great results were expected by Dr. Crosby, and by those who ignore all the lessons of experience, was probably not as liberal in its indulgence as this, but even it failed to fix the limit between moderation and intemperance ; and no wonder, for the task is an impossible one.

The Right Honorable John Bright, in an "Address to Professing Christians," in 1843, argued this point in a plain and impressive manner, when he said :

"To drink deeply—to be drunk—is a sin : this is not denied. At what point does the taking of strong drink become a sin ? The state in which a body is when not excited by intoxicating drink is its proper and natural state : drunkenness is the state farthest removed from it. The state of drunkenness is a state of sin : at what stage does it become sin ? We suppose a man perfectly sober who has not tasted anything which can intoxicate : one glass excites him, and to some extent disturbs the state of sobriety, and so far destroys it : another glass excites him still more : a third fires his eye, heats his blood, loosens his tongue, inflames his passions : a fourth increases all this : a fifth makes him foolish and partially insane : a sixth makes him savage : a seventh or eighth makes him stupid, a senseless, degraded mass ;—his reason is quenched, his faculties are for the time destroyed. Every noble and generous and holy principle within him withers, and the image of God is polluted and defiled. This is sin, awful sin ! for 'drunkards shall not inherit the kingdom of God.' But where does the sin begin ? At the first glass, at the first step towards complete intoxication, or at the sixth, or seventh, or eighth ? Is not every step from the natural state of the system towards the state of stupid intoxication an advance in sin, and a yielding to the unwearied tempter of the soul ? "

The experience of any and every person who has become intemperate is also a corroboration of the statement that moderation is indefinable. No man can tell where in his own career the line should be run that marks the distinction between his moderate and his excessive use of the intoxicating cup; while so great is the fascination and so thorough the delusion of the drinking habit that multitudes who have a general reputation of sottishness, still pride themselves on their temperate lives and their ability and success in keeping within the limits of moderation.

V. Add to all this that men affect others by the examples which they set, and that no example of sottishness is ever contagious, but is never other than repulsive and disgusting, and we cannot fail to be convinced that so-called moderate drinking is the most mischievous and immoral of all use of intoxicating beverages; the degree of its power to harm being determined by the respectableness and standing in society accorded to those who thus indulge. Our sons never will become drunkards, our daughters never will be in danger of becoming the wives of drunkards, when the only examples before them of the effects of drinking, are the disreputable who reel and stagger in their loathsome degradation. Respectable moderate drinking furnishes the chief and the most fascinating temptation, and before that they are in most imminent danger of falling.

The following incident illustrates the truth of the foregoing statement: At a social gathering of clergymen, the fanaticism of the plea for total abstinence was strongly reprobated, and the superior virtue of temperance or moderation was extolled. Among others, a clergyman of extensive learning and of large influence, was especially vohement in his plea for the moderate use of wine, and unsparing in his invectives against fanatical total abstinence. At the close of his speech, which was warmly applauded, a layman who was present was asked to say a few words, and responded as follows:

"It is not my purpose to answer the learned argument you have just listened to. My object is more humble, and I hope more practical. I once knew a father, in moderate circumstances, who was at much inconvenience to educate a beloved son at College. Here his son became dissipated; but after he had graduated and returned to his father, the influence of home acting upon a generous nature, actually reformed him. The father was overjoyed that his cherished hopes of other days were still to be realized. Several years passed, when the young man having completed his professional study, and being about to leave his father to establish himself in business, was invited to dine with a neighboring clergyman, distinguished for his hospitality and social qualities. At this dinner wine was introduced and offered to this young man, who refused; pressed upon him, and again refused. This was repeated, and the young man ridiculed for his peculiar abstinence. The young man was strong enough to overcome appetite, but he could not resist ridicule. He drank and fell; and from that moment became a confirmed drunkard, and long since has found a drunkard's grave. I am that father, and it was at the table of the clergyman who has just taken his seat that his hospitality ruined the son I shall never cease to mourn."

We conclude, therefore, that the arguments from Scripture, health, longevity and example, necessitate the position that any use as a beverage, of that which intoxicates, is an intemperate use; that temperance in regard to it is necessarily total abstinence; and that the definition given by Socrates, and nobly illustrated in the example set by his life, as according to Xenophon, "He was temperate by refusing on all occasions to prefer what is merely agreeable to what is best," * is the only just and consistent meaning of the word when employed, as in this case, to denote temperance in regard to the use of an ever dangerous and mis-

---

* Memorabilia, B. iv. c. viii. v. 11.

chievous poison. With the old philosopher agrees the great schoolman, St. Thomas Aquinas:

"The temperate man does not use in any measure things contrary to soundness or a good condition of life, for this would be a sin against temperance;"* 'and the modern philosopher Hobbes: "Temperance is the habit by which we abstain from all things that tend to our destruction; Intemperance the contrary vice."† To which may be added the following from the Jewish Catechism: "Q. What is Temperance? A. Temperance consists in abstaining from all that is forbidden and sinful, and in the wise and prudent use of what is good and lawful."‡

Continuing to employ the word Intemperance to denote the use of alcoholic intoxicants as a beverage, and so leaving wholly out of the account any consideration of the intoxication produced by the use of opium, and other poisons, of whatever name, we shall be aided in our view of the long continued history of this evil, and also of the extent of its prevalence, by noticing the great variety of substances which men have either fermented or distilled, in order to obtain the alcoholic poison. Of course the list does not claim to be exhaustive, but it is believed to be more full and accurate than has ever been given in any previous work of this character.

ALOE. *Agave Americana.*—The fermented sap is called Actli, Ponchra, Pulque. It is used in Mexico, Paraguay, Peru and Spain. From this a fiery spirit is distilled, called Brandy Chinguerite, Vino Mercal and Mexical. "*Pulque,*" says Humboldt, "smells like putrid flesh."

AMINATA MUSCARIA.—This is a mushroom, found in great abundance in various parts of the Russian empire. From it, by a disgusting process, an intense intoxicant is prepared. Morewood quotes Dr. Langsdorff, a Russian physician, as authority for the statement that

---

* Quærtio, cxl. De Temperantia.
† Quoted in "Bacchus Dethroned," p. 175.
‡ The Road to Faith, for the use of Jewish Elementary Schools. By Dr. Henri Loëb." Philadelphia, p. 48.

" The most extraordinary effect of the aminata is the change
it makes in the urine, by impregnating it with an intoxicating
quality, which continues to operate for a considerable time. A
man moderately intoxicated to-day, will by the next morning
have slept himself sober ; but, as is the custom, by drinking a
cup of his own urine, he will become more powerfully intoxicated than he was the day preceding. It is therefore not uncommon for confirmed drunkards to preserve their urine as a
precious liquor, lest a scarcity in the fungi should occur. This
inebriating property of the urine is capable of being imparted
to others, for every one who partakes of it, has his urine similarly affected. Thus with a very few amanitæ, a party of drunkards may keep up their debauch for a week. Dr Langsdorff
states, that by means of the second person taking the urine of the
first, the third that of the second, the intoxication may be pro-
pagated through five individuals. The relation of Strahlenberg,
that the rich lay up great stores of the amanitæ, and that the
poor, who cannot buy it, watch their banquets with wooden
bowls, in order to procure the liquor after the second process,
is fully confirmed by the statement of Langsdorff." *

ARA.—An intoxicating pepper plant, of Borneo. The
root is chewed, and on the spittle and masticated pulp, a
little water or cocoanut milk is poured, and from the ferment
that ensues a strong and quickly inebriating drink is procured, greatly delighted in by the natives. †

APPLES.—The expressed juice is commonly called cider,
but in Brazil, Kooi. It is made in the Barbary States,
Brazil, Canada, Chili, France, Germany, Great Britain,
Peru, Poland, Spain, Syria and the United States. Distilled cider produces brandy ; and from the portion of cider
which does not freeze when a large quantity is exposed to a
low temperature, Pomona wine is made by adding brandy,
in the proportion of one gallon of brandy to six gallons of
cider.

ARTICHOKE, (*Helianthus tuberosus.*)—Raised in Northern France, for distillation, and produces a strong spirit.

---

* History of Inebriating Liquors, by Samuel Morewood, edition of 1838, pp. 129, 130.
† Ibid, p. 313.

ANTS.—In Sweden a large species of black ant, which produces on distillation a resin, oil and acid, is used to give a special flavor and power to brandy. These ants are found in great abundance, making their hills at the roots of the fir tree.*

ALGOBARA.—A shrub like the acacia, from the pods of which the Peruvians make Chica.

ALMONDS.—From the fruit of the dwarf almond, the Russians distil a beverage.

AIPIMAKAKARA.—A species of Manioc, in Brazil. From the roots a kind of wine called Aipy, is prepared. The roots are first sliced and chewed by women, then put into a pot of water and boiled. The liquor, after fermentation, is drunk lukewarm.†

ANANAS.—From this wild fruit in Brazil, brandy is distilled.

ARRACHACA.—A vegetable cultivated in Paraguay. Produces a fermented drink.

BANANAS.—Ripe bananas are infused in water by the Peruvians, and from this ferment a drink is distilled. In Madagascar the fermented liquor is used.

BARBERRIES.—Wine, in Hungary.

BARLEY.—A fermented drink was made in Egypt, to which was given the several names, Ceres Vinum, Ceria, Ceroisa, Cœlia, Curmi, and Zythum. It was also an ingredient with curds, honey and melted butter, in making Sura, a powerful intoxicant concocted by the ancient Aryan races of India. The Greeks also had their "wine made from barley," described by Ovid as a strong drink. In Syria and throughout the Turkish Empire common drinks are brewed from barley, and called Bouza and Zythum. In Arabia the drink made from fermented barley is called Curmi; in Nubia, Bouza, Merin and Ombelbel, each name

---

* Consetts' Remarks on a Tour through Sweden.
† Morewood, p. 315.

3

denoting a certain degree of fermentation ; in Abyssinia it is callen Swoir, or Sowa, and when drugs are mixed with it for the purpose of producing more rapid intoxication, it is called Sava.   In various parts of Ethiopia, barley is the basis of a drink called Maiz, which is sometimes made very strong and distilled to brandy.   In Southern Africa barley is also made into beer and porter.   In Tartary the drink is called Bakscuni ; in Thibet, Chong ; and a powerful intoxicant is distilled from Chong called Arra ; in China, Tarasun ; in Russia, beer, quass, brandy ; in Holland, beer ; and distilled, is largely used in the manufacture of gin.   Austria, Spain, France, Germany, Denmark, beer ; Ireland, Curmi, Leann ; Great Britain, British North America and the United States, beer, ale, porter ; France, beer ; Norway and Sweden, beer, brandy ; Caucasus and Siberia, brandy ; Brazil, Kaviaraku ; Peru, Cluro, Neto, Sora.

Batata Root.—A fermented liquor in Brazil, called Vintro de Batatas.

Beets.—Experiments in extracting the saccharine properties of beets have been made in various portions of Europe and America, but nowhere so successfully as in France, where the fermented juice is manufactured into ale, and distilled, producing strong spirits.

Birch Sap.—A fermented drink is made in England, Japan, Norway and Siberia.

Cashew-Nut.—From the fermented juice brandy is distilled, in the West Indies.

Cassada.—This root of the Manioc plant is the base of a fermented drink in the East and West Indies, called in some places Piworree, in others Ouycon.   It is made by the females chewing the root and flowers, and spitting them into a wooden trough.   This, with water added, soon runs into fermentation and yields the intoxicant.   It is also made from the cassava bread, chewed and treated in the same manner.*

---

* Morewood, p. 319.

Cava.—Before the appearance and influence of the missionaries, the natives of the Friendly Islands produced an intoxicant from the root of the cava plant, a species of pepper, in the following manner :

"The root is scraped, cut into small pieces and distributed among the people to be chewed. In some places, says Kotzebue, only the old women do the chewing, but the young women spit on it to thin the paste. The chewing of each mouthful occupies about two minutes, and when thus masticated it is placed in a wooden bowl, where it is mixed with water by the men; then being strained and clear, about a pint is given to each person to drink." *

Capt. Cook states, in his account of his third voyage to the Friendly Islands :

"I have seen the natives drink it seven times before noon ; yet it is so disagreeable, or at least seems so, that the greatest part of them cannot swallow it without making wry faces, and shuddering afterwards."

Cebatha Berries.—From these the Jews of Arabia make a strong spirit.

Cherries.—In Turkey, a distilled liquor called Maraschino ; in Switzerland, also distilled, Kirschenwasser ; in Russia, mead, wine.

Cocoanut Milk.—In India and the East and West India Islands, a fermented drink is made, and is distilled to resemble Arrack. In Peru a fermented drink is obtained from boiling the leaves and the stems to which the nuts are attached.

Dates, and the juice of the Palm.—Wherever the date palm is found, its fruit is highly prized as food ; and is, as is also the sap of the numerous varieties of the palm tree, converted into a beverage. Unfermented, these drinks are sweet, delicious to the taste, and healthful; if however, the liquor is not deprived of its watery parts by evaporation in boiling, it soon ferments, and will produce intoxication. It is not certain, therefore, that in all

---

* Morewood, p. 250.

cases where palm or date wine are mentioned, an intoxicating drink is to be understood; but it is probable that all people who prepare it and do not prevent its fermentation, use it both while it is new, and after it has become intoxicating; and it is known that some who boil it afterwards mix water with it for the purpose of producing fermentation. As an intoxicant it is used or has been used in the following countries: In Syria it had the name of *Shechar*, a word which denotes all manufactured drinks, whether intoxicating or otherwise, except wine. The Greeks had their palm wine. A liquor distilled from dates by the Christians in Syria, is called Araki; the Egyptians distil a liquor called Arrack; the Nubians drink the fermented wine, and also distil it; the Abyssinians call their drink made of dates and meal, Amderku; the inhabitants of Fezzan prepare an intensely fiery spirit from dates, called Busa; on the western coast of Africa, five kinds of wine are manufactured from as many varieties of dates; on the Gold Coast there are four varieties; at Sierra Leone, there are three kinds. Livingstone found the natives of South Africa intoxicated on a liquor called Malova, which they manufactured from the juice of the palm oil-tree.

The Jews of Morocco distil brandy from dates; in Eastern Africa the fermented drink is used, and brandy is also distilled; in the Barbary States the drink is called date tree water; the Jews there also prepare a kind called Laghibi; and in other portions of Africa it is known as Ballo Cœcuta Congo, Embeth Kriska, Lugrus, Pali, Pardon or Bardon. In several of the states of farther India the palm juice is distilled into Arrack; in several of the East Indies the fermented juice is called Soura; in others, Talwagen and Vellipatty. In Java and Amboyna a spirit is distilled from Tyffering, the fermented fruit of a variety of the palm called Sagwire, so strong and fiery that the drinkers call it "hell water." In Manilla and Mindora from another species of palm is procured a drink called tuba; in China, cha; in Japan, brandy; in Surinam, wine.

ELDER BERRIES.—A fermented and also a distilled spirit is made from them in Hungary; in Tartary, arraki; in England, wine.

EPILOBIUM.—Ale, in Russia.

FIGS.—A distilled liquor called Mahayah, is extracted from figs by the Jews of Morocco; from damaged figs brandy is distilled in Portugal.

FLESH OF LAMBS.—In China, a beer or wine.

FLESH OF SHEEP.—Tartary, a beer or wine; Afghanistan, beer; China a distilled spirit called Kan-yang-tsyew.

GAGAHOGUHA.—A fruit of Southern Africa, from which the Caffres make wine.

GRAPES.—Wine in Palestine, Syria, India, Egypt, Greece, Rome, Arabia, Abyssinia, Barbary States, Persia, Peru, Chili, France, Germany, Italy; in short in all the countries where the grape grows. Pliny reckoned about one hundred and ninety-five sorts of wine in use in his day; Henderson, in his History of Ancient and Modern Wines, published in 1824, gives the names of seventy-eight varieties of the former, and three hundred and fifty-nine of the latter; Redding on Ancient and Modern Wines, published in 1851, enumerates eighty varieties of ancient, and · eleven hundred and seventy-nine of modern wines; the American edition of Dr. B. W. Richardson's Cantor Lectures on Alcohol, gives the names of forty-four varieties of ancient Roman wines, and ninety-five varieties of modern wines. It would be impossible to give an exact and exhaustive list.

GRAPE SKINS AND REFUSE.—In Arabia the Jews and Christians distil Arrack; Abyssinia, brandy; Eastern Africa, brandy; Germany, Troster, which mixed with ground barley or rye, makes a fermented drink.

HEMP.—In India the seeds are fermented for a beverage called Brug.

HONEY.—In Nubia, honey is diluted with water, boiled, and then fermented in the sun, when it is called Hydromel; the Abyssinians prepare it in a similar way, and also pro-

duce a very intoxicating liquor from potatoes and honey.
Fermented honey is said to be distilled to brandy in some
of the Barbary States.   The Tartars make a fermented
drink called Ball; in Caffaria, by fermenting honey with
the juice of a native root, an intoxicating mead is produced;
the Hottentot does the same by mixing it with a plant
called gli; German distillers employ it in making Rosolis,
as do also the Italians.   In Ireland, Eastern Africa and
Persia, it is called mead; in Southern Africa, Hydromel;
in Russia, Metheglin; in Madagascar, Toak; and a mead
which requires several years before it comes to perfection,
is made from it in Poland.

JIN-JIN-DI ROOT.—In Central Africa, a fermented
drink.

JUNIPER BERRIES.—In France, Juniper wine; in Holland and other countries where gin is manufactured, they
are a prominent ingredient in that liquor.

LEMON FLOWERS.—In China a drink is distilled from
the flowers of a species of lemon tree.

LOTUS BERRIES.—In Africa and China, wine.

MADLUCA FLOWERS.—Mentioned in the Institutes of
Menu, as producing one of the three inebriating liquors of
the Hindoos.

MAGUE.—A fruit in Peru, resembling the cherry, from
which a wine is made called Theka.

MAKKAHNYEYE.—A fruit resembling guava, growing
in Eastern Africa, from which the natives make a fermented
drink called Wocahyeye.

MALLE.—The berries of a tree in Peru, from which is
made a wine called Malle.

MANDIOCH POBIONE.—A root resembling a chestnut in
taste, found in Paraguay, from which is made a drink
called Mandebocre.

MANIOC.—At Sierra Leone a fermented drink; Peru,
Kiebla, and when distilled Puichin; Mexico, Masato.

MAIZE.—Egyptians, Bonza, Curmi; Arabs, Bonza;
Bournon, Sza; Badagary, Gear; Congo, Guallo; African

Slave Coast, Southern Africa, Whidah, beer; Eastern Africa, Epcahla; Mexico, Demaize, Pinole, Pulque; Chili, Paraguay, beer: Thibet, Chong; Russia, Siberia, brandy; Brazil, Kaviaraku; Surinam, Chiacor; United States, Prussia, Denmark; whiskey; Peru, Chica. Chica is made by pounding maize to a fine powder and placing it in a heap, around which a number of females sit, and chew the material into a kind of paste. After chewing it is rolled between the hands into round balls, which, being placed in the form of a pyramid, are baked in the fire, and then immersed in water, where they ferment and form the intoxicating draught.*

MIENGON.—A fruit in Tonquin, resembling the pomegranate, produces cider.

MILK.—Mare's milk is distilled by the Tartars and Calmucks, and nearly all the tribes of Central Asia, and when sufficient cannot be obtained, recourse is had to the milk of cows, camels, and sheep. It receives the names of Airen, Arjan, Caracosmus, Koumiss, Skhon, Vina and Yaouste. In Iceland, fermented milk is called String, and fermented whey, Syra; in Lapland it is Prima; Siberia, Koumiss; Afghanistan, Sihce.

MILLET.—In Egypt, a drink called Curmi; Dahomey, and other countries on the African Slave Coast, Pitto; Eastern Africa, beer, Huyembo or Puembo. Southern Africa, beer, Ballo, Pombie; Central Africa, Kissery, Otee; Circassia, Hautkups, Soar; Yantzokbl; China, Sew-hengtsow; Sau-tchoo; Tartary, Baksoum, Busa; Corea, wine; Russia, beer, Braga.

MOLLE.—A fruit in Chile, of the color and shape of pepper, a red wine called Huigan.

MOTHERWORT.—From the flowers the Japanese distil a drink called Sacki.

MULBERRIES.—In the Island of Chios, brandy; Russia, wine.

* Morewood, p. 293.

MUTILLAS.—A species of myrtle berry in Peru, from which is made Chica de Mutilla; and from the fruit of Myrtus luna, another species of myrtle, the Chilians make wine.

OATS.—Norway, Siberia, brandy; Russia, beer, Braga, Quass.

ORANGES.—China, cordial; Spain, **wine.**

PEACHES.—From the blossoms Sacki is distilled in Japan. From the fruit, brandy in America; wine in Russia.

PEARS.—The Jews of Morocco, brandy; Hungary, wine; Russia, Perry, wine; England, Perry.

PITAUGA.—A species of myrtle in Brazil; a spirit is distilled from its berries.

PERSIMMONS.—Southern parts of United States, beer, brandy.

PLANTAINS.—In Peru a fermented drink called Masato, which they often distil.

PLUMS.—Cacongo, Central Africa, Japan, England, wine; Hungary, Schliwowitza; Tartary, Arraki.

POMEGRANATE.—In Persia, wine.

POTATOES.—In Germany, Prussia, Hungary, Norway, Sweden, France, Great Britain, North America, fermented and distilled drinks.

PSAK.—Berries in Tartary resembling dates, Bursa.

RAISINS.—An intoxicant is made from fermented raisins, in Arabia; Syria, Eastern Africa, and by the Jews of Morocco, brandy; from damaged raisins brandy is distilled in Portugal.

RASPBERRIES.—In Russia, mead.

RHODODENDRON.—In Siberia the steeped leaves make what is called Intoxicating Tea.

RICE.—One of the three inebriating drinks made by the ancient Hindoos. Modern India, Phaur; China, wines which take different names from their respective colors, and from the lees they distil Sam-tchoo; Tarquin and Cochin China, wine and Arrack; Japan, a strong beer called Sacki, and Mooroo-facoo, Samtchoo and Sotschio, distilled drinks; Tartary, Caracina or Teracina; Thibet, a fermented drink, called Chong; and a distilled, Arra; Japan, its various distilled qualities are called Badek, Brom, Kiji, Sichew, Tanpo. Central Africa, Ballo; Thibet, Chong; Manilla, Pangati; Eastern Africa, Corea, wine; Russia, beer; Siam, Sumatra, Corea, Arrack. In the interior of Formosa the women take a quantity of rice and boil it till it becomes quite soft, and then bruise it into a sort of paste. Afterwards they take rice flower, which they chew, and put with their saliva into a vessel by itself, till they have a good quantity of it. This they use as leaven or yeast, and mixing it with the rice paste, work it together like bakers' dough. The whole is then put into a large vessel, and having water poured over it, is suffered to stand for two months. Meanwhile the liquor works up like new wine, and the longer it is preserved the stronger it becomes.*

RYE.—France, Prussia, Denmark, Siberia, Kamtschatka, Great Britain, America, whiskey; Holland, gin; Norway, brandy; Russia, beer, Kisslyschtschy, Quass.

SATER.—From this tree, resembling the cocoanut, a liquor called Araffer, is obtained in Madagascar.

SELLAH.—An African plant, beer in Abyssinia.

SINGIN ROOT.—Eastern Africa.

SLOE FRUIT.—In France a drink resembling whiskey; Tartary, Arraki.

SLOKA TRAVA.—A sweet grass in Kamtschatka, Raka.

STRAWBERRIES.—Russia, mead.

SUGAR.—From the dregs, one of the three inebriating liquors made by the ancient Hindoos.

SUGAR CANE.—Upper Egypt, India, East and West Indies, New South Wales, Mexico, Brazil, North America, rum; East and West Indies, Tongare, Chilang; Peru, Guarapo.

TEE ROOT.—Sandwich Islands, Y-wer-a.

TEFF PLANT.—Abyssinia, beer.

* Morewood, p. 235.

Tocurso.—Abyssinia, beer.

Vontaca.—Madagascar, wine.

Water Melons.—Russia, brandy.

Wheat.—India, Phaur ; Congo, Guallo ; Prussia, white beer ; Holland, beer; Caucasus, brandy.

Wines are also made from gooseberries, raspberries, strawberries, cherries, mulberries, blackberries, quinces, peaches, and mountain ash berries; and intoxicants have been produced from parsnips, beets. and turnips.

Fermented and unfermented Wines.—Henderson, in his *History of Ancient and Modern Wines*, says that: "The invention of wine is enveloped in the obscurity of the earliest ages of the world." This fact confronts all who attempt to investigate the subject, but it is not an insuperable barrier to a clear apprehension, statement and defence of the fact that a real distinction between unfermented and intoxicating wines can be established by indisputable proof. To a brief statement in this direction we devote a few pages of this work. So much has been written and published on the subject, both in this country and elsewhere, since 1834, that anything more than hints at the results reached would fill several large volumes.

1. First, then, we notice this, that although the first account given in well authenticated history, of the use of wine, in the case of Noah, shows us that an intoxicating agent was known by that name,* the earliest notice of any mode of preparing wine, that given in the dream of Pharaoh's butler, and the interpretation thereof by Joseph,† as clearly shows that an unintoxicating agent was also called wine. "I took the grapes," says the butler, "and pressed them into Pharaoh's cup, and I gave the cup into Pharaoh's hand." "Thou shalt," says Joseph, in his interpretation of the dream, "deliver Pharaoh's cup into his hand, after the former manner, when thou wast his butler." In comment-

---

* Genesis ix. 21.   † Gen. xl. 9-13.

ing on this passage the learned German, Rosenmuller, produces historic proof to show that no other kind of wine was allowed to be used by Egyptian kings.

2. A second significant fact is found in the institution and observance of the Jewish Passover. As originally instituted, (see Exodus xii. and xiii.) no mention is made of drink of any kind; but it is expressly declared, that " whosoever eateth that which is leavened, even that soul shall be cut off from the congregation of Israel."* The late Prof. Stuart, has well said, that:

" As the word translated *eating*, is, in cases innumerable, employed to include a partaking of all refreshments at a meal, that is, of the drinks as well as the food, the Rabbins, it would seem, interpreted the command just cited as extending to the *wine* as well as *bread*, of the Passover. * * * The Rabbins, therefore, in order to exclude every kind of fermentation from the Passover, taught the Jews to make a wine from raisins or dried grapes expressly for that occasion, and this was to be drunk before it had time to ferment. * * * That the custom is very ancient, that it is even now almost universal, and that it has been so for time whereof the memory of man runneth not to the contrary, I take to be facts that cannot be fairly controverted. * * * I am not able to find evidence to make me doubt that the custom among the Jews of excluding fermented wine as well as bread, is older than the Christian era." †

The Jews " are forbidden," says Allen (*Modern Judaism*, p. 394), " to drink any liquor made from grain, or that has passed through the process of fermentation. Their drink is either pure water, or *raisin wine* prepared by themselves."

So also Hyam Isaacs, (*Ceremonies of the Jews*, p. 98,) says:

"Their drink during the time of the feast is either fair water, or *raisin wine* prepared by themselves, but no kind of leaven must be mixed."

" Mr. A. C. Isaacs, says:—" I spent among my own people six and twenty years of my life, and, prior to becoming a convert from the Jewish to the Christian faith, I sustained among them the office of Hebrew teacher. I can therefore speak confidently on the subject of your inquiries. All the Jews with

---

* Exodus xii. 19. † " Bibliotheca Sacra," 1843, pp. 507, 508.

whom I have ever been acquainted, use *unintoxicating wine* at the Passover, a wine made expressly for the occasion, and generally by themselves. If it ever should be fermented, it is certainly unknown to them, and against their express *intention ;* but I never I knew it to exhibit any of the symptoms. " *

The following American testimonies may be added : The late Judge M. M. Noah, of New York, after describing the manner in which raisin-wine is prepared, says :

" This is the wine we use on the nights of the Passover, because it is free from fermentation, as we are strictly prohibited, not only from eating leavened bread, but from drinking fermented liquors." †

To the same effect, the late Rev. Isaac Lesser, of Philadelphia, a learned Rabbi, and translator of the Hebrew Scriptures, says :

" For religious purposes, we uniformly exclude Gentile wines from the ceremonies. Hence in countries where the vine is not cultivated, we resort to artificial wines, such as raisin wine, &c.; or even cider, lemonade, mead made of honey; but seldom on such occasions do we employ spirituously fermented liquors ; *and never, so far as my knowledge goes, on the Passover nights, when uniformly the unintoxicating preparations are used, if Jewish wine is not readily accessible.* This is not, however, on temperance principles, but because all fermented liquors, of which grain is the basis, are *leaven,* and therefore strictly prohibited on the Passover." ‡

3. It is also true, as no one denies, that the Bible makes a discrimination between unintoxicating and inebriating wines, by its commendation of some as beverages, and its as emphatic condemnation of others. Unfortunately for the common reader, the thirteen different terms employed by the sacred writers to designate these varieties of beverages, are almost always rendered in our English version by the one word wine; and hence arises misunderstanding and confusion.

---

* The above are all cited from Dr. F. R. Lees' Works, Vol. II. pp. 125, 170.

† *The Enquirer,* December, 1841, p. 32.

‡ Ibid, p. 29.

" One of the greatest faults of our otherwise admirable version of the Bible," says an able writer in the *Encyclopedia Britannica,* " is, that the translation of the same original word is often improperly varied at the expense of perspicuity ; while, on the other hand, ambiguity is sometimes occasioned by the rendering of two original words, in the same sentence, by only one English word ; which, however, is used in different meanings. Not only two, but thirteen different and distinct terms are translated by the word wine, either with or without the adjectives, ' new,' ' sweet,' ' mixed,' and ' strong.' If the first rule for a translation, as laid down by Dr. George Campbell, be correct,—that the translation should give a complete transcript of the idea of the original—the common version must, on this point, be deemed exceedingly defective."

A' minutely critical examination of this subject, necessitating the taking up of each of these Hebrew and Greek words, and examining its use in every instance where it is employed, would more than fill this volume, to the exclusion of everything else.* We content ourselves, therefore, with the summing up of the argument as given by Prof. Stuart in his Essay on the " Scripture View of the Wine Question : "

" Wherever I find declarations in the Scriptures respecting any matter which appear to be at variance with each other, I commence the process of inquiry by asking whether these declarations respect the same object in the same circumstances ? Wine and strong drink are a good, a blessing, a token of divine favor, and to be ranked with corn and oil. The same substances are also an evil. Their use is prohibited ; and woe is denounced on all who seek for them. Is there a contradiction here,—a paradox, incapable of any satisfactory solution ? Not at all. In the light of what has already been said, we may confidently say,—not at all. We have seen that these substances were employed by the Hebrews in two different states: the one was a fermented state, the other an unfermented one. The ferment-

---

* This work has been ably done by Dr. F. R. Lees, in his Works, especially in his " Commentary," and " Bible Wine Question ; " by Ritchie in his " Scripture Testimony against Wine ;" by Parsons, in his " Anti-Bacchus ; " by Miller, in his " Nephalism ;" and by others, to whom the reader is referred for more full information.

ed liquor was pregnant with alcohol, and would occasion inebriation, in a greater or less degree, in all ordinary circumstances; and even where not enough of it was drunk to make this
effect perceptible, it would tend to create a fictitious appetite
for alcohol, or to injure the delicate tissues of the human body.
The unfermented liquor was a delicious, nutritive, healthy beverage, well and properly ranked with corn and oil. It might be
kept in that state by due pains, for a long time, and even go on
improving by age. Is there any serious difficulty now in acquit-
ting the Scriptures of contradiction in respect to this subject?
I do not find any. I claim no right to interfere with the judgment of others; but for myself, I would say, that I can find no
other solution of the seeming paradox before us. I cannot regard the application of the distinction in question between the
fermented and unfermented liquors of the Hebrews, to the solution of declarations, seemingly of an opposite tenor, as any
forced or unnatural means of interpretation. It simply follows
suit with many other cases, where the same principle is concerned. Wine is a blessing, a comfort, a desirable good.
When, and in what state? Wine is a mocker, a curse, a thing
to be shunned. When, and in what state? Why, now, is not
the answer plain and open before us, after we have taken a
deliberate survey of such facts as have been presented? I can
only say, that to me it seems plain,—so plain, that no wayfaring man need to mistake it. My final conclusion is this: viz.,
that wherever the Scriptures speak of wine as a comfort, a blessing, or a libation to God, and rank it with such articles as
corn and oil,—they mean,—they can mean, only such wine as
contains no alcohol that could have a mischievous tendency;
that wherever they denounce it, prohibit it, and connect it
with drunkenness and revelling, they can mean only alcoholic
or intoxicating wine. I need not go into any minuteness of
specification or exemplification, for the understanding of my
readers will at once make the necessary discrimination and
application. If I take the position that God's word and works
entirely harmonize, I must take the position that the case befor·
us is such as I have represented it to be. Facts show that th·
ancients not only preserved wine unfermented, but regarded i
as of a higher flavor and finer quality than fermented wine.
Facts show that it was and might be drunk at pleasure without any inebriation whatever. On the other hand, facts show
that any considerable quantity of fermented wine did and would
produce inebriation; and, also, that a tendency toward it, or a
disturbance of the fine tissues of the physical system, was and
would be produced by even a small quantity of it, full surely,

if this was often drunk. What, then, is the difficulty in taking the position, that good and innocent wine is meant in all cases where it is commended and allowed, or that the alcoholic or intoxicating wine is meant in all cases of prohibition and denunciation? I cannot refuse to take this position without virtually impeaching the Scriptures of contradiction or inconsistency. I cannot admit that God has given liberty to persons in health to drink alcoholic wine, without admitting that his word and works are at variance. The law against such drinking, which he has enstamped on our nature, stands out prominently, read and assented to by all sober and thinking men. Is his word now at variance with this? Without reserve I am prepared to answer in the negative."

4. The facts referred to in the foregoing extract, that the ancients preserved wine in an unfermented state, and they preferred it to all other beverages, are attested by numerous witnesses.

Aristotle says of sweet wine: "It is wine in name, but not in effect; for the liquor does not intoxicate." "The wine of Arcadia, he says, "was so thick, that it was necessary to scrape it from the skin bottles in which it was contained, and to dissolve the scrapings in water."

Pliny says: "There is an intermediate thing between *dulcia* (sweets) and *vinum* (wine,) which the Greeks call *ucigleuces*."

Discorides ranks, in his 'Materia Medica,' 'boiled wine,' under the head of 'wine.' So also Pliny, Columella, and Theophrastus, pronounce that wine the best, which is nutritious and unintoxicating, a syrup which could have been prepared from the grape juice only before it had fermented, and which, to be used, must be diluted with water. Modern travellers and observers testify to the same thing. Rev. Henry Homes, Missionary in Constantinople, said in the " Bibliotheca Sacra," May, 1848 :

"Simple grape-juice, without the addition of any earth to neutralize the acidity, is boiled from four to five hours, so as to reduce it to one-fourth the quantity put in. * * * It, ordinarily, has not a particle of intoxicating quality, being used freely by both Mohammedans and Christians. Some which I have had on hand for two years has undergone no change. * * * In the manner of making and preserving it, it seems to correspond with the recipes and descriptions of certain drinks included by some of the ancients under the appellation, wine."

In 1845, Capt. Treatt wrote : "When on the south coast of Italy, last Christmas, I inquired particularly about the wines in common use, and found that those esteemed the best were sweet and unintoxicating. The boiled juice of the grape is in common use in Sicily. The Calabrians keep their intoxicating and unintoxicating wines in separate apartments. The bottles were generally marked. From inquiries I found that unfermented wines were esteemed the most. They were drank mixed with water. Great pains were taken, in the vintage season, to have a good stock of it laid by. The grape-juice was filtered two or three times, and then bottled, and some put in casks and buried in the earth. Some kept it in water (to prevent fermentation.)" *

Mr. Delavan wrote to " The New York Observer," in 1840 : "While I was in Italy I obtained an introduction to one of the largest wine manufacturers there, a gentleman of undoubted credit and character, and on whose statements I feel assured the utmost reliance can be placed. By him I was instructed in the whole process of wine-making, as far as it could be done by description, and from him I obtained the following important facts :

" First, That with a little care, the fruit of the vine may be kept in wine countries free from fermentation for several months, if undisturbed by transportation. Wine of this character, he exhibited to me in January last, several months after the vintage.

" Secondly, That the pure juice of the grape may be preserved free from fermentation for any length of time *by boiling*, through which the principle of fermentation is destroyed; and in this state, may be shipped to any country, and in any quantity, without its ever becoming intoxicating.

Thirdly, That in wine-producing countries unfermented wine may be made any day in the year. In proof of this the manufacturer referred to, informed me that he had then in his lofts (January) for the use of his table till the next vintage, a quantity of grapes sufficient to make one hundred gallons of wine ; that grapes could always be had at any time of year to make the desirable quantity ; and that there was nothing in the way of obtaining the fruit of the vine free from fermentation in wine countries, at any period. A large basket of grapes were

---

* Dr. Lees' Works, Vol. II. p. 144. See also many citations in his "Preliminary Discourse," in "Temperance Commentary."

sent to my lodgings which were as delicious and looked as fresh as if recently taken from the vines, though they had been picked for months. I had also twenty gallons made to order from these grapes, which was boiled before fermentation had taken place ; the greater part of which I have still by me in my cellar. As a further proof that new wine may be kept in its sweet and un-fermented state, I travelled with a few bottles of it in my carriage over 2,000 miles, and upon opening one of the bottles in Paris, I found it the same as when first put up." *

Subsequently, after keeping some of this wine in his cellar for years, he sent a bottle of it to Prof. Silliman, who, after subjecting it to chemical tests, reported that he could not find a particle of alcohol in it.

Malt Liquors.—Ale or beer, is an Egyptian invention. It was first made " by pouring hot water on barley and allowing the fluid to ferment. It has been said that they called this drink ' bouzy,' from Busiris, the name of a city which contained the tomb of the god Busiris or Osiris. So, says one of our quaint old authors, we get the term ' bouzy,' which we apply to a man who has taken a great deal of beer, and whom we call a bouzy fellow. The word beer probably comes from barley, or from the Hebrew word *bar*, corn." †

Browne, a modern traveller, says, that

" The Egyptians still make a fermented liquor of maize, millet, barley and rice, but it bears little resemblance to our ale. It is of a light color, and in hot seasons will not keep above a day." ‡

The ancient Britons also made beer, which, though for a time displaced by mead, the favorite Saxon beverage, became again, after the Norman conquest, the national beverage, and was especially brewed in large quantities at the monasteries. † From the law passed by the Plymouth Col-

---

* *The Enquirer*, December, 1841, pp. 29, 30.
† Dr. Richardson's Temperance Lesson Book, p. 45.
‡ Travels in Africa, Egypt, and Syria, p. 26.
§ Teetotaller's Companion, p. 20.

ony in 1636, prohibiting the retailing of beer, it is evident that it was among the earliest beverages prepared in the New World.

The manufacture of ale or beer at the present time, is by a similar process in all countries, which is thus described by Rev. James B. Dunn, who was at one time approached by some capitalists to take charge of a large brewing establishment, and who, in order to inform himself on the subject, visited a brewery, and ascertained how beer was made. He says :

"In the manufacture of good beer, as it is called, (I speak not now of the abominable adulterations, though they are common enough,) three things are necessary: malt, hops, and water. The water, though useful, is not food. The hop gives flavor, and helps to preserve the liquor, but it contains no feeding properties. To name its chemical constituents will suffice. These are volatile oil, resin—a bitter principle—tannin, malic acid, acetate, hydrochlorate, and sulphate of ammonia. The malt, then, is the only substance that can make the liquor feeding, either as it remains in the liquor, or as it may be converted into some other substance. Malt, we all know, is vegetated barley. Barley is food next in nutrition to wheat, and all we have to do is to ascertain how much of this feeding substance is found in the beer when men drink it. The brewing process will give us that; in tracing which we shall find, that at every step the object is, not to secure a feeding, but an intoxicating liquor; and that to obtain this, the feeding properties of the barley are sacrificed at every stage.

"In making a gallon of beer six pounds of barley are used, which, to commence with, is six pounds of nutritious food. In manufacturing this into beer, it has to undergo four processes, in every one of which it loses part of its nutriment. The first is malting, or sprouting. By this process the malters spoil the barley of one-fourth of its nutriment, just in the same way as wheat is spoiled if it gets wet and sprouts in the field. Every housekeeper knows that when potatoes or onions sprout they lose much of their nutritive properties. The next process is that of mashing, by which a saccharine solution is extracted from the grain, and here one-third of the barley is lost. Then follows the fermenting process, by which one-fourth of it is converted into alcohol. The fourth process is that of fining. People don't like thick or muddy beer; and as some thick matter cannot be prevented coming over in mashing, the liquor is put to settle, and these settlings are disposed of as ' barrel bottoms.' These bottoms are really parts of the barley, and here is another loss. Now, in this gallon of beer, how much of the barley is there left ? At the outset you had some six pounds, or ninety-six ounces. What is there now ? Less than ten ounces. The truth of this you can easily ascertain. Get a pint of ale or beer, and place it in a saucepan, then gently boil it over the fire. The fluid part will go—the solid part will remain. Thus every grain of solid matter can be obtained, and its properties and amount fully ascertained. Scientific men have frequently made the experiment, and by careful tests demonstrated that the average quantity of solid matter found in a gallon of malt liquor is less than ten ounces. So that in manufacturing ale or beer you actually lose very nearly eighty parts out of eighty-eight, and all that you obtain in the place of it is upward of three ounces of alcoholic poison, and which constitutes the strength of the liquor. What would you think of the man who should buy ninety-six ounces of wheat, making it sprout, drying it, pouring hot water upon it, giving a part to the pigs, and throwing a part down the gutter—should waste upward of eighty ounces, and should leave for himself and family only ten ounces ? What if he did this for the purpose of getting about four ounces of poison, which will injure his health, destroy his reason, and corrupt his heart ? Would you say that God sent the grain to be thus wasted, or would you call the poison which the ingenuity of this prodigal had extracted, ' a good creature of God ?' Much has been said of waste and extravagance, but we know of no instance or example that will bear any parallel with the prodigality that is practiced in converting barley into malt, and malt into beer. * * * * What, then, we ask, is there to support or to strengthen a man in a pint of ale or beer ? Its contents are fourteen ounces of water, part of one ounce of the extract of barley, and nearly an ounce of alcohol. The water and alcohol go immediately into the veins, and while the alcohol poisons, the water, if not needed, unnecessarily dilutes the blood, overcharges the vessels, and loads the kidneys and bladder ; while there remains less than an ounce of indigestible extract of malt, which has been grown, roasted, scalded, boiled, embittered, fermented, and drenched with water and alcohol till it seems unfit for the brute, far less the human stomach. Yet this is all that is left in the stomach to be digested." *

---

* Tract No. 26, National Temperance Society.

With this statement in regard to nutrition, agrees the judgment rendered by the great chemist, Baron Liebig :

"In the brewing of beer a separation takes place between the sanguigenous (nutritive) matters of the barley, and the starch. Of the former, those portions which dissolve in the wort, and are separated as *yeast* during the fermentation, *are lost for the purpose of nutrition.* We can prove, with mathematical certainty, that *as much flour as can lie on the point of a table knife, is more nutritious than eight quarts of the best Bavarian Beer ;* that a person who is able *daily* to consume that amount of beer, obtains from it, *in a whole year,* in the most favorable case, exactly the amount of nutritive constituents which is contained in a 5 lb. loaf of bread, or in 3 lb. of flesh."

Dr. E. Lankester says, " Beer contains but one per cent of nutritive matter, and is not a thing to be taken for nutrition at all." Professor Lyon Playfair says, " 100 parts of ordinary beer or porter contains 9½ parts of solid matter, of which only about one-half part consists of flesh-forming matters ; in other words, it takes 1.666 parts of ordinary beer to obtain one part of nourishing matter." *

A similar confession is made by the brewers and beer sellers of Great Britain. After a series of experiments which resulted in demonstrating that ale or beer can be made from sugar much cheaper than from malt, the sanction of government was sought for the substitution of sugar and molasses for malt ; and the question was debated to the public in the columns of the " *Morning Advertiser,*" the avowed organ of the brewers, distillers, and publicans. In the issue for October 30, 1846, appeared the following :

" With respect to the quality of beer made from sugar, all who have tried it declare that it possesses the same qualities as the beer from malt. . . . By some it may be supposed that the working man will lose a nutritive beverage, but this is a misapprehension of the subject. After fermentation *no albumen, or flesh-forming principle remains in the liquor,* which has now become vinous. . . . As to spirits, those [already] produced from sugar [as rum] are well known ; and no question can be raised

---

* Alcohol : Its Combinations, &c., by Col. J. G. Dudley, pp. 38, 39.

in regard to comparative properties of nutrition, since *all kinds are equally deficient.*" *

No wonder then that Martin Luther said :

"The man who first brewed beer was a pest for Germany. Food must be dear in all our land, for the horses eat up all our oats, and the peasants drink up all our barley in the form of beer. I have survived the end of genuine beer, for it has now become small beer in every sense ; and I have prayed to God that he might destroy the whole beer-brewing business; and the first beer-brewer I have often cursed. There is enough barley destroyed in the breweries to feed all Germany."

The percentage of alcohol in ale or beer differs in the several varieties. Dr. Edward Smith, in his work on " Foods," says : (p. 412.)

"It bears a relation to the amount of saccharine matter which was fermented in the brewing. Brande in his day found 4.20 per cent. of alcohol (specific gravity 0.825,) in porter ; 8.88 per cent. in ale ; and 6.80 per cent. in brown stout. At the present day there may be 10 per cent. in the strong East India pale ale, and 15 to 20 per cent. in many old home-brewed ales, stored for private use ; but usually the amount varies from 5 to 7 per cent. in moderately good ales, and may be only 1 to 3 per cent. in small beer. Hence, one pint of strong home-brewed ale may contain as much alcohol as is found in several bottles of good claret wine; but as a general expression, a pint of good ale is equal in that respect to a bottle of fairly good claret."

Dr. Richardson says : " Some specimens of ales and stouts contain as much as ten per cent. of alcohol, and in very strong old ale that quantity may be exceeded. There is, however, a great deal of trickery played with the ale which is commonly sold in retail, so that it is difficult to arrive at any correct standard. I had once the duty of determining the quantity of alcohol in an immense number of specimens of ales vended from the London public-houses during the dinner hours of working-men. In many of these samples the alcohol presented did not exceed five per cent. and in a few instances it was actually as low as four per cent. The reason of this was, that at the particular time of the day named, the fresh beer in the casks was especially diluted with water, containing a little treacle and salt, so as to reduce the strength and increase the profit." †

---

* Teetotaller's Companion, pp. 444, 445.
† Temperance Lesson Book, pp. 92, 93.

According to Dr. Bence Jones, from tests made with the Alcoholometer, "New bitter ale contains 6 to 12 per cent. of alcohol; porter, 6 to 7 per cent.; stout, 5 to 7 per cent." *

Prof. Wood, of the Harvard Medical School, finds in "Boston lager beer, from 5½ to 6 per cent. of Alcohol." †

Mr. Henry H. Rueter, in his pamphlet entitled " Argument in Favor of Discriminating Legislation regarding the sale of Fermented and Distilled Liquors—addressed to the Joint Special Committee of the Massachusetts Legislature on the sale of Intoxicating Liquors," claims the percentage of alcohol in different malt liquors to be as follows:

"Ottawa" Beer....................contains	2.00	per cent.
Average German Lager-beer.......... "	3.80	"
Common Massachusetts Lager-beer... "	4.00	"
Common Massachusetts Ale.......... "	4.10	"
Average Massachusetts Ale.......... "	5.20	"
Home-brewed "Hop-beer" (made by farmers and private families from molasses and hops)................ "	5.50	"
Strong Massachusetts Lager-beer..... "	5.80	"
Common Cider...................... "	6.10	"
London Porter (imported)........... "	6.10	"
Strong "Stock" Ale .... "	6.30	"
London Ale (imported).............. "	6.80	"
Dublin Porter (imported)............ "	7.00	"
Edinburgh Ale (imported)... .... "	7.50	"

Let it not be forgotten that these large or small amounts of alcohol in these various beverages are none of them a natural production, but are invariably the result,—Baron Liebig being authority,—of "fermentation, putrefaction and decay." "These," he says, "are processes of *decomposition*, and their ultimate results are to *reconvert* the elements of organic bodies into that state in which they exist, *before they participate in the processes of life.*" "Fermentation," it is stated in Turner's Chemistry, edited by Liebig,

---

" is nothing else but the putrefaction of a substance containing no nitrogen." The formation of alcohol," says the great French chemist, A. F. Fourcroy, " takes place at the expense of the destruction of a vegetable principle : thus spirituous fermentation is a commencement of the destruction of principles formed by vegetation." " Nature," says Count Chaptal, " never forms spirituous liquors ; she rots the grape upon the branch, but it is art which converts the juice into wine." " Alcohol," said Dr. E. Turner, " is the intoxicating ingredient of all spirituous and vinous liquors. It does not exist ready formed in plants, but is a product of the vinous fermentation." *

This product, alcohol, is, by the agreement of scientific observers,—whether they are teetotalers or otherwise, or whatever their theories in regard to what becomes of alcohol after it is taken into the system,—a *Narcotico-Acrid Poison.* On this subject there are no more competent authorities than Orfila, Christisson, Dr. Taylor, Pereira, Professor Binz, Dr. Lallemand, Perrin, Dr. Willard Parker, Dr. Richardson, Professor Parks, Professor Duroy, Dumorel, Magnus, Dunglison, Dr. Edmunds, Professor Davis, Powell, Dermarquay, Wetherbee, Burns, and Dickenson, all of whom are agreed as to the character of the poison.

Distilled Liquors.—Dr. B. W. Richardson, in summing up the results of his researches into the history of Alcohol, says that there are these " five points " to be remembered :

" 1. The fluid containing alcohol that was first known was the fermented fluid obtained from fruits by fermentation, and called wine.

" 2. The wine was distilled, and thereby a fine spirit was obtained, which was called the spirit of the wine.

" 3. When the spirit of wine was discovered, it was treated in different ways, by which spirits of different tastes, colors, and strengths were obtained, and called by different names, such as whiskey, brandy, rum and gin.

---

* Cited in Text-Book of Temperance, pp. 29, 30, 33. See Dr. Hargreaves' " Alcohol, what it Is, and what it Does."

"4. Sugar and other substances than fruits were made to yield spirit by fermentation.

"5. At last the pure spirit, from whatever source it was got, was called *alcohol.*" *

These facts are significant in various ways, but chiefly in this : their showing that alcohol is not created in the act of distillation, but exists already in the fermented article, be it wine, beer, or whatever name may be given it ; and that distillation simply separates the alcohol from the other substances with which it is mixed in these fermented beverages ; and no more alcohol can be obtained by this process of distillation than was already in the fermented article before the distilling took place.

The first experiments in distilling wine, are said to have been made in the eleventh century of the Christian era, by an Arabian chemist, named Albucasis. He called it "The spirit of wine," and for a long time its use was confined to the laboratory of the chemist, being used to preserve animal substances from decay, and to dissolve oils, resins, gums and balsams, which water would not change. Subsequently it was employed as a medicine ; and afterwards as a beverage, to be used in health. The names then given it, were *vinum adustum,* burnt wine ; *spiritus ardens,* strong spirits ; as well as *spiritus vini,* spirit of wine ; and later *aqua vitæ,* water of life. According to Mr. Stanford, designated by Dr. Richardson as " a very learned scholar :"

" *Aqua vitæ* was used as a drink as early as the year 1260 of our present era. The Arabians, he thinks, taught the use of it to the Spaniards, and the Spaniards to the monks of Ireland. It thus came into use in Ireland, and obtained the Irish name by which it is still known in one form, 'whiskey.' In the old Erse, or Irish tongue, it was called *usize-biatha,* which means *aqua vitæ.* In time this term was shortened into *usque-baugh,* and this again was shortened into *usige,* from which comes the word whiskey. Sometimes in Ireland this same strong drink is called *potheen, or poteen.* This word, *poitin,* means a small pot or still, the vessel from which the liquor was distilled, and pooteen was, perhaps, derived from the Latin word *potio*, a drink." *

The name *Alcohol* was given to distilled spirit some time in the latter part of the seventeenth century. Dr. Richardson finds its mention in a chemical work by Nicholas Lemert, published in 1698.

"From Lemert's description, it appears that Alcohol was a term intended to describe something exceedingly refined or subtile. He uses the word sometimes as a verb, and explains that when any substance is beaten into a very fine powder, so that it is impalpable, *i. e.*, when it cannot be felt rough to the touch, it is alcoholized. The same word, he adds, is employed to describe a very fine, pure spirit, and so the spirit of wine well rectified is called the alcohol of wine."

"Other scholars have tried to trace out the origin of the word itself, and the most accepted explanation on this point is that the word is Arabic, A'l-ka-hol, meaning a very fine essence or a powder used by the women of the East to tinge their hair and the margins of the eyelids. Afterwards, as described by Lemert, it was applied to all refined substances distilled by the heat of the fire." †

Another, and quite different conjecture has been offered respecting the etymology of the word as applied to an intoxicant, its present exclusive significance, by the author of "Neuces Philosophicæ" and quoted by Dr. Lees in his " Chemical History of Alcohol." ‡

"It is an Eastern superstition," says Dr. Edward Johnson, the author referred to, " to suppose that the earth is infested with evil spirits called *gouls*—and this word is spelt in several different ways, as *goul*, *ghoul*, and *ghole*. They were supposed to frequent burying grounds, and to prey upon dead bodies. They were also supposed to assume different shapes, and sometimes to enter the body, and *to possess it, as it were, with a devil.* When anything fearful was heard or seen, it was a common expression to exclaim, ' The ghole! the ghole!' And when the

---

* Ibid, p. 60, which see, for the origin of names of other distilled liquors. See also this subject treated *in extenso*, in Dr. Hargreaves' "Alcohol, What it Is and what it Does."

† Temperance Lesson Book, p. 64.

‡ Works, Vol. II. pp. 85, 86.

Arabian chemists first discovered alcohol, and observed the effect (intoxication) which it produced on the first person who took it, it seems very natural that they should suppose him to be possessed by an evil spirit, and that alcohol was, in fact, only one of the forms which the *ghole* had assumed, in order to enter and torture the human body. And, frightened at what they beheld, it was very natural that they should exclaim : '*al ghole, al ghole !*'—and it seems very probable that a fluid capable of producing such extraordinary effects should continue, for some time, to be supposed to be an evil spirit in disguise, as it were ; and when this notion was laid aside, the great evil which this liquid-devil was observed to work amongst men, would still be likely to cause it to *retain its name of Al ghole,* or the *evil spirit.* Our manner of spelling it is no objection to this etymology, for *alghole* might easily be corrupted into alkohol ; thus, *alghole, alghol, algohol, alkohol.* In this word, as in almost every other, the *name* defines the nature of the thing."

Dr. Lees makes note of the singular coincidence of this definition with the name suggested to Shakespeare, as he observed the effects of drinking in his day :

"O thou invisible spirit of wine, if thou hast no name to be known by, let us call thee *Devil.*"

When we consider that all intoxication, literally poisoning,—for that is just what the word means,—prior to the twelfth century, was produced by alcohol in fermented drinks, that this covers all the drunkenness which was known in ancient times, including that described and denounced in the Scriptures, we are better prepared to notice the fearful rapidity with which intemperance increased after distillation became common ; and also how it is, since so many drink, not because they love the taste of the liquors, but because they desire the delirium the narcotic induces,—oblivion which is the end and object of drinking,—that it is so inevitably sure that the lighter intoxicants will fail to satisfy when ardent spirits are within reach.

The following (determined by Dr. Bence Jones) is the percentage of alcohol contained in samples of the liquors named, as given by the Alcoholometer :

Port Wine, 20 to 23.
Sherry, 15 to 24.
Madeira, 19.
Champagne, 14.
Burgundy, 10 to 13.
Rhine Wine, 9 to 13.
Claret, 9 to 11.
Moselle, 8 to 9.

Rum, 72 to 77.
Whiskey, 59.
Brandy, 50 to 53.
Genoa, (Gin) 49.
Bitter Ale, (new) 6 to 12.
Porter, 6 to 7.
Stout, 5 to 7.
Cider, 5 to 7. *

Adulterations of Liquors.—Poisonous as alcohol is of itself, other poisons are often mixed with alcoholic liquors. Sometimes the object is to cheapen the beverages, and at others to produce a more quickly inebriating drink. The practice is an old one: "How can wine prove innoxious," exclaims Pliny, "when it is mixed with so many destructive ingredients?" * An ordinance of the French Police, bearing date of 1696, mentions the adulteration of wine with litharge (vitrified lead).† Often, say the Committee of Convocation of the Province of Canterbury, in their report: "These adulterations arise out of the competition among rival dealers, and frequently supply the only margin of profit by which the trafficker is enabled to keep possession of his house as the tenant of some brewer or distiller." "Vegetation," says a competent authority, "has been exhausted, and the bowels of the earth ransacked, to supply trash for this purpose. So unblushingly are these frauds practised, and so boldly are they avowed, that there are books published, called 'Publicans' Guides,' and 'Licensed Victuallers' Directors,' in which the most infamous receipts imaginable are given to swindle their customers. § "The following deceptions," says Tovey, "are often practised: Aroma is added to give the appearance of age to young wines. Wine is sweetened with cane sugar, or with other fruit than that of the grape. Coloring ingredients are added to imitate deeper colored wines. Water is added to

* Dr. Lees' Text Book of Temperance, pp. 46, 47.
† Hist. Nat. xiv. 20.
‡ Henderson's History of Wines, p. 339.
§ Redding on Ancient and Modern Wines, p. 358.

strong wine to increase the quantity.   Spirit is added to weak wine to increase the strength." *

Addison said, a long time ago, in the "*Tattler*," No. 131: "There is in the city a certain fraternity of chemical operators, who work under ground, in holes, caverns and dark retirements, to conceal their mysteries from the eye and observation of mankind.   These subterranean philosophers are daily employed in the transmutation of liquors, and by the power of magical drugs and incantations, raising under the streets of London, the choicest products of the hills and valleys of France.   They can squeeze Bordeaux out of the sloe, and draw Champagne from an apple.   Virgil, in that remarkable prophecy,—'The ripening grape shall hang on every thorn,'—seems to have hinted at this art, which can turn a plantation of northern hedges into a vineyard.   These adepts are known among one another by the name of wine brewers; and, I am afraid, do great injury, not only to her majesty's customs, but to the bodies of many of her good subjects."

Much later Charles Dickens said, in his "Household Words," "Henceforth, let no one boast of his fruity port, of his tawny, or of his full-bodied.   Those small strong-smelling bottles, on the dusty shelves of an analytical chemist's laboratory, will rise up in judgment against him; butyric ether, acetic acid, and that deadly cognac oil, will stand out against him, accusing witnesses of his simplicity and ignorance.   Henceforth, the mystery of wine-making is at an end; but wine itself is a myth, a shadow, a very Eurydice of life.   There is no such thing, we verily believe, as honest, grape-juice now remaining — nothing but a compound of vile, poisonous drugs, and impurely obtained alcohol; all our beautiful Anacreontics are fables like the rest, for wine hath died out from the world, and *the laboratory is now the vineyard.*"

"A German newspaper," says Samuelson, "recently gave an account of a prosecution in Berlin, in which it was stated that one large store which had been inspected contained only artificial wines, into the manufacture of which the juice of the grape had never entered, although the names borne by the labels of the bottles were those of well-known wines." †

Says a recent number of the Parisian: "The wine crop of 1879 was about twenty-five million hectolitres, or thirty million hec-

---

* Wine and Wine Countries, by Charles Tovey, p. 6.
† History of Drink, p. 90.

tolitres below the average of the last ten years. The annual consumption in France is forty to forty-five million hectolitres. Everybody expected a rise in the price of wine, and some conscientious dealers laid in a stock from abroad. The rise in price, however, never came, and the market remained well supplied. The reason was that the natural deficit was compensated for by artifical means. Wine was manufactured out of dry grapes. All the raisins to be found in the eastern ports were bought up and wine manufacturers sprang up all over the country. Around Paris alone there are seven steam power wine manufactories. The cost of a cask of raisin wine is about fifty francs, and it was sold at one hundred francs, thus giving a profit of a hundred per centum. But the competition has now become such that the price of raisins has risen from twelve francs to seventy-five francs the one hundred kilogrammes.

" The consequence is that raisins have been abandoned, and wine is now manufactured out of glucose, a sugary matter obtained from the potato, out of the residues of molasses, out of rotten apples, dried prunes, dates, figs, and all kinds of refuse fruit, and even out of beetroot. These abominable liquids are colored artificially, and mixed more or less with Spanish wines or white wine. The adulteration and manufacture of wine has attained such vast proportions that the principal dealers, who had taken measures to supply the market really with harvest wine from foreign countries, have taken steps to put a stop to the gigantic fraud. The imposture has reached such a pitch that not one-third of the wine drunk at Paris is real grape wine."

Dr. Hiram Cox, a distinguished chemist of Cincinnati, was directed by the Legislature of Ohio to analyze and examine the liquors in that market. He says :

"I was appointed to the office of Chemical Inspector on the 19th of March, 1855. Since then I have made over six hundred inspections of stores, and lots of liquors, of every variety, and now positively assert that over ninety per cent. of all that I have analyzed were adulterated with the most pernicious and poisonous ingredients!" " I called at a grocery store one day where liquor was being sold. A couple of Irishmen came in while I was there, and called for some whiskey. The first one drank, and the moment he drank, the tears flowed freely. while he, at the same time, caught his breath like one suffocating or strangling. When he could speak, he said to his companion, 'Och, Michael, by the powers! but this is warming to the stoomach, sure!' Michael drank, and went through like contortions, with the remark, 'Troth, and wouldn't it be foin on a coold frosty morning, Timothy?' After they had drank I asked the proprietor to pour me out a little in a tumbler. I went to my office, got my instruments, and examined it. I found it seventeen per cent. alcoholic spirits, when it should have been fifty, and the difference in percentage was made up by sulphuric acid, red pepper, pellitory, caustic potash, brucine, and one of the salts of nux vomica (strychnine). One pint of such liquor (at one time) would kill the strongest man." *

"A druggist in Cincinnati, sent to New York for two hogsheads of seignette brandy, so as to supply the physicians with the very best article for medical purposes. One cask was dark seignette, the other pale seignette. Dr. Cox tested them; poured some into a tumbler; sunk a polished steel blade into it, and let it remain there fifteen minutes. At the end of that time the steel blade had 'turned the brandy black as ink. The steel spatula itself corroded, and when dried left a thick coating of rust, which, when wiped off, left a copper coat (on the spatula) almost as thick as if it had been plated with copper.' Dr. Cox warned the druggist not to sell it, and advised him not to pay for it. The New York man sued the druggist for his pay. At the trial, Dr. Cox analyzed the stuff in the presence of the court and jury. In one cask he found 'sulphuric acid, nitric acid, nitric ether, prussic acid, Guiana pepper, and abundance of fusil oil. He pronounced it base, common whiskey. Not one drop of wine.' In the other cask he found 'the same adulterations as the first but in greater abundance, with the addition of catechu. This is most villanous.' The jury decided that the liquor was worthless, and the New York man left town without his pay." †

"Dr. Draper, Professor of Chemistry in the Medical College of the University of New York, some time since analyzed thirty-six samples of brandy, whiskey, etc., mostly taken from the bars of first class hotels and restaurants in the city of New York, where liquors are retailed at the highest prices, and supposed by the drinkers to be pure; and he found only four samples that did not contain fusil oil and coloring matter of some sort. In the lower and second class bars he found not only fusil oil, but cayenne pepper, salt and other substances. These mixtures were sold for

---

* Alcohol, its Nature and Effects. By Dr. Charles A. Story. pp. 252, 253.

† Ibid. pp. 377, 378.

pure full-proof liquors, when the analysis showed but about thirty-two per cent. on the average. It should have given fifty per cent. of alcohol, or in fact thirty-six per cent. below proof of spirits; but the deleterious chemicals with which the liquor was adulterated would produce the effects of intoxication, and the drinker was deceived; for instead of solacing himself with pure liquor, he was impregnating his system with compound poisons.

"Fusil oil, or amylic alcohol, as it is called in chemistry, is one of the products of distillation obtained from all substances containing starch—like corn, potatoes, wheat, etc.—and more or less is found in all these alcohols, according to the method of distillation and rectifying. Dunglison, one the highest authorities, says it is an acrid poison and destroys the mucous membrane of the stomach. It is nearly worthless by itself, and is produced as the last product of distillation, and if mixed with the ethylic alcohol, it greatly reduces the cost of liquor. Much more water can be put in the liquor where the amylic alcohol or fusil oil is allowed to remain in it, and is not removed by the process of rectifying. In fact some of the manufacturers of imitation liquors recommend adding it to inferior liquors, in order to 'reduce them' or 'lengthen them out,' as they term it. But it means, to enable them to add more water and still keep up the intoxicating quality of their liquor.

"The immense amount of whiskey made in this country furnishes the basis for most, if not all, of the imitation liquors and wines, and the presence of so much fusil oil invariably found in them is due, first, to the fact, that the distillers have discovered methods by which they can get a much larger quantity of alcohol out of a given quantity of grain than formerly. By adding blue vitriol and an extra quantity of yeast to their mash, they hasten the process of distillation by inducing a fermentation in about twenty-four hours that formerly required seventy-two. By thus artificially hastening the distillation, more fusil oil and other impurities are mingled with the whiskey, and its dangerous and deleterious qualities are greatly increased. And, farther, to get the largest amount of alcohol possible out of a given quantity of grain, they carry the process of distillation to the farthest possible extent, thus getting into the last portion of the product, most, if not all the amylic alcohol or fusil oil.

"The only way to purify these whiskeys and get rid of the fusil oil is by rectifying. When this is done, the whiskey is sold at various degrees of strength, under the different names of French spirits, pure spirits, or Cologne spirits, and these are used for making the imitation liquors; and the reason why so much fusil oil is commonly found in the counterfeit brandies, rum and wines is, that the whiskey which is used in making them has not been properly rectified, and possibly not rectified at all. If a large portion of the fusil oil remains in the whiskey, the stronger it will be for intoxicating purposes, and the manufacturer can increase his profits by putting in more water, and mixing in some kind of drug to make it bear a bead; and the drinker, when he feels the intoxicating effects coming on, will be satisfied that he has been furnished good, pure, strong liquor, when in fact it is many degrees below unadulterated alcoholic drinks.

"Whenever there is a failure of the grape crop in France, there is always a large demand for raw whiskey from that market, which comes back to us in due time, mixed, and which is sold here in the shape of pure French brandies and wines. We also import large quantities of cognac oils and liquor essences, flavoring matters and other drugs to be used by American manufacturers of counterfeit liquors. Many of these preparations are made in this country, and to add double-refined rascality to villany, some of them are adulterated, so that the compounder of these mixtures does not know himself exactly what quality of devil's broth he is brewing.

"The amount of adulterated liquors is enormous; and with a few exceptions, the entire liquor traffic of the world is not only a fraud, but, (perhaps without all of the dealers being aware of the fact) it also amounts to a system of drugging and poisoning.

"The business of making adulterated liquors has been so simplified that any novice who knows enough to make a punch or a cocktail can learn in a short time how to make any kind of liquor that will pass muster with nine-tenths of the drinking community. The oils and essences are within the reach of any dealer, wholesale or retail, and, with the chemical preparations, he can procure the directions for making a large or small quantity in a short time.

"Many books have been published in England and this country, giving instruction on this subject. The dealers in these articles observe secresy and caution. In some of their circulars they say to their customers that 'goods ordered to be forwarded by express and collected for on delivery, are sent with the amount only on the collection bill, *giving no indication of the nature of the articles*, and a detailed bill of items sent by mail.' They also say, for the purpose of encouraging the compounders in this country, that, 'The wine growers of Europe make use of compound ethers and oils to convert the grain spirit into brandy of superior quality, and that the liquors prepared with their flavors mix with the foreign in most economical proportions.'

"If the oils, essences, and other chemical preparations, are wanted for converting corn whiskey into any other kind of liquor, they can easily be obtained. You can procure brandy oil enough to change eight barrels of corn whiskey into eight barrels of French brandy for sixteen dollars, and enough chemicals to convert sixteen barrels into old Holland gin, London cordial gin, Old Tom gin, or Schnapps, for twelve dollars; to make old Bourbon, malt, Monongahela, rye or wheat whiskey, enough of these chemical compounds can be purchased for eight dollars to make four barrels; and to make four barrels of Irish or Scotch whiskey, the chemical materials can be procured for ten dollars. Then there is the cost of the coloring matter, and what the dealers call "age and body preparation." By using these drugs new whiskey is converted into any kind of liquor, of any age or color, in a short time. Some of these materials are known to be deadly poisons. The more highly the imitation liquor can be charged with the cheap poisonous drugs, to supply the intoxicating properties of alcohol, the more water can be added, thus reducing the cost, and keeping up the intoxicating power of the liquor. These preparations can be procured in any quantity. A small retailer can purchase a small quantity, sufficient to convert a gallon or two of whiskey into brandy, gin, or rum, as his daily wants may require, but they are generally used for larger quantities.

"In addition to the foregoing there are an immense number of receipts for making all kinds of intoxicating liquors. From various authentic sources I have procured a large number of these, which have been made use of at different times, or are in use now. For the benefit of moderate drinkers I will give a few; and as cider is generally considered a very wholesome beverage, they can always procure a sufficient quantity of it, even in those years when the apple crop fails.

" To make sweet apple cider :
20 pounds of brown sugar,
1 *pound of cider flavor*,
20 gallons of water,
1 pint of good brewers' yeast.
Add to each ten gallons of this mixture one quart of rectified spirits.

" To make Cognac brandy :
40 gallons of French spirits,
¼ pound extract of chicory,
1 pound of green tea,
¼ pound black currant leaves,

5

1 quart of burnt sugar or lime water,
A small quantity of simple syrup, to soften and give it
age, caramel or burnt sugar to color.

"Before the war, when real French brandy could be imported
for $2.50 per gallon, and corn whiskey was cheap, this imitation
of brandy could be made for less than 37½ cents per gallon.

"To make French brandy that can be sold for Cognac, Sazarac,
or Martell's, by varying the coloring:

97 gallons pure spirits,
7 pounds red argolls,
3 pounds acetic ether,
3 gallons wine vinegar,
7 pounds of bruised raisins,
1 ounce bruised bitter almonds.

"Distil this mixture, and add oak shavings, catechu and caramel to color, then throw in a few bits of old Russia leather, to
give the flavor of age.

"To make an imitation of pure old Monongahela whiskey:

40 gallons high-proof corn whiskey,
3 gallons tincture Guinea pepper,
40 gallons water,
1 quart tincture pellatory,
2 ounces acetic ether,
1½ gallons strong tea.

"This will produce from the forty gallons of corn whiskey,
about eighty-four gallons of what will be sold for pure old
Monongahela. The fusil oil not being rectified out of the
whiskey, the intoxicating quantity will be superior; and, aided
by the tincture of pellatory, disguised with pepper, ether, and
strong tea, enables the dealer to add largely of water, and also
to use cheap whiskey.

"To make Holland gin:

15 gallons proof spirit,
1 gallon gin essence,
1 quart white syrup.

"Mix thoroughly, and filter if necessary. This is simple, but
there is no gin in it, so it is a pure fabrication, as almost every
one of the imitation liquors are.

"Real imported Holland gin sometimes has sugar of lead
added to it, to give it a peculiar roughness and flavor, *which is
much esteemed* by some gin drinkers.

"The following is a favorite receipt for making a very high
flavored Holland gin, which is much admired by some gin drinkers, and it is no wonder so many have Bright's disease of the
kidneys:

80 gallons French spirits,
1 pint oil of turpentine,

3 ounces oil of juniper,
1 drachm essential oil of almonds. (*This is almost prussic acid.*)
2 ounces creosote. (*This is a deadly poison, for which no antidote is known.*)
Simple syrup enough to soften, and give the appearance of age.

" To make a very rich flavored French brandy.

100 gallons pure spirits,
2 quarts acetic ether,
4 ounces cassia buds,
2 ounces bitter almonds,
6 ounces orris root,
1 ounce cloves,
2 quarts white wine vinegar,
1 pound catechu,
2 gallons Jamaica rum,
1½ ounces cayenne pepper,
1 quart caramel for coloring.

Let it stand two weeks, occasionally stirring it.

" To make old London cordial gin of the highest quality:

"Ninety gallons of gin; oil of almonds one drachm; oils of cassia, nutmeg and lemon, of each two drachms; oils of Juniper, caraway, and coriander seed, of each three drachms; essence of orris root, four ounces; orange flower water, three pints; lump sugar, fifty-six to sixty pounds. The oils and essences must be dissolved in a quart of spirits of wine, and the sugar in three or four gallons of water. The essences must be added gradually to the gin, until the requisite flavor is produced, when the dissolved sugar must be mixed along with sufficient quantity of soft water, holding four ounces of alum in solution, to make up one hundred gallons. When the whole is perfectly mixed, two ounces salts of tartar, dissolved in two or three quarts of water, must be added, and the liquor again well rummaged or stirred up, after which it must be tightly bunged down, and allowed to repose. In a week or ten days it will have become brilliant, and ready for sale, or racking off and bottling. Many persons use this pretty freely for the benefit, as they suppose, of their kidneys; and such a compound or compounds must produce an effect not only on their kidneys, but also on every organ of the body.

" To make brandy which can be sold for pale or dark brandy:

40 gallons pure spirits (*common proof,*)
1 drachm Cognac oil, (*This is a deadly poison.*)
1 pint spirits of raisins,
1 pint spirits of prunes,
1 drachm tannin powder,
1 ounce acetic ether,
3 drops oil of neroli, dissolved in 90 per cent. alcohol.

" Color to make dark or light brandy, according to the market you are preparing it for. Fusil oil is found in nearly all the imitation brandies, showing that the whiskey used for the basis of them has been very imperfectly rectified.

" When real brandy is first distilled from wine, it is quite colorless, but after being kept some time in oak casks it becomes of a pale amber color, the color being derived from the wood. Very dark brandies owe their color to caramel, or burnt sugar. The characteristic taste of brandy is due to the presence of a volatile oil obtained from the skin of the grape.

"To make old Bourbon whiskey :

40 gallons pure rectified spirits,
½ pint of brandy coloring,
¼ pint of concentrated essence of Bourbon,
1 pound age and body preparation.

"Absinthe is one of the most deadly poisons, nevertheless they make a counterfeit absinthe as follows :

2 ounces of essence of absinthe,
4 ounces green coloring,
1 gallon of simple syrup,
4 gallons of rectified spirits.

"Here we have about 5 gallons of absinthe cordial, which contains 2 ounces of deadly poison, and 2 gallons of *pure* alcohol.

" To make Santa Cruz rum :

45 gallons N. E. rum,
5 gallons Santa Cruz rum,
4 drachms vanilla essence.

"To make Jamaica rum :

60 gallons proof spirit,
1 pound rum essence.

" This is simple and easy, but when we think we are drinking good old Jamaica rum, we are served with corn whiskey.

" Wines are as universally and as badly adulterated as the distilled liquors. In fact, prepared chemicals can be found in the stores of men who deal in these articles to make every kind of wine, with directions how to mix them. Whiskey is used as the basis for nearly all wines, and upon chemical analysis fusil oil is almost always found in counterfeit wines. What was sold by one of our respectable New York hotels for fine old port wine was analyzed, and found to contain 25 per cent. of alcohol, some *fusil oil*, extracts of cherry and elderberry, and some kind of coloring matter. This is a fine medicine to give sick persons to strengthen them.

" Receipts for making Madeira wine :
    20 pounds of figs, mashed up,
    50 pounds raisins,
    20 ounces linden or tilla flowers, with the leaves on,
    3 drachms of Turkish rhubarb,
    10 grains of cloves,
    3 gallons of sugar syrup.

" Infuse the above for ten days in 30 gallons of spirits, then add 90 gallons of water, and filter, and you have nearly 130 gallons of what is sold for pure old Madeira wine, without a drop of grape juice in it, but, upon analysis, fusil oil will sometimes be found.

" To make sherry wine :

100 pounds sugar,	200 gallons water,
40 gallons spirits,	70 gallons sherry wine.

" Color according to the kind of sherry you wish to imitate. Agitate and stir this mixture up for several days, and we have 230 gallons of what is sold for pure old sherry wine.

" A portion of the so-called champagne wines consumed in this country is composed of the expressed juice of turnips, apples, and other vegetables, to which sufficient sugar of lead is added to produce the necessary sweetness and astringency. The terrible headaches and depression of spirits that follow fashionable champagne suppers are attributable to the united poisons of lead and alcohol.

" Logwood is the great coloring matter for wines. Blackberries, elderberries, and bilberries are also used. Wines are adulterated with distilled spirits, lime salts, tannin, alum, lead, copper, cider, perry, etc. Port wine, as sold in the market, when not entirely counterfeit, is usually a mixture of pure port, or Marsala, Bordeaux, and Cape wines with brandy. Inferior port is still more highly adulterated with logwood, elderberries, catechu, prune juice, sandalwood, and alum.

" Many people suppose if they go to the Custom-house, and buy liquors in bond, under Custom-house lock and key, they will get them pure; but in this they are mistaken, for the liquors are as badly drugged in other countries as they are here. Professor Parkes gives an analysis of between forty and fifty of the different kinds of wines made in Europe. He says it has been stated that the fermentation of the grape, when properly done, cannot yield more than 17 per cent. of alcohol, and that anything beyond this has been added; and that some of the finest wines do not yield more than from 6 to 10 per cent. He found, upon analyzing the port, sherry, and Madeira wines in London, that the port ran from 16¾ to 23¼ per cent. alcohol ; the sherry from 16 to 25, and the Madeira from 16¾ per cent. to 22, and champagnes from 5½ per cent. to 13. The other wines averaged from 6¾ to 19 per cent. Mulder on 'Wine' (p. 186) quotes Guijal to the effect that pure port never contains more than 12¾ per cent., but Mulder doubts this. Dr. Gorman stated before a Parliamentary Committee that pure sherry never contains more than 12 per cent. of alcohol, and that from 6 to 8 gallons of alcohol is added to every 108 gallons of sherry. Some port used in the Queen's establishment contained but 16⅓ per cent., the highest was 18½; and the sherry only 16, and the clarets from 6¾ to 7 per cent. of alcohol. These were the purest wines to be found in London. Upon comparison we should find that the foreign wines in our market would show a much larger per cent., and as the corn whiskey they obtain from the United States is the cheapest form of alcohol they can procure, it is used for this purpose; and when not perfectly rectified, fusil oil will be found in the foreign wines.

"Thudichum and Dupré (on 'Wine,' p. 682) state that natural wine may contain 9, while the maximum limit is 16 per cent. of alcohol (of weight in volume.) They also state that a pipe of 115 gallons of port wine has never less than three gallons of brandy added to it, and the rich port wines have from thirteen to fifteen gallons added.

"I have not space to say more of the adulterations of wines and ardent spirits, but it is a system of fraud and deception the world over. Ales, porter and beers are as badly adulterated in this country and England as other liquors are.

"Professor Gallatin, of the chemical department of the Cooper Institute, has analyzed many samples of the best ales from the largest breweries in New York and vicinity, and others of the best reputation, and found none free from adulteration. He did not find as deadly drugs as the English brewers are said to use, or as the English books recommend, but he found that salt, alum, and lime are extensively used. 'The substances added to give 'head' to beer are alum, salt, and ferrous sulphate.' The effect of these adulterations on the consumer is very injurious. The cumulative effect of alum is to produce a general derangement of the digestive organs, and the diseases which grow out of it.

"The English works recommend coculus indicus, sweet flag-root, grains of paradise, alum, capsicum, absinthe, nutgalls, potash, and several other drugs. Dr. Beck, in his work on 'Adulterations,' asserts that they use strychine, opium, and hyosciamus, all deadly poisons. Keeping new ales until they are old is quite expensive, and they are converted into old ales cheaply and in a short time by adding oil of vitriol (sulphuric acid,) and the new ale acquires almost immediately the flavor of hard old ale, so much admired by beer drunkards.

" Coculus indicus is largely imported into England, ostensibly for tanners' use, although it is never used by them, but finds its way into the brewers' hands, in spite of a severe law against its use, and is used by them to give greater intoxicating effect to their beer, and by adding water they reduce its cost and retain the intoxicating properties. It is also imported into this country, and it is said that it is used by some of the American brewers for the same purpose. It is obtained from the seed or fruit of a shrub growing in the East Indies, and is imported in various sized packages. The trade-mark is ' B. E.,' meaning Black Extract. It is an acrid-narcotic poison. Dr. Taylor, one of the highest authorities on the subject of poisons, experimented with coculus indicus, and killed a rabbit with two drachms in two hours ; three drachms killed one in an hour, half an ounce in a quarter of a hour, and one ounce killed one in four minutes. It is also sometimes called fishberry, as the fishermen use it to cast into the water, and all the fish within reach of its influence become paralyzed, and float on the surface of the water, where they can be easily taken. Its poisonous effects more nearly resemble those of alcohol than any other known substance.

" The excessive use of malt liquors produces softening of the brain, and many other diseases. When it is adulterated its effects are always injurious, *and it is now so generally adulterated that the only safety is in letting it alone.*" *

Glycerine is now used to a large extent by the brewers, both for the purpose of preventing rapid fermentation, reducing the bitter taste of the " old or poor hops " sometimes employed, and making a " sweet, full beer."

Says an authority on this subject : " Glycerine is present in all fermented liquors, which fact was established by Pasteur, in 1859. An addition, therefore, of glycerine to beer will not be necessary, except in especial cases. Pasteur first used glycerine for the improvement of wine, in which respect its action was found so excellent that the attention of the brewers was called to its properties, and its use has since been considered with considerable favor by many brewers. * * * * Analyses made of various beers of Saxony, Bohemia and Bavaria, showed that Erlanger beer contained the largest percentage of glycer-

---

* Alcohol : Its Combinations, Adulterations and Physical Effects. By Col. J. G. Dudley, pp. 21-38.

ine. As is well known, glycerine, although it possesses a pure
sweet taste, is not capable of undergoing fermentation, and by its
addition to the beer the same acquires a sweet, full taste, and destroys the bitter taste which the beer acquires if a great amount
of hops has been used. * * * The amount of glycerine to
be used varies with the amount of hops which has been used,
from ¼ to 1 gallon for every 100 gallons of beer. Its careful use
not only improves its taste but also its keeping qualities. As
regards the amount of hops used in brewing, it will be observed
that a certain amount must at least be used in order to insure
the keeping qualities of the beer. If, however, old or poor hops
are to be used, the amount must be increased, whereby the bitterness as well as the keeping qualities are correspondingly increased. In order to neutralize this increased bitterness, an addition of glycerine offers the most ready means, and at the same
time most beneficial remedy. In regard to expenses, it will be
observed that by the rational use of glycerine they are not
materially increased, as the expense of the glycerine may be
covered by using correspondingly, less malt and hops, without
detracting from the quality of the article. It may be assumed
that one pound of glycerine represents within the beer the properties of three pounds of malt, and in using glycerine a corresponding reduction in the use of malt may be made. The
increased expense caused by the use of glycerine may also be
covered in another way, if the amount of malt is not to be diminished. Such beer may be sold at an increased price, as the
preference which it will find among consumers, who generally like a full beer which is not too bitter, will no doubt sustain
this course. Glycerine is a colorless, syrupy liquid, of a sweet
taste, and easily soluble in water. It is prepared from fatty
substances, which consist of glycerine and fatty acids. It is
therefore obtained in large quantities, as a by-product in the
manufacture of soaps and candles, and has for many years been
allowed to go to waste, but of late has been utilized for a great
many purposes." * Delicious soap-grease!

---

* "The Western Brewer," for Sept. 15, 1880, pp. 934–5.

CHAPTER II.

History of Intemperance, and its Political, Moral and Religious Effects, in China, India, Persia, Egypt, Greece, Rome; with the Jews and Contemporary Nations mentioned in the Old Testament, in Germany, Great Britain, and the United States.

INTEMPERANCE has been declared by an American Statesman to be "The gigantic crime of the age, and the great source of danger to our republic."* Unfortunately it is no recent evil, nor are its dangers less imminent in any country. † Yet it is impossible to give a full account of its extent, or to trace its origin in every instance of its existence in various ages and climes, since in some localities it was no doubt practised at a period prior to the begining of authentic history; in others the inferences with regard to its use are far fetched and inconclusive; and in still others the traditions are mere surmises, or unwarranted declarations, too recent in their origin to be of any value as in· timations of what was done in the distant past.

(A). The statement in Genesis ix. 20, 21, "And Noah began to be an husbandman, and planted a vineyard, and drank of the wine, and was drunken," etc., is regarded by able critics, as referring, not to some new thing in the way

---

* U. S. Senator Morrill.

† "Drunkenness," says the *Westminster Review*, " is the curse of England—a curse so great that it far eclipses every other calamity under which we suffer. It is impossible to exaggerate the evils of drunkenness."

of the culture of the vine, but to the revival of general husbandry, the vine having been cultivated before the flood. The criticism is both reasonable and just, and granting its correctness, the first mention of wine in the Bible does not pretend to take us back to its origin. As we shall conclusively see, still further on, its first mention among other people is traditional, long before it appears in authentic history.

(B). Somewhere from three thousand to seven thousand years ago—a conveniently wide margin of difference—there existed, according to modern authority,* a race of partially civilized men, who built their dwellings on piles driven into the beds of lakes in Switzerland. They practised agriculture, and were familiar with, if they did not cultivate, grapes, apples, pears, plums, cherries and barley, as charred and dried apples, and pears, stones of grapes and the other fruits, and whole ears of barley have been discovered among the traces and remains of their dwellings. It has been surmised by some that because fermented drinks can be made from these products of the soil, therefore the Lake Dwellers manufactured and used intoxicants. The inference is worthless.

(C). An ignorant confounding of the Indians of North America with the native wild men of the southern part of the continent—families wholly different and distinct—and attributing to the former customs which it is by no means certain that the latter ever established, has led to the unwarranted charge that our Aborigines were addicted to intemperance before their intercourse with the whites. The assertion is contradicted by all authentic history, and by every reliable tradition in regard to their primitive habits. There is no proof whatever, that they knew anything of any kind of intoxicants before the arrival of Europeans. There is no better authority on this subject than Rev. John

---

* Keller's Lake Dwellings, p. 344.

Heckewelder, Moravian Missionary to the Indians. He says:

" Of the manner in which they have acquired the vice of Intemperance, I presume there can be no doubt. They charge us in the most positive manner with being the first who made them acquainted with ardent spirits, and what is worse, with having exerted all the means in our power to induce them to drink to excess. It is very certain that the processes of distillation and fermentation are entirely unknown to the Indians, and that they have among them no intoxicating liquors but such as they receive from us. The Mexicans have their *Pulque*, and other indigenous beverages of an inebriating nature, but the North American Indians, before their intercourse with us commenced, had absolutely nothing of the kind." *

History is sufficiently full, however, of positive evidence of the existence of the vice of Intemperance at very early periods in human experience, and of its extent over vast portions of time and space, to furnish us with abundant data for the purposes of this chapter: a view of intemperance among nations before and after the birth of Christ, and its effects on religious and social life and on the State. Our real difficulty in this work will be to condense the great amount of historic material in hand, into the space in which it must be limited here, and still give a just view of the subject.

I. CHINA.—We begin with the people who claim to be the most ancient nation on the earth, the Chinese. It is impossible to deny their claim, and there is much to show their great age and their early civilization. Their great philosopher, Confucius, and his eminent disciple Mencius, the former flourishing in the fifth century before Christ, and the latter two centuries later, were not only eminent as teachers of their nation, for which they are greatly reverenced, but

---

* History, Manners, and Customs of the Indian Nations, etc. By Rev. John Heckewelder, chap. xxxvi. See also numerous authorities cited in Halkett's Historical Notes, respecting the Indians of North America, chap. viii.

also rendered most signal service by editing and perfecting two great works of historic value: "The Shoo-King, or History," and "The She-King, or book of ancient Poetry," a series of writings handed down through many generations, together with the commentaries written thereon by the ancient wise men.   From these books we learn that intemperance was frequently putting the Empire in danger, and that stringent measures for its suppression were often employed. The earliest account in the "Shoo-King," is the following, in the year 2187 B. C.

"T'ae-k'ang occupied the throne like a personator of the dead. By idleness and dissipation he extinguished his virtue, till the black-haired people all began to waver in their allegiance."

His five brothers visit him to endeavor to bring him back to virtue.   They call to his mind the counsels of the ancients.   "The second said :

'It is in the lessons :—
When the palace is a wild of lust,
And the country a wild for hunting:
When wine is sweet, and music the delight,
When there are lofty roofs and carved walls,—
The existence of any one of these things,
Has never been but the prelude to ruin.'" *

Another account dates 2154 or 2127, B. C., in the reign of Chung-k'ang :

"He and Ho had neglected the duties of their office, and were sunk in wine in their private cities, and the prince of Yin received the imperial charge to go and punish them." He and Ho were Ministers of the Board of Astronomy, but through their licentious indulgences unfitted themselves for their duties, and in consequence, the people, dependent on them for knowledge of the times and seasons, received no light and guidance. An eclipse comes on them unawares, and the Astronomers are too much intoxicated to notice it. The prince of Yin assembles his troops, and thus addresses them: "Ah! ye, all my troops, these are the well-counselled instructions of the sage founder of our dynasty, clearly verified in their power to give security

* Shoo-King, Bk. iii. ch. i. 6.

and stability to the State : 'The former kings were carefully attentive to the warnings of Heaven, and their ministers observed
the regular laws of their offices. All the officers, moreover,
watchfully did their duty to assist the government, and the
sovereign became entirely intelligent.' Every year in the first
month of spring, the herald, with his wooden-tongued bell goes
along the roads, proclaiming, ' Ye officers able to direct, be prepared with your admonitions. Ye workmen engaged in mechanical affairs, remonstrate on the subject of your business! If any
of you disrespectfully neglect this requirement, the country has
regular punishments for you.' Now here are He and Ho. They
have entirely subverted their virtue, and are sunk and lost in
wine. They have violated the duties of their office, and left
their posts. They have been the first to allow the regulations
of Heaven to get into disorder, putting far from them their proper business. On the first day of the last month of autumn, the
sun and moon did not meet harmoniously in Fang. The blind
musicians beat their drums; the inferior officers and common
people bustled and ran about. He and Ho, however, as if they
were mere personators of the dead in their offices, heard nothing
and knew nothing;—so stupidly went they astray from their
duty in the matter of the heavenly appearances, and rendering
themselves liable to the death appointed by the former kings.
The statutes of government say, when they anticipate the time,
let them be put to death without mercy ; when they are behind
the time, let them be put to death without mercy." *

Again, 1122 B. C. The Emperor Chow becomes dissolute, "being lost and maddened with wine." His pernicious
example is so generally followed that the Viscount of Wei
finds it impossible to rule in his Principality. He therefore
seeks the Grand and Junior Tutors and inquires what can
be done. They give him no help. The dynasty is too
corrupt to be changed, and nothing but its overthrow can
be looked for. The Viscount is advised to flee and save
his life, while the Grand Tutor resolves to stay and share
in the death which may come to all who are in the government :

"King's son, Heaven in anger is sending down calamities,
and wasting the country of Yin. Thence has come about that

lost and maddened condition through wine. He has no reverence for things which he ought to reverence, but does despite to the aged elders, the old official fathers. Now the people of Yin will even steal away the pure and perfect victims devoted to the spirits of heaven and earth; and their conduct is connived at; and though they proceed to eat the victims they suffer no punishment. On the other hand, when I look down and survey the people of Yin, the methods of government to them are hateful exactions, which call forth outrages and hatred, and this without ceasing. Such crime equally belongs to all in authority, and multitudes are starving with none to whom to appeal. Now is the time of Shang's calamity; I will arise and share in its ruin. When ruin overtakes Shang, I will not be the servant of another dynasty. But I tell you, O King's son, to go away, as being the course for you. Formerly I injured you by what I said, but if you do not go forth now, our sacrifices will entirely perish. Let us rest quietly in our several parts and present ourselves to the former kings. I do not think of making my escape.*

In " the She-King," are many allusions to the habits of the people. The following are descriptions of the ways of the settlers in Pin, under King-lëuz, B. C. 1496–1325 :

> " In the tenth [month] they reap the rice,
> And make the spirits for the spring,
> For the benefit of the bushy eyebrows."

This is interpreted to mean that the spirits distilled from rice cut down in the tenth month, would be ready for use in the spring ; and that the use of spirits was restricted to the aged. In another poem, however, allusion is made to the custom of drinking healths :

> " In the tenth month they sweep their stack-sites,
> The two bottles of wine are enjoyed,
> And they say, ' Let us kill our lambs and sheep,
> And go to the wall of our prince,
> There raise the cup of rhinoceros horn,
> And wish him long life, that he may live forever." †

Mencius says (about 300 B. C.): "There are five things which are said in the common practice of the age to be unfilial. * * *

---

* Shoo-King, Bk. xi. 1- 9.
† She-King, Part I., Bk. xv., Ode I.

The second is gambling and chess-playing and being fond of wine, without attending to the nourishment of his parents." *

Of the extent of drinking in modern China it is impossible to speak with accuracy. The art of distillation was known and practised there somewhere between the tenth and sixteenth centuries, and both fermented and spirituous liquors are imported into the country. But travellers do not agree in their statements and opinions as to the extent of intoxication among the people; some insisting that instances of it are very rare, and others declaring that it is scarcely less prevalent than among Europeans, although those who imbibe are cautious how they exhibit themselves to the public while under the influence of liquor. Of the influences which have produced this changed condition, we shall speak elsewhere.

II. INDIA.—For our knowledge of the drinking customs of ancient India, we are indebted to the writings of the Brahmans, called the Rig-Veda, or Sacred Books, which were brought together about 400 years B. C., but are supposed to have been composed in a remote antiquity, the nearest date of which to our own time, is 1200 B. C.† These books contain the Ancient Hymns, as recited or sung by the priests when engaged in their official duties. Their religious ceremonies were chiefly sacrificial, and the principal sacrifice was called " Soma," after an intoxicating drink made from the juice of the creeping plant *Asclepias acida.* This plant, after being cleaned and macerated in water, was pressed between stones, and the juice, strained through ram's wool, was mixed with malt and clarified butter, and then fermented. The sacrifice was made by pouring the fermented liquid on the sacred fire, where it was supposed to be drank by the gods. Sometimes it was be-

---

* Bk. iv., Pt. II., Ch. xxx.

† According to Dr. Haug, Superintendent of Sanskrit Studies in the Poona College, Bombay, the oldest hymn in the Rig-Veda is to be placed between 2000 and 2400 B. C.

lieved to be miraculously transformed into the god himself, and so is occasionally addressed as a person. Indra was the god to whom the Soma was most frequently offered, and unless he was intoxicated with it nothing was expected from him, while all his great exploits were said to be due to his "exhilaration with Soma." "Indra delights in it from his birth : lord of bay horses, we wake thee up with sacrifices : acknowledge our praises in the exhilaration of the Soma beverage." "Be exhilarated by the Soma. The Soma is effused, the sweet juices are poured into the vessels; this propitiates Indra." "Indra comes daily seeking for the offerer of the libation. The pleasant beverage that thou, Indra, hast quaffed in former days thou still desirest to drink of daily : gratified in heart and mind, and wishing our good, drink, Indra, the Soma that is placed before thee. As soon as born, Indra, thou hast drunk the Soma for thine invigoration. I proclaim the ancient exploits of Indra, the recent deeds that Maghavan has achieved : when indeed he had overcome the divine illusion, thenceforth the Soma became his exclusive beverage." "Indra verily is the chief drinker of the Soma among gods and men, the drinker of the effused libation, the acceptor of all kinds of offerings; whom others pursue with offerings of milk and curds as hunters chase a deer with nets and snares, and harass with inappropriate praises." "When thou hast expelled the mighty Ahi from the firmament, then the fires blazed, the sun shone forth, the ambrosial Soma destined for Indra flowed out, and thou, Indra, didst manifest thy manhood." *

James Samuelson, in his "History of Drink," p. 38, refers to Langlois' translation of the Rig-Veda as authority for asserting that, "Just as in one of the Hebrew Psalms every verse ends with the words, 'For his mercy endureth forever,' so in one hymn to Indra, each verse concludes as follows : 'In the intoxication which Soma has caused him, see what Indra has accomplished.' "

---

* Rig-Veda Sanhita, translated by H. H. Wilson, M. A., etc. Edited by E. B. Cowell, M. A., pp. 67, 72, 195, 219, 22?.

Sometimes another god is associated with Indra in the Soma sacrifice: "May the prayers that are repeated to you, reach you, Indra and Vishnu; may the praises that are chaunted reach you; you are the generators of all praises, pitchers recipient of the Soma libation." "Indra and Vishnu, agreeable of aspect, drink of this sweet Soma, fill with it your bellies; may the inebriating beverage reach you: hear my prayers, my invocation." "Indra and Varuna, observant of holy duties, drinkers of the Soma juice, drink this exhilarating effused libation; sitting on the sacred grass, be exhilarated by the draught." "The prompt effuser of the libation offers the Soma to Indra and to Vayu to drink at the sacrifices, at which devout priests, according to their functions, bring to you two the first portion of the Soma." "Come with gracious minds, Indra and Agni, to this our Soma libation: Ye are never regardless of us, therefore I propitiate you with constant sacrificial viands. Utter destroyers of Vritra, exhilarated by the Soma, you who are worshipped with hymns and prayers and songs, come hither, destroy with your fatal weapons the mortal who is malignant, ignorant, strong, rapacious, destroy him like a water jar, with your weapons." *

Other gods, as Mitra, the Maruts, Aryaman, the Aswins, and Sakra, are called upon in the Soma festivity, and all are said to be endowed with wonderful capacity for containing and enjoying the beverage, to owe their power to it, and to be expected to grant favors to mortals only as they are well supplied with Soma.

"That you may drink the sacrificial beverage, you come promptly upon this my invitation." "When the stone, seeking to propitiate you two divinities, is raised aloft, and loudly sounds, expressing for you the Soma juice, then the pious worshipper brings you back, beautiful divinities, by his oblation." "The divine Soma juices, flowing like water, self-renowned at religious assemblies, support Indra and Varuna." "Praise together Indra, the showerer of benefits when Soma is effused."

* Ibid, pp. 16, 17, 95, 140, 185, 187, 189.

"Come hither, Indra, be exhilarated by the wonderful liba- tory affluence, and with thy fellow-topers, the Maruts, fill with the Soma juices thy vast belly, capacious as a lake." "Indra, drink this effused libation till thy belly is full." "The pota- tions of Soma contend in thy interior for thine exhilaration like the ebriety caused by wine : thy worshippers praise thee, filled full of Soma like the udder of a cow with milk." "Quickly, priest, pour forth the Soma, for Indra is thirsty; verily he has harnessed his vigorous steeds, the slayer of Vritra has arrived. Pour out, priests, the Soma libations to Indra, in his chariot : the stones, placed upon their bases, are beheld effusing the Soma for the sacrifice of the offerer." "Indra, when the Soma juices are effused, sanctifies the offerer and the praiser." "Thou, Indra, the most excellent drinker of the Soma [or it may mean, says Cowell, "thou who on drinking the Soma becomest preeminent,"] destroyest the adverse assembly that offers no libations." *

That the people, as well as the gods, were partakers of the Soma, is evident from their so definite descriptions of its effects ; their desire that the gods may not simply partake of it as offered by them, but may sit down on the grass and participate with the offerers in their libations; from their declaration that Indra is the chief drinker among gods and men ; and also from one of their prayers to Soma personi- fied :

"Where wishes and desires are, where the bowl of the bright Soma is, where there is food and rejoicing, there make me im- mortal." † "When the meal was prepared, they strewed the eating place with sacred grass, and invited the gods to take their seats and drink their fill. They then poured a portion of their food on the sacred fire, which was personified as a divine messenger who carried the sacrifice to the several deities; and when this was done the family apparently sat down and feasted on the remainder." ‡

The Soma sacrifices are now very rarely offered in any part of India. Their disuse will be accounted for in our

---

* Ibid. pp. 4, 148, 173, 211, 216, 218, 220, 232, 271, 279.
† Max Müller's Chips from a German Workshop, I., p. 46.
‡ History of Indra from the earliest ages, by J. Talboys Wheeler, Vol. III. p. 17.

sketch of the History of Efforts to suppress Intemperance. 'Dr. Haug, desiring to avail himself of all possible aids to the correct understanding of the Vedas, says :

"Seeing the great difficulties, nay, impossibility of attaining to anything like a real understanding of the sacrificial art from all the numerous books I had collected, I made the greatest efforts to obtain oral information from some of those few Brahmans who are known by the name of 'Srotriyas' or 'Srautes,' and who alone are the possessors of the sacrificial mysteries as they descended from the remotest times. The task was no easy one, and no European scholar in this country before me ever succeeded in it. This is not to be wondered at; for the proper knowledge of the ritual is everywhere in India now rapidly dying out, and in many parts, chiefly in those under British rule, it has already died out."

Müller continues: " Dr. Haug succeeded, however, at last, in procuring the assistance of a real Doctor of Divinity, who had not only performed the minor Vedic sacrifices, such as the full and new moon offerings, but had officiated at some of the great Soma sacrifices, now very rarely to be seen in any part of India. He was induced, we are sorry to say, by very mercenary considerations, to perform the principal ceremonies in a secluded part of Dr. Haug's premises. This lasted five days, and the same assistance was afterwards rendered by the same worthy and some of his brethren whenever Dr. Haug was in any doubt as to the proper meaning of the ceremonial treatises which give the outlines of the Vedic sacrifices. Dr. Haug was actually allowed to taste that sacred beverage, the Soma, which gives health, wisdom, inspiration, nay immortality, to those who receive it from the hands of a twice-born priest. Yet, after describing its preparation, all that Dr. Haug has to say of it is this: "The sap of the plant now used at Poona appears whitish, has a very stringent taste, is bitter, but not sour ; it is a very nasty drink, and has some intoxicating effect. I tasted it several times, but it was impossible for me to drink more than some tea-spoonsful." *

The drink of the common people of ancient India is also mentioned in the Vedas. It is called Sura, and in the earliest ages it was made from a tall native grass, curds, honey, melted butter, barley, and water ; later, rice, black pepper,

* Chips from a German Workshop, Vol. I., pp. 103, 104.

barley, lemon juice, ginger and hot water, entered into its composition. It was in very general use, was much more intoxicating than Soma, and in one of the hymns is confessed to be the cause of sinful debasement: "It is our condition that is the cause of our sinning; it is intoxication." "Sura, literally wine," adds Mr. Cowell.* Even after severe penalties were attached to intemperance, and all use of the bowl was denounced by the Laws of Manu, the vice prevailed to such an extent that—

"Palastya, an ancient sage, enumerates no less than twelve different kinds of liquor besides Soma; and the preparation of those drinks from the grape, from honey, sugar, dates, the palm, pepper, rice, cocoa-nut, etc., has been described with considerable minuteness. Besides these home-made drinks, large quantities of foreign wines were imported into India two thousand years ago, and met with a ready sale throughout the country." †

Morewood states that when, in 640, A. D., the trade of India was transferred from the Egyptians to the Saracens, and the Mussulmans would carry on no commerce in wine, the Indians manufactured their intoxicants from various substances, the chief of them being the fermented juice of the palmin tree, and called Tari. ‡

Modern travellers tell us that intoxication in India to-day, is chiefly among the lowest castes and the half-castes, the higher orders very generally abstaining from all inebriating drinks. The common arrack, distilled from rice, is used most; although the very lowest and besotted drink a fiery compound called Pariah arrack, the distilled juice of the palm and the thorn apple, a powerful narcotic. Rousselet describes an annual debauchery in the Spring season, when under the guise of religion, all classes in India give themselves up to beastly drunkenness.

"The carnival," he says, "lasts several days, during which the most licentious debauchery and disorder reign throughout

---

* Wilson's Rig-Veda, p. 175.
† The History of Drink, by James Samuelson, p. 42.
‡ Morewood on Inebriating Liquors, p. 71.

every class of society. It is the regular saturnalia of India. Persons of the greatest respectability, without regard to rank or age, are not ashamed to take part in the orgies which mark this season of the year." " Troops of men and women, wreathed with flowers, and drunk with bang, crowd the streets, carrying sacks full of a bright red vegetable powder. With this they assail the passers-by, covering them with clouds of dust, which soon dye their clothes a startling color." "Never have I seen so revolting a spectacle. Groups of native wretches dead drunk were wallowing in the gutters, and at every step the most disgusting debauchery was exhibited with unblushing effrontery." *

III. PERSIA.—The beginnings of the history of the people who established the Persian Empire are involved in no little obscurity, and much that is accepted as true in regard to the place from which the settlers came, as also the causes of their emigration, is conjectural. But it seems well proven that they brought a religion with them having in form many features in common with the religion of India, though in spirit containing much and aiming at much in wide contrast with the Brahminical writings. Their leader and most renowned man was, he claimed, and they believed, favored with a revelation from the Supreme God, Ormazd, who directed that they should call themselves the Mazday-asnas, the people of Ormazd. The book containing the revelation, the Zend-Avesta, which means Zend-translation, sometimes commentary, Avesta—sacred † writings is made up of the Sacred Law, Invocations and Hymns; the former, being much the larger portion, is put into the form of a conversation between Zarathustra, whom the modern Persians call Zerdusht, and whom we know by the Greek translation, as Zoroaster, and Ormazd, the former asking and the latter answering questions.

When Zoroaster flourished is unknown. " Eudoxus declares," says Pliny, " that this Zoroaster lived six thousand

* Quoted in Samuelson's History of Drink, p. 5.
† Essays on the Sacred Language, etc., of the Parsis, By Martin Haug, Ph. D., p. 120.

years before the death of Plato. So also, Aristotle. Hermippus, who wrote with the utmost care on the whole art, and commented on two million verses composed by Zoroaster, and prepared indexes of his works, reports that Azonaces was the teacher by whom he was instructed, and that he lived 5000 years before the Trojan war." [*]

Bunsen suggests " that the date of Zoroaster fixed by Aristotle, cannot be said to be so very irrational; but he adds : " At the present stage of the inquiry, the question whether this date is set too high cannot be answered either in the negative or affirmative." [†]

Spiegel, one of the translators of the Zend-Avesta, considers Zoroaster as a neighbor and cotemporary of Abraham; Rapp, in his Religion of the Persians, concludes, after a thorough comparison of ancient writers, that Zoroaster lived B. C. 1200 or 1300; while Prof. Whitney, of New Haven, places him at least B. C. 1000. The range is therefore a wide one, from 1000 to 6350 B. C.

Like the Brahmins, the Mazdayasnas offered sacrifices of an intoxicating beverage to their gods. This drink they called Haoma, or Homa, and sometimes Parahoma, and as was also the case with the Brahmins, the name was given to the tree or plant, and to the god, as well as to the beverage itself. It is supposed, and not without good reason, that Soma and Homa are identical, as the initial S of Sanscrit is always represented by H in Zend, an indication that in remote antiquity the ancestors of the two people were one family. The Hom or Homa tree is said " to grow on the tops of mountains in Gilân, Shirvân, Mazenderân, and according to Antequil, the Parsees of India still send from time to time one of their priests to Kirmân for cuttings." [‡] It is often praised in the Avesta for the golden color of the liquid. A White Homa, a mystical plant, sometimes called

---

[*] Nat. Hist., Bk. xxx. ch. 1.
[†] Egypt's Place in Universal History, Vol. iii. p. 471.
[‡] Spiegel's Arvesta, ii. lxxii.

Gaokerena, possessing even greater virtues than the real one, inasmuch as whoever tastes of it becomes immortal, is also mentioned." *

In the ninth Yasna, Spiegel's translation, the god Haoma appears to Zarathustra, "At the time of the morning dawn, as he was purifying the fire and reciting the Gâthâs," and calls upon him : "Praise me with songs of praise." Zarathustra having complied, enters into conversation with the god, asking, "Who first, O Haoma, prepared thee in the corporeal world ? What holiness thereby became his share ? What wish was bestowed on him ?" An answer being given, as also to the inquiry who the second and the third were that had prepared him in the corporeal world, he questions in the same form in regard to the fourth man, and is answered :

"Pourúshâspa has prepared me as the fourth man in the corporeal world ; this holiness became thereby his portion, this wish was fulfilled to him :

"That thou wert born to him, thou pure Zarathustra, in the dwelling of Pourushâspa, created against the Daevas, devoted to the belief in Ahura.

"The renowned in Airyana-vairya, which spreads itself abroad four-fold.

"Afterwards the other prayer with mighty voice.

"Thou madest that all the Daevas hid themselves in the earth, O Zarathustra, which before were going about on the earth in the shape of men.

"Thou, the mightiest, strongest, most active, swiftest, the most victorious among the heavenly beings.

"When answered Zarathustra : Adoration to the Haoma !

"Good is Haoma, well-created is Haoma, rightly created is Haoma.

"Well-created and health-bringing.

"Gifted with good body, rightly acting.

"Victorious, golden, with moist stalks.

"He is very good when one eats him, and the surest for the soul.

"Thy wisdom, O Golden, praise I ;

"Thy powers, thy victory,

---

* Vendidad, xx. 17.

"Thy healthfulness, thy healing power,
"Thy furtherance, thy increase.
"Thy powers in the whole body, thy greatness in the whole form.
"Praise that I may go about the world as Ruler, paining the tormenters, smiting the Drujas;
"That I may torment all the torments, the tormenting Daevas and men."

The fourteenth Vispered is a prayer or ascription to be recited by the priests while preparing the Homa; and among the necessary utensils of the priest, "the cup for the Homa" is mentioned.* It is offered up "for satisfaction to the good waters created by Mazda;" "for satisfaction to the Fravashi (the soul) of the holy Zarathustra," and for "all departed souls;" "for praise to Ahura-Mazda" (Ormazd); "in prayer for strength to those who fight the demons;" in propitiation of all the gods;† at the sacrifice for the purification of the land, of the killer of a dog, and of the licentious.‡ It is the only thing incapable of defilement by being brought in contact with the unclean: "The prepared Haoma has neither dissolution nor death; not even when it is brought to a dead body." And it is one of the mightiest weapons with which to fight the demons.§ The laity are to offer it to all the genii of the waters, the stars, and the cattle, and are to praise it continually.|| Homa curses the person, the dwelling and the posterity of those who do not prepare it, or who hinder others in their preparation of it.¶

From the foregoing quotations, and they are fair specimens of what is said of Homa in the Zend-Avesta, it appears that the offerings as well as the praises of the beverage, were made to the good god and his assistants; but Plutarch conveys the idea that the offerings were made to the evil god and the demons, for the purpose of averting

---

* Vendidad, xiv. 31.   † Yasna, vii. x. xii. xxiii. xxiv. lxvii.
‡ Vendidad, ix. xiii. xviii.   § Vendidad, vi. xix.; Yasna, x.
|| Kordash-Avesta, xxi.–xxvi. xl.   ¶ Yasna, xi.

their wrath : " They beat a certain plant called Homomi, in a mortar, and call up Pluto and the dark ; and then mix it with the blood of a sacrificed wolf, and convey it to a certain place where the sun never shines, and there cast it away." * But whatever the intent with which the Homa was sacrificially used, it is evident that, unlike the use of the Soma, it was not to be employed either by priests or by laity for the purpose of producing their intoxication ; it was wholly for the gods. Drunkenness was supposed to be the work of Ahriman, the god of darkness and evil, and therefore was forbidden by Ormazd.

Two other intoxicants were known to the people, and in spite of injunctions to the contrary, were employed in producing drunkenness. The one, Hura, identical probably in quality as in name with the Brahmin Sura, and the other, Banga, sometimes denounced in the Vendidad as producing abortions, and sometimes represented as one of the three demons who are ever hostile to man.†

Later in the history of Persia, the vine seems to have been cultivated, intoxicants became more common, and intemperance increased. Sir James Malcom, in his History of Persia, quotes from the MSS. of Moullah Ackber, to the effect that Jem Sheed, the founder of Persepolis, was passionately fond of grapes, and desiring to have some always easy of access, concealed a large quantity in a vault. Great was his surprise on visiting his treasure to find that much of the mass had been crushed, and that the escaped juice was so acid that he believed it to be poisonous. Not knowing what might yet come of it, he filled some vessels with the liquid and stored them in his own apartment, labelling them " Poison." A favorite concubine suffering from nervous debility, meditated suicide as the only relief from her malady, and seeing the vessels of poison, opened one and swallowed its contents. Stupefied by the draught, she fell

* Plutarch's Morals, Vol. iv. Art. Isis and Osiris.
† Vendidad, xv. xix.

into a sleep, and on waking was delighted to find herself
free from pain. Charmed with the sensations experienced,
she continued her experiments till she had drunk up all the
monarch's poison. Confessing the theft, and describing
the delightful effects which it produced, and her thorough
restoration to health by its use, Jem Sheed caused large
quantities of grapes to be gathered and left in a bruised
condition in larger vessels, and soon his entire court sung
the praises of the Zeher-e-koosh, or " the delightful poison,"
as they named it.

Hafiz, the favorite poet of the Persians, thus sings the
praises of wine :

That poignant liquor which the zealot calls the mother of sins,
is pleasanter and sweeter to me than the kisses of a maiden.

" The only friends who are free from care are a goblet of wine
and a book of odes.

" The tulip is acquainted with the faithlessness of the world ;
for from the time that it blows till it dies, it holds the cup in its
hand.

" Give me wine! wine that shall subdue the strongest ; that I
may for a time forget the cares and troubles of the world.

" The roses have come, nor can anything afford so much
pleasure as a goblet of wine.

" The enjoyments of life are vain ; bring wine, for the trap-
pings of the world are perishable." *

Herodotus, who wrote about 450 B. C., says that " the
Persians are much addicted to wine. They are used to
debate the most important affairs when intoxicated ; but
whatever they have determined on in such deliberations, is,
on the following day, when they are sober, proposed to
them by the master of the house where they have met to
consult ; and if they approve of it when sober also, then
they adopt it ; if not, they reject it ; and whatever they
have first resolved on when sober, they reconsider when intoxicated." He also relates that when Cyrus, about 538
B. C., made war upon the Massagetæ, of Central Asia, he

---

* Morewood on Inebriating Liquors, p. 61.

made a feint of deserting his camp, leaving in it flowing goblets of wine, which tempting the enemy to excess, Cyrus attacked them and gained a victory; also that Cambyses, son and successor of Cyrus, sent, among other gifts, a cask of palm wine to the king of Ethiopia.*

Although, as will be shown hereafter, the rules of the Mohammedan religion are acknowledged by the modern Persians, their earlier successes as conquerors of Babylon, their union with the Medes, a people of luxurious habits, addicted them to intemperance; and leading them to the cultivation of the vine for the purpose of obtaining wine, their ancient Empire was overthrown, they having become in two hundred years from the conquest of Babylon, the most drunken nation on the earth, entailing on their descendants a love and practice of inebriation. They are much less strict Mohammedans than are other nations that have adopted the creed of the Prophet, and their wines are celebrated for their abundance, strength and flavor. Sir J. Chardin, who travelled extensively among them, states that "as much as a horse can carry of their best wines can be purchased for twelve shillings, and the more common sorts do not cost more than half that money." Attending an entertainment at the house of a royal prince, he describes their manner of drinking as follows:

"The prince's nearest relations, selecting about eight in number, were first presented with vessels of wine, which they drank standing up. The same bowls being filled again, were carried to the next persons, and so on, until the health had been drunk round. After this, the next health was drunk in larger cups, for it was the custom of the country to drink the healths of great personages in large vessels. This was done on purpose to make their guests more effectually drunk. This desired climax would soon be attained, when we consider the size of their glasses. The first glasses used were of the common sort, but the last contained about a pint and a half of wine." †

---

* Herodotus, i. 133, 211. iii. 20.
† Sir J. Chardin's Travels, pp. 228, 229.

Tavernier, another traveller, bears witness to the same excess in Armenian Persia. No man who gives an entertainment considers that he has shown true hospitality till he has made his guests so drunk that they cannot find their way out of the room. The more they reel and stagger about, the less reason has he to regret the expense of the feast. He also says of the Persian Georgians, that their use of stimulants is so common that on entering the dining-room each guest is presented with a half-glassful of aqua vitæ, to excite his appetite. Wine, though the native drink of the country, soon fails there, as elsewhere, to satisfy the toper. "They love the strongest drinks best, for which reason, both men and women drink more aqua vitæ than wine. It is also observable that at the women's festivals, there is more wine and aqua vitæ drunk than at the men's."* A drink prepared from herbs, and made more intoxicating by an infusion of hemp seed, was in use at that time. It was called Bengueh, and was no doubt similar to if not identical with the Banga of the Avesta. More recent travellers speak of the great quantities of arrack, a fiery and rapid intoxicant, consumed in Persia. Rev. J. H. Shedd, writing from Oroomiah, Persia, says of the Protestant missionaries in that province:

"We have never found wine an ally to the temperance cause, though it flows around us almost as cheap and abundant as water. During the wine season beastly drunkenness is too common to excite comment. I have been in large villages on a feast-day, when it was nearly impossible to find a sober man in the place. The corruption of morals, the degradation of mind, the midnight carousals, the losses from riotous living, from idleness, quarrelling, and crime, are too enormous to be exaggerated. The wine-weddings, with their train of evils, are the enemy of the Christian peasant, and the source of debt and misery that often crush him, and break up his home. Many acquire the passion for stimulants, and pass from wine to *arrack*, a rum distilled from raisins. Thus wine is a Mocker, and multitudes are in the road to ruin through the curse of Strong Drink.

---

* Persian Travels, vol. i. p. 243.

Among the nominal Christians of Persia, and many other places of the East, the worst destroyer of the soul and obstacle to the Gospel is wine, and the attendant intemperance." *

"A Moslem prince lately asked me," says Arthur Arnold, "why I drank wine? It does not make you drunk. I take arrack." † Many excuse themselves for drinking ardent spirits, on the ground that it is the use of wine alone that Mohammed prohibits.

The Parsees of India, not over 110,000 in number, according to their great champion, Framjee,‡ claim to be the descendants, in point of faith, of the ancient Zoroastrians. Concerning them, Framjee, while conceding that such as can afford it, drink large quantities of wine at supper, denies that they drink intoxicants during the daytime. No doubt, then, the nights are long, and faithfully devoted to drinking, else how could their "826 tavern keepers, and 5,227 liquor sellers, distillers, and palm-wine drawers," which he enumerates, (and against whom he places but "417 bakers and confectioners,") find patronage?

IV. EGYPT.—The testimony of ancient writers in regard to the intoxicating drinks made and used by the Egyptians, is conflicting, and therefore of little worth. Herodotus says: "They use wine made of barley, for they have no vines in that country." § But previous to this, speaking of the advantages enjoyed by the priests, he says: "Sacred food is cooked for them, and a great quantity of beef and geese is allowed each of them every day, and wine from the grape is given them."|| His statement in regard to "no vines in the country," must therefore refer to that part of the country

---

* Quoted by Rev Dawson Burns, in Christendom and the Drink Curse, pp. 219, 220.

† Through Persia by Caravan, Vol ii. p. 322.

‡ The Parsees, by Desabhoy Framjee. Quoted by Samuelson, p. 57.

§ Bk. ii. 77.

|| Ibid. Bk. ii. 37.

" which is sown with corn," the part specified at the commencement of the paragraph. He also identifies Osiris with the Greek Bacchus, and claims that the Egyptians also made them identical. Plutarch's testimony is to the same effect. * The latter says of the use of wine among the Egyptians, that their kings, being also priests—

" Began first to drink it in the reign of King Psammeticus, but before that time they were not used to drink wine at all, no, nor to pour it forth in sacrifice, as a thing they thought anyway grateful to the gods, but as the blood of those who in ancient times waged war against the gods, from whom, falling down from heaven, and mixing with the earth, they conceived vines to have first sprung; which is the reason, say they, that drunkenness renders men beside themselves and mad, they being, as it were, gorged with the blood of their ancestors." †

But Homer, who flourished about 1000 B. C., and 400 years before Psammeticus, refers the invention of drugged wines to Egypt :

" But Helen now on new device did stand,
  Infusing straight a medicine to their wine,
  That, drowning cares and angers, did decline
  All thought of ill. Who drank her cup could shed
  All that day not a tear, no not if dead
  That day his father and his mother were,
  Not if his brother, child, or chiefest dear,
  He should see murder'd then before his face.
  *   *   *   *   *   *   *
  And this juice to her Polydama gave,
  The wife of Thoon, an Egyptian born." ‡

And we find from Genesis xl. 11–13, that the vine supplied grapes for the king's table in the time of Joseph, 1876 B. C., and that it was the duty of the butler to press the grapes into Pharaoh's cup, and then deliver the cup into Pharaoh's hand. This, it is true, produced an unintoxicating beverage, but it shows that the vine was not then regarded with contempt.

---

* Morals, Vol. iv.   Article, Isis and Osiris, p. 79.
† Ibid. p. 71.
‡ Odyssey, Bk. iv.

Hellanicus the historian, about 400 B. C., says that

"The vine was first discovered in Plinthina, a city of Egypt; on which account Dion, the Academic philosopher, calls the Egyptians fond of wine and fond of drinking; and also, that as subsidiary to wine, in the case of those who, on account of their poverty, could not get wine, there was introduced a custom of drinking beer made of barley; and moreover, that those who drank this beer were so pleased with it that they sung and danced, and did everything like men drunk with wine." *

It also appears from the Monuments, that the cultivation of grapes and the art of wine-making were well understood in Egypt from the time of the Pyramids, according to Bunsen, 3229 years B. C., or according to Lepsius, 3426 B. C.† And although Herodotus and Plutarch and other authorities differ as to the extent of the use of wine in the sacrificial rites, it is evident, from the delineations on the most ancient frescoes, that no restrictions were put upon its use by men or women in social and private life. Wilkinson gives several illustrations of these pictures,‡ in some of which servants are carrying their insensible masters home from a drinking frolic, while the female attendants on their wives and daughters, are represented as supporting them as they sit at the feast, unable without such help to prevent themselves from falling on those seated beside or behind them, and often so sick in their debauch as to be unable to conduct themselves with decency. So ambitious were they in their gross rivalries as to who should imbibe the most, that various articles of stimulating food were placed on their tables, intended to create thirst and otherwise excite the palate. So great was the consumption of wine, that, in the time of Herodotus, large importations were received twice a year from Phœnicia and Greece.

"Egyptian beer was made from barley; but as hops were unknown, they were obliged to have recourse to other plants, in

---

* Athenæus, Bk. i. chap. 61.
† Egypt, Past and Present, by Joseph P. Thompson, p. 349.
‡ "A Popular Account of the Ancient Egyptians, by Sir J. Gardner Wilkinson." Vol. I. pp. 52, 53.

order to give it a grateful flavor; and the lupin, the skirret (*Sium sisarum*), and the root of an Assyrian plant were used by them for that purpose.

"Besides beer, the Egyptians had what Pliny calls factitious, or artificial wine, extracted from various fruits, as figs, myaxas, pomegranates, as well as herbs, some of which were selected for their medicinal properties. The Greeks and Latins comprehended every kind of beverage made by the process of fermentation, under the same general name, and beer was designated as barley-wine; but by the use of the name zythos, they show that the Egyptians distinguished it by its own peculiar appellation." *

V. GREECE AND ROME.—Entering now, as our field of observation, those classic regions with whose customs all are more or less acquainted, the literature of the people being in part an enforced study in the curriculum of a liberal education, we have to do with two of the most wonderful nations that have ever flourished on the earth.

The Greeks have an early and long extended history, and in addition thereto, a mythical period, valuable at least for this, that it acquaints us with the customs which must have prevailed when the writers of that period flourished.

The most valuable work illustrative of the domestic and social life and manners of the Greeks, and also throwing much light on the manners and customs of other ancient peoples, is the Deipnosophistæ, (Banquet of the Learned), by Athenæus, a rhetorician , and encyclopædian compiler, who lived in the beginning of the third century of the present era. His work is in the form of a dialogue between above twenty eminent lawyers, poets, and representatives of the various learned professions, who are supposed to meet at a banquet given by a rich citizen of Rome, where each draws upon his learning to discourse of feasts in general, and to enlarge on the great variety of subjects naturally suggested by talking of the customs of the ancients. They profess to deal chiefly with facts, and refer profusely to the authorities for their statements.

----

* Ibid, p. 54.

Of course, they have much to say of wine, for although the Greeks made intoxicating drinks from figs, roots and the palm, as also from barley, and their mixed drinks were almost without number, their chief beverage was wine made from grapes. The origin of the wine, is, in many fanciful ways, attributed to the gods, chiefly, but not exclusively, to Bacchus. He discovered, rather than created it. Homer frequently enumerates vineyards among the possessions of his heroes ; but probably because so little is really known of the origin of the vine, many fables were originated to account for it, both among the poets and the common people. Deucalion, who is famed in fable as having peopled the earth after the flood, had a son Orestheus, who owned a dog, which in lieu of giving birth to pups, brought forth a small piece of wood, which being buried, sprung up a vine loaded with grapes. Orestheus having shortly after, a son, named him Œneus, from the vine, for that was the ancient name for the vine. Athenæus gives as his authority for this fable, Hecatæus, who wrote about 450 B. C. He also quotes Nicander, 146 B. C., as authority that wine, oinos, has its name from Œneus :

> " Œneus poured the juice divine
> In hollow cups, and called it wine."

And a still earlier authority for the story, is found in Melanippides, about 450 B. C., who said : "'Twas Œneus, master, gave his name to wine." *

Others suppose a spot near Olympia to have first produced it, in proof of which, a miracle was said to be wrought there annually during the Dionysiac (Bacchic) festival ; and still others that the Greeks brought it from the shores of the Red Sea.† Plato, 400 B. C., after ordaining in his Second Book of Laws that boys should never taste wine at all, and men of thirty years of age should drink sparingly, if at all, but that those who are forty may feast at large

---

banquets and invoke the gods, especially Bacchus, since he gave wine as an antidote against the austerity of old age, adds:

"But there is a report and story told that this god was once deprived of his mind and senses by his mother-in-law, Juno; on which account he sent Bacchic frenzy, and all sorts of frantic rage, among men, out of revenge for the treatment which he had received; on which account also, he gave wine to men." *

This story is certainly the best borne out by facts of any of the numerous ancient fables of the origin of wine, for whatever transient joy its use may impart to its users, it is sure to be revealed as an enemy at the last. There is a Grecian legend not mentioned indeed by Athenæus, but quite old enough to be worthy of as much regard as is paid to such as we have already cited, that contains some suggestive thoughts, so well does it portray the present as well as the past consequences of wine drinking. It runs in this wise:

"When Bacchus was a boy he journeyed through Hellas to go to Naxia; and, as the way was very long, he grew tired, and sat down upon a stone to rest. As he sat there, with his eyes upon the ground, he saw a little plant spring up between his feet, and was so much pleased with it that he determined to take it with him and plant it in Naxia. He took it up and carried it away with him; but, as the sun was very hot, he feared it might wither before he reached his destination. He found a bird's skeleton, into which he thrust it, and went on. But in his hand the plant sprouted so fast that it started out of the bones above and below. This gave him fresh fear of its withering, and he cast about for a remedy. He found a lion's bone, which was thicker than the bird's skeleton, and he stuck the skeleton, with the plant in it, into the bone of the lion. Ere long, however, the plant grew out of the lion's bone likewise. Then he found the bone of an ass, larger still than that of the lion. So he put the lion's, containing the bird's skeleton and the plant, into the ass's bone, and thus he made his way to Naxia. When about to set the plant, he found that the roots had entwined themselves around the bird's skeleton and the lion's bone and the ass's bone; and, as he could not take it out without damaging the

* Ibid, B. x. ch. 55.

roots, he planted it as it was, and it came up speedily, and bore, to his great joy, the most delicious grapes, from which he made the first wine, and gave it to men to drink. But, behold a miracle! When men drank of it, they first sang like birds; next, after drinking a little more, they became vigorous and gallant like lions; but, when they drank more still, they began to behave like asses."

Bacchus, as we have already seen, had different names in different countries. His Greek name was Dionysus, and the story of his origin is both wonderful and ridiculous. His common name, Bacchus, is from a Greek word which means " to revel," and the other names given him by the Greeks denoted other peculiarities by which he was distinguished.* He is represented in the ancient paintings with a red face, a bloated body, carried in a chariot sometimes drawn by tigers and lions, sometimes by other animals, and having as a guard a drunken band of satyrs, demons and nymphs that preside over the wine presses. He is often followed by Silenus, his foster-father, who drinks with him from the same cup, and is almost always intoxicated, as he is described in the sixth eclogue of Virgil.

Bacchus was a great traveller, and wherever he went he taught the culture of the vine and the mode of making wine. He also taught certain mysteries, which were chiefly followed by the women, who from the effects which the rites had on them, were called Thyades and Maenades, names which denote madness and folly. At Sparta he was worshipped under the name of Sukites, because, says Sosibios, he was supposed to be the discoverer of the fig. Sophocles called him the "many named," because in the Orphic hymns alone, more than forty of his appellations are met with. Not only do we glean from the representations of him during his supposed existence on the earth that like the gods of India and Persia he delights in intemperance, bnt the festivals instituted in his honor after his death be-

---

* Tooke, in his Pantheon of the Heathen Gods, gives him nineteen other names.

came so riotous and dissolute that at last the arm of the law was invoked for their suppression.

In the mythic writings of Homer we find that both the higher and the lower orders of demi-gods and men are similarly affected by the use of wines; that the wines of one country are exported to other lands, and wherever used produce intoxication. For example when Ulysses reaches the " outlawed Cyclops' land," he finds

> "A race
> Of proud-lived loiterers, that never sow,
> Nor put a plant in earth, nor use a plow,
> But trust the gods for all things; and their earth,
> Unsown, unplow'd, gives every offspring birth
> That other lands have; wheat, and barley, vines
> That bear in goodly grapes delicious wines;
> And Jove sends showers for all."

Obtaining provisions from a neighboring island abounding in goats, the crews of the "twelve ships in the fleet," made a feast, at which they had abundance of wine.

> " Even till the sun was set,
> We sat and feasted, pleasant wine and meat
> Plenteously taking; for we had not spent
> Our ruddy wine a ship-board, supplement
> Of large sort each man to his vessel drew,
> When we the sacred city overthrew
> That held the Cicons."

The day after the feast, Ulysses took twelve of his friends and went on shore to visit the great Cyclop Polyphemus, and, as a present, carried,

> "A goat-skin flagon of wine, black and strong,"

which he had obtained, " In Thracian Ismans."*

The Cyclop having received them in a barbarous manner, and devoured six of the crew, he is made drunk by the wine, and Ulysses with his remaining comrades escapes.

---

* Odyssey, Bk. ix. 167-174; 239-245.

In the drunken riot in which the suitors of Penelope engage, Ulysses slays Antinous "As he was lifting up the
bowl;" and in the conversation which precedes this tragedy,
it is made known that the long war terminating in the
destruction of the Centaurs, was occasioned by a drunken
frolic.*

Homer also makes Agamemnon say,

> " Disastrous folly led me thus astray,
> Or wine's excess, or madness sent from Jove."

And Achilles, thus to reproach Agamemnon:

> "Tyrant, with sense and courage quelled by wine." †

The writers in the Historic period, even such of them as
praise wine, also bear witness that it is a mocker, and that
the most fearful consequences follow its use. " Pittacus,
one of the Seven Wise Men of Greece," 612 B. C., "recommended Periander of Priene not to get drunk, ' so that,'
says he, ' it may not be discovered what sort of a person
you really are, and that you are not what you pretend
to be!'

> ' For brass may be a mirror for the face—
> Wine for the mind.'"

" On which account they were wise men who invented the
proverb : ' Wine has no rudder.' Accordingly, Xenophon
the son of Gryllus, (when once at the table of Dionysius,
tryrant of Sicily, the cupbearer was compelling the guests to
drink), addressed the tyrant himself by name, and said,
' Why, O Dionysius, does not also the confectioner, who is
a skilful man in his way, and one who understands a great
many different recipes for dressing things, compel us also,
when we are at a banquet, to eat even when we do not
wish to; but why, on the contrary, does he spread the
table for us in an orderly manner, in silence?' And Sophocles, 450 B. C., in one of his satiric dramas, says:

> ' To be compelled to drink is quite as hard
> As to be forced to bear with thirst.'

From which also is derived the saying :

'Wine makes an old man dance against his will.'

And Sthenelus the poet, 400 B. C., said very well:

'Wine can bring e'en the wise to acts of folly.'*

"Panyasis the epic poet, 490 B. C., allots the first cup of wine to the Graces, the Hours, and Bacchus ; the second to Venus, and again to Bacchus ; the third to insolence and destruction.† Euripides, 480 B. C., says, 'Drinking is sire of blows and violence.'‡ Epicharmus, 470 B. C., says,

> 'A. Sacrifices feasts produce,
> Drinking then from feasts proceeds.
> B. Such rotation has its use.
> A. Then the drinking riot breeds;
> Then on riot and confusion
> Follow law and prosecution ;
> Law brings sentence, sentence chains;
> Chains brings wounds and ulcerous § pains.'

Enpolis, 446 B. C., says :

> 'He who first invented wine,
> Made poor man a greater sinner.'" ‖

Æschylus, 490 B. C., represents the Greeks as frequently so drunk as to break their drinking cups and other utensils about each other's heads. And Sophocles says, in his Banquet of the Greeks :

> "He in his anger threw too well
> The vessel with an evil smell
> Against my head, and filled the room
> With something not much like perfume ;
> So that I swear I nearly fainted
> With the foul steam the vessel vented." ¶

Antiphanes, 408 B. C., in his Æolus, speaking of a temptation to do a base thing, which came to Macareus, says that he

---

*Ibid, Bk. x. 31. †Ibid, Bk. ii. 3. ‡Ibid, Bk. ii. 4.
§ Ibid, Bk. ii 3. ‖Ibid, Bk. i. 30. ¶Ibid, Bk. i. 30.

> "For a while
> Repressed the evil thought, and checked himself;
> But after some short time he wine admitted
> To be his general, under whose lead
> Audacity takes the place of prudent counsel,
> And so by might his purpose he accomplished." *

Critias, 400 B. C., in his Elegies, speaks of the effects of wine :

> " After draughts like this, the tongue gets loose,
> And turns to most unseemly conversation ;
> They make the body weak; they throw a mist
> Over the eyes; and make forgetfulness
> Eat recollection out of the full heart.
> For fierce, immoderate draughts of heady wine
> Give momentary pleasure, but engender
> A long-enduring pain which follows it." †

Pytheas, 380 B. C., says on the same theme : " You see the demagogues of the present day, Demosthenes and Denades, how very differently they live. For the one is a water-drinker, and devotes his nights to contemplation, as they say ; and the other is a debauchee, and is drunk every day, and comes like a great pot-bellied fellow, as he is, into our assemblies." ‡

Eubulus, 375 B. C., after extolling water, and saying that it never produces bad effects, while " wine obscures and clouds the mind," makes even Bacchus say :

> " Let them three parts of wine all duly season
> With nine of water, who'd preserve their reason.
> The first gives health, the second sweet desires,
> The third tranquillity and sleep inspires.
> These are the wholesome draughts which will men please,
> Who from the banquet-house return in peace.
> From a fourth measure insolence proceeds ;
> Uproar a fifth ; a sixth wild license breeds;
> A seventh brings black eyes and livid bruises;
> The eighth the constable introduces ;
> Black gall and hatred lurk the ninth beneath ;
> The tenth is madness, arms, and fearful death.

---

* Ibid, Bk. x. 62.   † Ibid, Bk. x. 41.   ‡ Ibid, Bk. ii. 23.

For too much wine poured in one little vessel
Trips up all those who seek with it to wrestle." *

And Alexis, 350 B. C., testifies, in his Ulysses Weaving:

"For many a banquet which endures too long,
  And many and daily feasts, are wont t'engender
Insult and mockery; and those kind of jests
Give far more pain than they do raise amusement,
For such are the first ground of evil-speaking;
And if you once begin t'attack your neighbor,
You quickly do receive back all you bring,
And then abuse and quarrels surely follow;
Then blows and drunken riot. For this is
The natural course of things, and needs no prophet." †

And in his Phrygian, he says:

"If now men only did their headaches get
  Before they get so drunk, I'm sure that no one
Would ever drink more than a moderate quantity:
But now we hope to 'scape the penalty
Of our intemperance, and so discard
Restraint, and drink unmixed cups of wine." ‡

Wisely then, does he ask:

  "Is not, then, drunkenness the greatest evil,
  And most injurious to the human race?" §

Diphilus, about 340 B. C., says of Bacchus:

"You make the lowly-hearted proud,
  And bid the gloomy laugh aloud;
You fill the feeble man with daring,
And cowards strut and bray past bearing." ‖

Crobylus, 324 B. C., in his Female Deserter, says of the
use of wine, that men

        "Can have
No pleasure in it, surely; how should it,
When it deprives a living man of power
To think as he should think? and yet is thought
The greatest blessing that is given to men."

---

* Ibid, Bk. ii. 19.   † Ibid, Bk. x. 17.   ‡ Ibid, Bk. x. 34.
    § Ibid, Bk. x. 61.   ‖ Ibid, Bk. ii. 2.

And he significantly asks

"What pleasure, prithee tell me, can there be
In getting always drunk? in, while still living,
Yourself depriving thus of all your senses;
The greatest good which nature e'er has given?"*

And Callimachus, 260 B. C., testifies that,

"Wine is like fire when 'tis to men applied,
Or like the storm that sweeps the Libyan tide;
The furious wind the lowest depth can reach,
And wine robs man of knowledge, sense, and speech."

The Greeks at their feasts, puzzled and amused each other with enigmas, conundrums, riddles, and such like mysteries as pleased their wits. Some of these were nonsensical, and some were wonderfully ingenious and acute. The penalty for not guessing or otherwise discovering the correct answers, was to be compelled to empty at one draught, the largest cups or goblets of wine. Their drinking cups, or as they were sometimes called, vases, were often formed from the large horns of the Molasian and Poemian oxen. Small cups were in bad repute. There was one bowl, which, on account of its enormous size, was called the Elephant.

"A. If this hold not enough, see the boy comes
      Bearing the Elephant.
"B.   Immortal Gods!
      What thing is that?
"A.   A double fountained cup,
      The workmanship of Alcon: it contains
      Only three gallons."†

The practice of drinking wine from the horns of bulls and oxen, has been regarded by some as suggesting to artists the idea of representing Bacchus with horns, and the epithet of the Bull Dionysus. At Cyzicos he was worshipped under the form of a bull.‡

* Ibid, Bk. x. 34, 61.
† Ibid, Bk. xi. 35.
‡ Boeckh. Pub. Econ. of Athens, Vol. ii. p. 254.

There was a peculiar kind of cup called Grammateion, from the letters of gold chased on it. Alexis thus speaks it:

"A. But let me first describe the cup ; 'twas round,
       Old, broken-eared, and precious small besides,
       Having, indeed, some letters on't.
 B.   Yes, letters ;
       Eleven, and all of gold, forming the name
       Of Saviour Zeus.
 A.   Tush ! no, some other god." *

Athenæus relates that at the marriage of Caranus, Proteas drank upwards of a gallon of wine at a draught, exclaiming—

          "Most joy is in his soul,
           Who drains the largest bowl ;"

and was immediately presented by Caranus with the immense goblet which he had drained. Other capacious goblets were then produced, and the host declared that every man should claim as his own property the bowl whose contents he could despatch. Nine valiant drinkers at once started to their feet, vieing with each other as to who should empty his goblet first, while one poor wight, whose capacity was not equal to such a venture, sat down and burst into tears, because he must go away cupless. The bridegroom, unwilling that any grief should mingle in the feast, graciously presented him with an empty cup. †

Among the inscriptions to be found on monuments in different cities, as preserved by Polemon, 150 B. C., are the following :

    " This is the monument of that great drinker,
      Arcadion; and his two loving sons,
      Dorcon and Charmylus, have placed it here,
      At this the entrance of his native city :
      And know, traveller, the man did die
      From drinking strong wine in too large a cup."

---

* Athenæus, Bk. xi. 30.   † Ibid, Bk. iv. 4.

'Twice was this cup, full of the strongest wine,
Drain'd by the thirsty Erasixenus,
And then in turn it carried him away."

From the large number of enigmas, puzzles and conundrums preserved by Athenæus, as invented at the feasts of
the Greeks, the following are selected as specimens of the
best :

" A.   It is not mortal, nor immortal either,
But as it were, compounded of the two,
So that it neither lives the life of man,
Nor yet of God, but is incessantly
New-born again, and then again deprived
Of this its present life ; invisible,
Yet it is known and recognized by all.
B.   You always do delight, O lady, in riddles.
A.   No, I am speaking plain and simple things.
B.   What child then is there which has such a nature ?
A   'Tis sleep, my girl, victor of human toils."

" A.   There is a thing which speaks, yet has no tongue:
A female of the same name as the male ;
A steward of the winds, which it holds fast :
Rough, and yet sometimes smooth ; full of dark voices,
Scarce to be understood by learned men ;
Producing harmony after harmony :
'Tis one thing, and yet many ; e'en if wounded
'Tis still invulnerable and unhurt ;
B.   What can that be ?
A.   Why, don't you know, Callistratus ?
It is a bellows.
B.   You are joking now.
A.   No ; don't it speak, although it has no tongue ?
Has it not but one name with many people ?
I'st not unhurt, though with a wound in the centre ?
Is it not sometimes rough, and sometimes smooth ?
Is it not, too, a guardian of much wind ?"

" I know a thing which, while it's young, is heavy,
But when its old, though void of wings, can fly
With lightest motion, out of sight o' th' earth."

The answer is, the thistledown ; for it,

" While it is young stands solid in its seed,
But when it loses that, is light and flies,
Blown about every way by playful children."

" S. There is a female thing which holds her young
Safely beneath her bosom ; they, though mute,
Cease not to utter a loud sounding voice
Across the swelling sea, and o'er the land,
Speaking to every mortal that they choose ;
But those who present are can nothing hear,
Still they have some sensation of faint sound.

" B. The female thing you speak of is a city,
The children whom it nourishes, orators ;
They, crying out, bring from across the sea,
From Asia and from Thrace, all sorts of presents ;
The people still is near them while they feed on it,
And pour reproaches ceaselessly around,
While it nor sees nor hears aught that they do.

S. But how, my father, tell me, in God's name,
Can you e'er say an orator is mute,
Unless, indeed, he's been three times convicted ?

B. And yet I thought that I did understand
The riddle rightly. Tell me then yourself.

S. The female thing you speak of is a letter,
The young she bears about her is the writing :
They're mute themselves, yet speak to those afar off
Whene'er they please. And yet a bystander,
However near he may be, hears no sound
From him who has received and reads the letter."

" Of all the things the genial earth produces,
Or the deep sea, there is no single one,
Nor any man or other animal,
Whose growth at all can correspond to this :
For when it first is born its size is greatest ;
At middle age 'tis scarcely visible,
So small it's grown ; but when 'tis old and hastens
Nigh to its end, it then becomes again
Greater than all the objects that surround it."

The answer is, a Shadow.

" What is the strongest of all things ? " " Iron," said
one, " for with that material men dig and cut all other
things." " No," said the second, "the blacksmith is the
strongest, for he makes the iron into any shape and for
any purpose that he chooses." " You are both wrong,"
said the third, "love can subdue even a blacksmith ; therefore love is the strongest of all things."

The following refers to Night and Day:

> " There are two sisters, one of whom brings forth
> The other, and in turn becomes its daughter." *

Music and song were accompaniments of the feasts, and a burning bowl called Oidos, or " the cup of song," rewarded him whose skill pleased the drinkers. † The young drank to their mistresses, sometimes taking as many cups as there were letters in her name ; sometimes restraining their appetite by taking a glass to each of the three Graces ; but when in for a frolic they chose the Muses for their patrons, and honored their mistresses' names with three times three. Hence, it is said, the custom so well observed in political circles, of honoring candidates for office, with cheers. ‡

Aristophanes, 430 B. C. represents the women of Athens as extravagantly given to the use of wine ; so much as to pawn their wardrobes to procure it, and manufacturing counterfeit keys to their husbands' wine cellars. § Phalæcus, 320 B. C., in his Epigrams, mentions a woman who was a notorious drinker :

> "Cleo bestow'd this splendid gift on Bacchus,
> The tunic, fringed with gold and saffron hues,
> Which long she wore herself ; so great she was
> At feasts and revelry : there was no man
> Who could at all contend with her in drinking."

Alexis speaks of a certain woman, as " Zopyra that wine-cask."

And Antiphanes, in his Female Bacchanalians, makes the sweeping assertion :

> "I'm sure
> He is a wretched man who ever marries

* Ibid, Bk. x. Chapters, 71, 73, 74, 75.
† Ibid, Bk. xi. 110.
‡ St. John. Manners and Customs of Ancient Greece, Vol. ii. p. 192.
§ Aristoph. Lysistrata, p. 18.

Except among the Scythians; for their country
Is the sole land which does not bear the vine." *

Æschylus, 480 B. C., is said to have been the first
person who introduced the appearance of drunken people
into a tragedy. "But the fact is," says Athenæus, "that
the practices which the tragedian himself used to indulge
in, he attributed to his heroes; at all events he used to
write his tragedies when he was drunk ; on which account
Sophocles used to reproach him, and say to him, 'O
Æschylus, even if you do what you ought, at all events
you do so without knowing it.'" †

The excuses offered twenty-five hundred years ago, for
drinking, were identical with those often so glibly offered
now. Alexus, the lyric poet, who flourished in 612 B. C.,
thus offers them :

"In winter cold

Let's drive away
The wintry cold, and heap up fire,
And mingle with unsparing hand
The honied cup, and wreathe our brows
With fragrant garlands of the season.

"In Spring :

Now does the flowery spring return,
And shed its gifts all o'er the land.

"In Summer :

Now it behooves a man to soak his lungs
In most cool wine ; for the fierce dog-star rages,
And all things thirst with the excessive heat.

"In Misfortunes :

By grieving
We shall not do ourselves much good.
Come to me, Bacchus ; you are ever
The best of remedies, who bring
Us wine and joyous drunkenness.

---

* Athenæus, Bk. x. 56, 57.   † Ibid, Bk. x. 33.

" In Joy:

> Now is the time to get well drunk,
> Now e'en in spite of self to drink." *

As in Greece, so in Italy, the first mention of wine is in the fables, and not in the history, of the people. Mezentius, a mythical king of the Etruscans or Tyrrhenians said to have been cotemporary with Æneas, of Troy, by whom he was slain in battle, granted peace to the people whom he conquered, on condition that they should annually pay as a ransom, all the wine that was produced in their country. Virgil's Æneid gives the story of his " unutterable barbarities," and of the drunkenness of his followers.

Pliny, who wrote about 65 B. C., states in his Natural History, that wine was well known in Rome from the earliest period in its history, about 700 B. C., as Mecenius slew his wife because she had tasted the intoxicating draught. The early laws and usages in relation to drinking by women, we shall have occasion to notice hereafter, but it is certain that in later periods of the history of the city men and women got drunk together in most licentious carousals. At the time of issuing the decree of Numa, about 650 B. C., wine was a scarcity, as its use was forbidden as a libation to the gods, and for sprinkling on the funeral pyre; and for a long time after it was so scarce that milk was employed instead : and when in 319 B. C., Papivius was about to engage in a decisive combat with the Samnites, he vowed, as the choicest and rarest gift that he could offer, to sacrifice " a small cup of wine to Jupiter, if he should grant him the victory."

Shortly after this, however, the production of native wines greatly increased, and there were large importations from Greece, so large that it is mentioned to the praise of Cato, another general, about 220 B. C., that he set him-self against the growing luxury of the times, by refusing to partake of any better wines than were served out to

---

* Ibid. Bk. x. 35.

the men of his command.   Little more than a century later, according to Varro, the illustrious Lucullus, on his return from a successful campaign in Asia, distributed about 60,000 gallons of Greek wine among the people, as a gift.   A little later, about 50 B. C., Hortensius, the rival orator of Cicero, left to his heir 10,000 casks of Chian wine in the cellar of one of his country residences.   It was at this time, according to Pliny, that Cæsar placed upon his table, at a banquet, Falernian, Chian, Lesbon and Mamertine wines, "the first occasion," says Pliny, "on which four kinds of wine were served at table."

From this time there is rapid degeneracy, the excess and debauchery of both men and women becoming offensively licentious and disgusting.*   Seneca complains: "The weak and delicate complexion of the women is not changed, but their manners are no longer the same ; they value themselves on carrying excess of wine to as great a height as the most robust men ; like them they pass whole nights at table, and with a full glass of unmixed wine in their hands, they glory in vieing with them, and, if they can, in overcoming them."   Drinking for wagers became frequent, and in order that men might in their competition for the prize, overfill themselves with wine, Pliny relates that some drank hemlock before going to their cups, that frightened by the thoughts of death, they might even force down wine as an antidote ; others after having filled their stomachs resorted to emetics, in order that the drinking might be renewed ; others betook them to the hot-baths, from which they were carried out half-dead ; and both sexes, without leaving the table, outraged all decency, by their beastly condition.†

Caius Piso was famed for this latter indecency, it being said of him that he would set for two days and nights drinking without intermission, or even stirring from the ta-

---

* Pliny. Natural History, Bk. xxxvi. chap. 21.
† Ibid, Bk. iv. 28.

ble. * Seemingly incredible stories are told of the capacity
of some of these old topers, as for example, of Torquatus,
who was knighted by Tiberius with the title of Tricongius,
or the three-gallons knight, for drinking three gallons of
wine at a draught, and without taking breath; and Tergilla,
who boasted that he ordinarily drank two gallons at a
draught; and later the Emperor Maximin, who, it was said,
could drink six gallons of wine without committing any
debauch! Although the Roman gallon (Congius) was
little less than six pints, our measure, these amounts are
simply enormous, and must, if admitted as at all real,
necessitate the conclusion that the wines were much weaker than those of the present day. It is known, indeed, that
they were largely diluted, some with three, some five, seven,
and even nine parts of water.

But the outward physical results of such dissipation
were the same as now, for Pliny describes their blotched
skin, purple nose, bleared eyes, and their " sleep agita-
ted by furies; " while they deprive their victim of reason, and " drive him to frenzy and the commission of a
thousand crimes." Wine at last became so common, says
the same authority, that it was even given to the beasts of
burden. No wonder that he exclaimed, " By Hercules,
pleasure has now begun to live, and life, so called, has
ceased to be." †

Athenæus confirms the foregoing account of dissipation
in the description which he gives of a feast at a Roman
mansion. From the pleasurable first excitement of the
wine, hosts and guests pass to the most debased sottishness;
the slaves being compelled to participate, that they may
be at no advantage over their masters, and at last, host,
guests, men, women and slaves, mingle in the wildest riot
and confusion.

The feast of the Saturnalia, marked by all the folly and licentiousness which intoxication produces, was extended from

---

.he day, its original limit, to three, and finally to seven days, by Caligula and Claudius. Wines were furnished at public expense, and under Vitellius, drunken feasts were held for three days on the battle-fields, while yet the dead lay un-buried. In the city itself, the people kept the same feast with riot and debauchery. "The whole city," says Taci-tus, " seemed to be inflamed with frantic rage, and at the same time intoxicated with drunken pleasures. * * * Thrice had Rome seen enraged armies under her walls, but the unnatural security and inhuman indifference that now prevailed, were beyond all example." *

The same writer describes the surprise and capture of the city of Terracina, by the Roman troops under the com-mand of the Emperor's brother, in consequence of the in-temperance of the garrison, commanded by Julianus and Appolinaris, " two men immersed in sloth and luxury; and by their vices, more like common gladiators than superior officers."

"No sentinels stationed, no night watch, to prevent a sudden alarm, and no care taken to guard the works, they passed both night and day in drunken jollity. The windings of that delight-ful coast resounded with notes of joy, and the soldiers were spread about the country to provide for the pleasures of the two commanders, who never thought of war except when it be-came the subject of discourse over the bottle." †

The Romans, under the advice of a renegade, surprised the city, slaughtered the drunken troops and put one of the commanders to death in a barbarous and ignominious manner.

In modern Rome, intemperance still prevails, as it also does in other parts of Italy. E. C. Delavan, Esq., late of Albany, N. Y., visited Rome in 1839, from whence he wrote home that Cardinal Acton, the supreme judge of the city, assured him that nearly all the crime in Rome origi-nated in the use of wine. The Judge directed him to a part of the city, that would compare well with the Five

---

* Histories, B. iii. sect. 83.   † Ibid, sect. 76.

Points in New York. "I visited that district," says Mr. Delavan, and there I saw men, women, and children sitting in rows, swilling away at wine, making up in quantity what was wanting in strength; and such was the character of the inmates of those dens that my guide urged my immediate departure, as I valued my life."

The same year, Horatio Greenough, the American sculptor, wrote from Florence, "Many of the more thinking and prudent Italians abstain from the use of wine; several of the most eminent of the medical men are notoriously opposed to its use, and declare it a poison. When I assure you that one-fifth, and sometimes one-fourth of the earnings of the laborers are expended in wine, you may form some idea as to its probable influence on their thrift and health." Hon. George P. Marsh, United States Minister to Italy, wrote from Rome to the Centennial Temperance Conference at Philadelphia, in 1876:

"It is undoubtedly true in Italy, as in most other European countries, that a very considerable proportion of the crimes accompanied with violence, originate in intoxication, and the police reports show a large and, I am sorry to say, an increasing number of such cases. The days of idleness, miscalled *religious* festivals, in Italy, are devoted by vast numbers of the lower classes to drinking, gambling, and other immoralities: and until public opinion shall become enlightened enough to suppress these occasions of vice, I should not expect much result from efforts of philanthropists in the way of temperance reform."

VI. THE JEWS.—In giving attention to the fact of Intemperance among the Jewish people, as it is manifest in the Histories and Prophecies contained in the Old Testament, the question in regard to the distinction between the intoxicating and unintoxicating wines mentioned in the Bible, need not here be considered, since its examination is reserved for another part of the work; and our immediate object is solely to set forth the fact that there was intemperance among the people to whom a special mission was assigned by Jehovah, and that its consequences were disastrous to them, as they uniformly are to all other nations.

The illustrations selected are not chronologically arranged, but given as they stand in the order in which the books comprising the Jewish Scriptures are placed before us in our English Bibles.

The account of the drunkenness and disgrace of Noah, as recorded in Genesis ix. 20-25, is the first mention of drunkenness in the Bible. Whether this was Noah's first experience in producing and partaking of wine, or an indulgence common to him before the deluge, but now for the first time, either through ignorance or carelessness, an indulgence after fermentation had commenced; or a use of that which he knew would intoxicate;—each of these theories or surmises having advocates among those who seek to interpret the account;—we are not likely to know. In either case, it is evident that Noah thus fell into sin, and furnished an occasion for the sin of his son, and probably of his grandson also; and that in consequence of it a heritage of sorrow and bondage was the portion of the descendants of Ham in the line of Canaan.

The description in Genesis xix. 32-38, of the incestuous conduct of Lot while senseless and unconscious under the influence of the intoxicating draught of the "wine," that "is the poison of dragons, and the cruel venom of asps," is a fitting first-mention of the licentiousness which has been such a constant accompaniment of drunkenness. What retributive consequences were entailed on his sin we are not informed, as no further mention is made of him in the sacred record. We simply see him "saved indeed from the conflagration of Sodom, but an outcast, widowed, homeless, hopeless, without children or grandchildren, save the authors and the heirs of his shame."

In Exodus xxxii. we have a sad account of the irreligion and licentiousness into which the children of Israel fell, on the occasion of the feast in which they indulged, when, in the language of the record, they "sat down to eat and to drink, and rose up to play." The significance of that description being, that, inflamed with wine, they committed all sorts of sexual uncleanness.* This also is the sense of the word "naked," in the 25th verse.† As the immediate consequence swift death came upon thousands, and the entire nation were discomfitted and distressed by the lengthening out of their wanderings in the wilderness.

Leviticus x. 1–11 gives an account of the sin of Nadab and Abihu, committed while they ministered at the altar; and of a command imposed by Jehovah immediately after, and probably on account of that sin. The offence consisted in the offering of an incense kindled by "strange fire," and the incitement to the offence seems to be more than implied in the command to Aaron: "Do not drink wine nor strong drink, thou, nor thy sons with thee, when ye go into the tabernacle of the congregation, lest ye die; it shall be a statute forever throughout your generations. That ye may put difference between holy and unholy, and between clean and unclean." It is the general opinion of the Jewish Commentators that the inebriation of Nadab and Abihu caused them to use the "strange fire;" and in this agree many eminent Christian critics.‡

In the time of the Judges, drunkenness seems to have become so common a vice in Israel as to have involved even the women of the nation in its shame. In 1 Samuel, i. 14, Hannah is unjustly accused by Eli, because in her prayers no words issue from her moving lips, of being "drunken." No surprise is expressed by this rebuke, that she should presume to present herself in the temple of Jehovah in this plight, for Hannah's answer intimates that such women did frequent that place, and the statement in

---

* It is the same word as is rendered "mock" in the false charge made by Potiphar's wife against Joseph, when she accused him of a licentious attempt. Genesis xxxix. 14–17.

† See "The Speaker's Commentary," in loco.

‡ See Prof. Geo. Bush's Commentary on Leviticus x. 9: also "the Speaker's Commentary," on vs. 1, same chapter.

the 22nd verse of the second chapter, shows that even lewd women were permitted there; and that these " daughters of Belial " had for their associates the sons of the Judge and High Priest Eli, who were so debauched as to be called the " Sons of Belial." Their intemperate habits not only involved them in licentiousness, and so incurred severe judgments on themselves, but also made them so negligent of their duties as custodians of the Ark of the Covenant, as to suffer the populace to take it from their keeping, to carry it, as a battle-flag, into their fight with the Philistines, by .whom it was captured, and in consequence the hand of Jehovah lay heavy on the nation.

When, under the strong rule of Saul, and the wise government of David, the nation regained its position and also the divine favor, drunkenness soon worked mischief in the royal household, and in producing discord and rebellion. Amnon cruelly ruined his half-sister. Her brother, Absalom, nursed vengeance in his heart, and on the first favorable opportunity had Amnon made drunk with wine and slew him: 2 Samuel xiii. Then came alienation, rebellion, and distress. No wonder that David employed the severest terms in reprobating the use of wine, and that, when endeavoring to set forth the most expressive idea of the judgments of the Almighty, he makes choice of the figure of an inebriating cup in the hand of Jehovah, which, as he pours it out upon the nations, spreads terror and desolation wherever its falls:

" In the hand of the Lord there is a cup, and the wine is red ; it is full of mixture, and he poureth out of the same; but the dregs thereof, all the wicked of the earth shall wring them out, and drink them." Psalm lxxv. 8.

Solomon, out of a deeper and more varied experience even than that of David, wise above all others while serving and obeying God, and the most besotted of all fools in his idolatry, luxury and licentiousness, tells us, after his vain " seeking in his heart to give himself to wine," what are the characteristics of the deceptive draught, and what consequences fatal to prosperity, happiness, and moral purity, follow its use :

"Wine is a mocker, strong drink is raging : and whosoever is deceived thereby is not wise." "Who," he exclaims, "Who hath woe? who hath sorrow? who hath contentions? who hath babbling? who hath wounds without cause? Who hath redness of eyes? They that tarry long at the wine; they that go to seek mixed wine."

Then he lifts up his voice in warning, shows how the momentary gratification of drinking is followed by the most sorrowful and bitter results ; how passion is given the mastery, God driven from the heart, life is put in fearful peril, and the power of the will so benumbed that still again and again the victim of drink rushes on to his indulgence and incurs repeated loss, misery and pollution :

"Look not thou upon the wine when it is red, when it giveth his color in the cup, * when it moveth itself aright. At the last it biteth like a serpent, and stingeth like an adder. Thine eyes shall behold strange women, and thine heart shall utter perverse things. Yea, thou shalt be as he that lieth down in the midst of the sea, or as he that lieth upon the top of a mast. They have stricken me, shalt thou say, and I was not sick; they have beaten me, and I felt it not; when shall I awake? I will seek it yet again."

The advice given to Lemuel, was it not based on what Solomon had found true in his own experience? "It is not for kings, O Lemuel, it is not for kings to drink wine; nor for princes strong drink : lest they drink and forget the law, and pervert the judgment of any of the afflicted." Proverbs xx. 1; xxiii. 29-35; xxxi. 4. Elah, one of the kings of Israel, became the victim of a conspiracy, and was slain by the "captain of half his chariots," while "drinking himself drunk in the house of Arza, steward of his house in Tirzah." 1 Kings xvi. 9.

---

* "When it giveth his color. Literally, 'its eye,' the clear brightness, or the beaded bubbles on which the wine drinker looks with complacency."—*The Speaker's Commentary.* Prov. xxiii. 31.

These pernicious personal examples of priests, princes and Kings, and other mighty and so-called noble men, could not fail to bear their fruit in infecting the nation at large with this fearful evil. Isaiah, who flourished just before the Babylonian Captivity, describes in terse and vigorous words the immoral condition of the masses, and ascribes their predicted ruin to their intemperate habits:

"Woe unto them that rise up early in the morning, that they may follow strong drink; that continue until night till wine inflame them! And the harp, and the viol, the tabret, and pipe, and wine, are in their feasts: but they regard not the work of the Lord, neither consider the operation of his hands. Therefore my people are gone into captivity, because they have no knowledge: and their honorable men are famished, and their multitude dried up with thirst. Therefore hell hath enlarged herself, and opened her mouth without measure: and their glory and their multitude, and their pomp, and he that rejoiceth, shall descend into it." "Woe unto them that are mighty to drink wine, and men of strength to mingle strong drink; which justify the wicked for reward, and take away the righteousness of the righteous from him! Therefore as the fire devoureth the stubble, and the flame consumeth the chaff, so their root shall be as rottenness, and their blossom shall go up as dust: because they have cast away the law of the Lord of Hosts, and despised the word of the Holy One of Israel." Isaiah v. 11-13; 22-24.

In equally severe terms does the same prophet announce the divine judgment on the people for their disobedience to the command of Jehovah, that during the seige of the city by the Persians, they shall humble themselves before Him, and by penitence obtain His favor and help; instead of which they become so lost to a sense of their obligations to God as to mock Him by feasting, drinking and riot:

"In that day did the Lord God of Hosts call to weeping, and to mourning, and to baldness, and to girding with sackcloth: and behold joy and gladness, slaying oxen, and killing sheep, eating flesh, and drinking wine: let us eat and drink; for to-morrow we shall die. And it was revealed in mine ears by the Lord of Hosts, Surely this iniquity shall not be purged from you till ye die, saith the Lord God of Hosts." Chapter xxii. 12-14.

So also when Isaiah denounces woe upon Samaria, it is because of intoxication, taught and encouraged by the unfaithful priests and prophets :

"Woe to the crown of pride, to the drunkards of Ephraim, whose glorious beauty is a fading flower, which are on the head of the fat valleys of them that are overcome with wine! Behold, the Lord hath a mighty and strong one, which as a tempest of hail and a destroying storm, as a flood of mighty waters overflowing, shall cast down to the earth with the hand. The crown of pride, the drunkards of Ephraim shall be trodden under feet. * * * * * They have erred through wine, and through strong drink are out of the way; the priest and the prophet have erred through strong drink ; they err in vision, they stumble in judgment. For all tables are full of vomit and filthiness, so that there is no place clean." Chap. xxviii. 1–8.

And once more, speaking of the general demoralization of the people, its cause is said to be the unfaithfulness of the besotted who are placed in power, who, under the figure of a blind and dumb watchman, and a stupid and foolish shepherd, invited, instead of preventing the encroachments of the devouring beasts :

"All ye beasts of the field, come to devour, yea, all ye beasts in the forest. His watchmen are blind : they are all ignorant, they are all dumb dogs that cannot bark; sleeping, lying down, loving to slumber. Yea, they are greedy dogs which can never have enough, and they are shepherds that cannot understand: they all look to their own way, every one for his gain, for his quarter: Come ye, say they, I will fetch wine and we will fill ourselves with strong drink ; and to-morrow shall be as this day and much more abundant."—lvi. 9–12.

In the days of Hosea, complaint is made that Israel "looks to other gods, and loves flagons of wine; Hosea iii. 1; and in vii. 7, that the princes have debauched the king by catering to his lowest passions, and so have unfitted him to rule in righteousness: "In the days of our king the princes have made him sick with bottles of wine;* he stretched out his hand with scorners."

---

* "Bottles of wine."—Literally, "poison of wine." The same word that is translated " poison," in Deut. xxxii. 33.

In the woes denounced on the people, by Joel, i. 5, the class doubtless esteemed by him the most guilty is thus addressed: "Awake ye drunkards and weep; howl all ye drinkers of wine."

Among the evils of which Amos complains, ii. 6-12, are these, that the tribe of Judah frequent the heathen banquets, drink their strong wines, and compel the Nazarites to break their pledge of total abstinence.

"They lay themselves down upon clothes laid to pledge by every altar, and they drink the wine of the condemned in the house of their God. * * * * And I raised up your sons for prophets, and of your young men for Nazarites. Is it not even thus, O ye children of Israel? saith the Lord. But ye gave the Nazarites wine to drink; and commanded the prophets, saying, Prophesy not."

So in the sixth verse of the sixth chapter, "drinking wine in bowls," is among the offences charged against those who "are at ease in Zion."

By the prophet Micah, it is declared, ii. 11, that the people have become so corrupt that true prophets were rejected, and only those received who encouraged them in their dissipation: "If a man walking in the spirit and falsehood do lie, saying, I will prophesy unto thee of wine and of strong drink; he shall even be the prophet of this people."

Habakkuk, foretelling the ruin of Judea by the Chaldeans, assigns as a reason because Nebuchadnezzar, the king, "transgresseth by wine;" and he denounces woe   •
upon him who, having made some of the people drunken, is thereby able to discover the extreme weakness of the Jewish nation, and so to encourage the warfare that results in its overthrow: "Woe unto him that giveth his neighbor drink, that puttest thy bottle to him, and makest him drunken also, that thou mayest look on their nakedness." ii. 5, 15.

VII. CONTEMPORARY NATIONS.—Special mention has been made of the Egyptians and the Persians, in another place. A brief mention of the facts indicating tho prevalence of intemperance among other heathen nations with whom the Jews were brought in contact, will be given here.

The Philistines, who had conquered and oppressed Israel, after several ineffectual attempts to capture Samson, whom God raised up to begin the work of the deliverance of his nation, at last succeeded ; and when Samson had in a measure regained his strength, of which at the time of his capture he had been deprived, the Philistines, at "a great sacrifice unto Dagon their god," made themselves drunken with wine, and had Samson brought out of his prison-house for their diversion. Judges xvi. 25. "They brought him to their feast," say Josephus, "that they might insult him . in their cups." *

The Amalekites, who had " smitten Ziklag, and burnt it with fire," and "had taken great spoil," were pursued by David, who found them in such a drunken and riotous condition that they were scattered far and wide, and so fell an easy prey to his avenging army. 1 Samuel xxx. 16.

The Syrians, under Benhadad, besieged Samaria, and made demand for the immediate surrender of its inhabitants and all their treasure. They came with a great army, far outnumbering the besieged. Ahab, under direction of the prophet, went out of the city to give them battle, and " slew the Syrians with a great slaughter." The reason of their success against such great odds, was, that " Benhadad was drinking himself drunk in the pavilions, he and the kings, the thirty and two kings that helped him." 1 Kings xx. 16.

The Babylonians were inordinate drinkers. In Daniel v. we have an account of a feast made by Belshazzar the king, during which the sacred vessels taken from tho temple at Jerusalem were brought in for the use of "the king, his princes, his wives, and his concubines, that they might

* Antiquities. Book v. c. 8.

drink therein." And while they thus "drank wine," the handwriting on the wall appeared. During the night, Cyrus and the Persian troops entered the city, and "in that night was Belshazzar the king of the Chaldeans slain." The deep-seated determination of this people to gratify their basest passions, and possess to themselves at any cost, of the intoxicating cup, is set forth in horrid detail by the prophet Joel, iii. 3: "They have given a boy for a harlot, and sold a girl for wine, that they might drink."

In the Apocryphal Book of Esdras, the following is attributed to one of the body-guard of Darius, king of the Medes and Persians:

"O ye men, how exceeding strong is wine! it causeth all men to err that drink it: it maketh the mind of the king, and of the fatherless child, to be all one: of the bondman and of the freeman, of the poor man and of the rich: it turneth also every thought into jollity and mirth, so that a man remembereth neither sorrow nor debt: and it maketh every heart rich, so that a man remembereth neither king nor governor; and it maketh to speak all things by talents; and when they are in their cups, they forget their love both to friends and brethren, and a little after draw out swords: but when they are from the wine, they remember not what they have done." Book I. c. iii. vs. 18-23.

This language, spoken and recorded before the time of Christ, although it is not possible to fix on the precise date, can hardly be improved upon in any description of the effects of intoxicants that might be attempted now, so invariable is the bewilderment, self-deception, oblivion of all true relationships, standing and duties, the treachery to friends, and violence even unto death, as are accompaniments of drunkenness in all periods of its history.

VIII. GERMANY.—For our earliest information of the Germans, we are indebted to Roman writers, Pliny and Tacitus. They bring them to our notice as they appeared about the time of the commencement of our Christian era. From the first we find them noted for their indulgence in strong drink, a habit so firmly fixed. that it overcame their vigor and enterprise, and their natural adaptation for successful offensive or defensive warfare, much more effectually than could the assaults of any enemy in arms.

"The liquor commonly drunk by them," says Tacitus, "is prepared from barley or wheat; which, being fermented, is brought somewhat to resemble wine. Those who reside on the banks of the Rhine use wine itself. Their diet is simple—wild fruits, fresh venison, and curdled milk. They satisfy their appetite without deserts or splendid appendages. The same abstinence is not observed with regard to the bottle; for if you will indulge them in drunkenness to the extent of their desires, you may as effectually conquer them by this vice, as with arms." *

Pliny says: "The Western nations produced their inebriating liquors from steeped grain. Moreover, those liquors are made use of most, and not diluted as is the custom with wine. Hercules seemed only to produce fruit from the earth; whilst, alas! the wonderful shrewdness of our vices has shown us in what manner even water may be made to administer to them." †

Tacitus, speaking of their custom of keeping their beds till late in the morning, says, that after bathing and having their breakfast—

"They, being armed, proceed to business; but as often to parties of conviviality, where they spend whole days and nights in drinking, without any disgrace being attached to it. At these feasts, when the guests are intoxicated, frequent quarrels arise, which terminate not only in abuse, but in blood. The subjects of debate at these feasts are the reconcilement of enemies, forming family-alliances, the election of chiefs, and lastly, peace and war. The German thinks the soul is never more open to sincerity, nor the heart more alive to deeds of heroism, than under the influence of the bottle; for then, being naturally free from artifice and disguise, they open the inmost recesses of their minds; and the opinions which are thus broached they again canvass the next day. There is safety and reason attached to both modes; for they consult when they are not well able to dissemble, and debate when they are not likely to err." ‡

It is related by the same author, in his "Historical

---

* Tacitus on Germany, xxiii.   † Natural History, B. xiv. 22.
‡ On Germany, xxii.

Annals," * that on the occasion of a war between the Romans and the Marsians, a German tribe, a notable victory was gained by the former, under Germanicus, one of their famous generals, on account of the intemperance of the latter. "The scouts brought intelligence that the approaching night was a festival, to be celebrated by the barbarians with joy and revelry." The advancing army surrounded their foes on every side.

"The barbarians were sunk in sleep and wine, some stretched on the beds, others at full length under the tables; all in full security, without a guard, without posts, and without a sentinel on duty. No appearance of war was seen; nor could that be called a peace, which was only the effect of savage riot, the languor of a debauch. Germanicus, to spread the slaughter as wide as possible, divided his men into four battalions. The country, fifty miles round, was laid waste with fire and sword; no compassion for sex or age; no distinction of places, holy or profane; nothing was sacred. In the general ruin the Temple of Tanfau, which was held by the inhabitants in the highest veneration, was levelled to the ground. Dreadful as the slaughter was, it did not cost a drop of Roman blood. Not so much as a wound was received. The attack was made on the barbarians sunk in sleep, dispersed in flight, unarmed, and incapable of resistance."

Mead was also a favorite drink among the ancient Germans, and according to Henderson, it was customary to drink it for thirty days after a marriage. † Hence, probably the familiar expression, the Honey-moon. Cider seems also to have been known, as Tatian, a writer of the second century, makes frequent allusion to it as a common drink.

Early in the history of the nation, all classes, and both sexes often drank to great excess, until so alarming was the evil that measures were resorted to as early as the eighth century, which will be more particularly mentioned further on, to interpose the arm of legal authority against the vice.

The author of the article on Germany, in the American

---

* Book I. Sections 50, 51.
† History of Inebriating Liquors, p. 466.

Encyclopædia, remarks that " Popular movements in favor of liberty in Germany have often been defeated by the excessive drinking of the people."

A popular song of the Middle Ages, as sung by the students, according to the Jus Potandi,—drinking code quoted by Samuelson, in his History of Drink, * gives an alarming picture of the extent of drunkenness at that period :

> "Bibit hera, bibit herus,
> Bibit miles, bibit clerus,
> Bibit ille, bibit illa,
> Bibit servus cum ancilla,
> Bibit velox, bibit piger,
> Bibit albus, bibit niger,
> Bibit constans, bibit vagus,
> Bibit rudis, bibit nagus.
>
> Bibit pauper et ægrotus,
> Bibit excul et ignotus,
> Bibit puer, bibit canus,
> Bibit præsul et decanus,
> Bibit soror, bibit frater,
> Bibit anus, bibit mater,
> Bibit iste, bibit ille,
> Bibit centum, bibit mille."

Further quotations from the same curious works,—which, whether a genuine collection of rules really enforced, or only a satire on the besotted condition of the people, is unknown, and perhaps is of no consequence, since in either

---

* Samuelson, p. 107 : "A literal rendering of the above, without any attempt at versification, is : The mistress drinks, the master drinks, the soldier drinks, the clergy drinks, the man drinks, the woman drinks, the man-servant together with the maid servant drinks, the active drinks, the lazy drinks, the white drinks, the black drinks, the constant drinks, the fickle drinks, the learned drinks, the boor drinks, the poor and sick drink, the exile and stranger drink, the young and old drink, the dancer and dean drink, the sister and brother drink, the wife and mother drink, this one drinks, that one drinks, hundreds drink, thousands drink."

case it reveals in lively coloring what is no doubt true of
the age in which it was written ;—reveals the fact that men
not only boasted to their neighbors how well they had
succeeded in making all their guests drink the night before,
and how long some had shown themselves tougher headed
than the others, but even that fathers made it part of their
special care and boast to train their lads to drink. "Now
let us see," said the fond parent to his little son, "let us see
what you can do. Bring him a half-measure;" and later
on, "Bring him a measure."

Hans Sachs, "the national poet," is also quoted by Mr.
Samuelson, as giving "an account of a drunken tournament
which he had witnessed, where twelve ' beer heroes' succeeded in drinking from ' pots and cans' a tun of beer in six
hours ! "

After the drinking-code was established and accepted,
"there was," we are told, "no promiscuous hobnobbing,
and caste was duly respected then as now.

"Nobles were not permitted to drink with tradespeople, but
they might raise their glass to a student, and he in like manner
might condescend to notice a tradesman, for there was no knowing of what advantage such a recognition might be to a student.
A case is cited were a merchant (pedlar, we presume,) actually
gave a poor ' studiosus' a pair of beautiful silk stockings the
morning after a carouse,* for which he had expressed a longing
during the entertainment. Young maidens were permitted to
drink platonically with virtuous young men, · but they are
warned in droll and not very modest terms against 'pseudo-
prophêtes,' who are ' lupi rapaces' in sheep's clothing, and the
evils of drinking ' sisterhood ' with such ravening wolves are
duly and circumstantially set forth in the code. One clause is
devoted specially to the expressions in vogue amongst ladies,
who may find it necessary, whilst at table, to protect themselves against the too gross familiarity of their gallant neighbors.

---

* Disraeli, in his " Curiosities of Literature," Part I. p. 198, is
authority for saying that, "According to Blount's Glossographia,
carouse is a corruption of two old German words, *gar* signify-
ing all, and *audz*, out: so that **to drink** *garauz* **is to drink all**
out: hence carouse.

"As a rule, guests might not pledge persons who were present, unless it were a sweetheart, and that toast must be drunk 'ad unguem'—that is to say, in a bumper—the drinkers afterwards reversing their goblets and ringing them on their thumbnail, to show that not a drop was left therein. This has been a common drinking custom in several countries. Toasts were drunk in various ways; sometimes one man drank from two glasses at once; at others, when virtuous young ladies sat by the side of respectable young men, they were allowed to drink simultaneously from the same goblet; and it was deplored that such a mode of drinking could not become more general, on account of the wild behavior of the youth of the period. Regular penalties were inflicted for sneezing and coughing into the goblets, and for certain other offences against decency and propriety, which, although they seem to have been everyday occurrences at those carousals, are unfit to be spoken of in genteel society. When new-comers arrived, the goblet was offered to them, with sundry compliments and orations, and to refuse to drink was a mortal offence, usually followed by a bloody encounter. When a guest found it difficult to keep pace with the company, or could not empty his goblet at a draught, he might avail himself of the aid of any *young* lady who sat by his side, but *old* ladies were not allowed to render assistance under such circumstances, for they were too fond of their liquor themselves.

"When men became riotous, gentle means were first to be employed to quiet them; if they still persisted, warnings followed; and should they then remain contumacious, they were to be well thrashed and sent home 'as cheaply as possible.' Table and window breaking were severely punished, and certain acts of indecency, if practised before ladies, were to be resented by seizing the offender and pitching him neck-and-crop into the street.

"Should the reader be desirous of studying this remarkable code (whatever view he may take of its authenticity as a serious production,) he will find it composed in mediæval German, interspersed with Latin and Greek phrases, as though it had been collated by some learned ecclesiastic, which is more than probable—that is to say, by some drunken hanger-on at a monastery; and he will see how the German youth of by-gone days studied as 'vini et cerevisiæ candidatus,' and eventually graduated in the courts of Bacchus. But if he imagines that the picture is overdrawn, we should recommend him to consult the historical records, and he will find that no language can adequately portray the state of morals in her

9

many in those days, at least so far as drunkenness was concerned." *

Aug. de Thou, in his Memoirs, liv. 11, describes scenes which he himself witnessed.

"There is before Mulhausen, a large place or square, where, during the fair, assemble a prodigious number of people of both sexes, and of all ages; there one may see wives supporting their husbands, daughters their fathers, tottering upon their horses or asses, a true image of a Bacchanal. The public houses are full of drinkers, where the young women who wait, pour wine into goblets, out of a large bottle with a long neck, without spilling a drop. They press you to drink, with pleasantries the most agreeable in the world. People drink here continually, and return, at all hours, to do the same thing over again."

Quite as strong is the testimony of De Rohan in his Voyages, published in 1646:

"I am well satisfied that the mathematicians of our time, can nowhere find out the perpetual motion, so well as here, where the goblets of the Germans are an evident demonstration of its possibility. They think that they cannot make good cheer, nor permit friendship or fraternity, as they call it, with any, without giving the glass brimful of wine, to seal it for perpetuity." p. 27.

Thackeray, in his Lectures on the Four Georges,† says of life among the German gentry in 1600:

"Every morning at seven, the squires shall have their morning soup, along with which, and dinner, they shall be served with their under-drink—every morning, except Friday morning, when there was sermon, and no drink. Every evening they shall have their beer, and at night their sleep-drink."

All accounts unequivocally agree that both the higher and the lower classes, the religious and the indifferent alike, were debauched by drink. The excesses and cruelties resulting therefrom in the so-called noble families, were beastly and fiendish, and worse than all were gloried in, and their remembrance paraded and perpetuated by means

---

* The History of Drink. By James Samuelson, pp. 107, 111.
† Lecture I.

of family records kept and handed down from generation to
generation; not only of the exploits of the men but also of
the women, in their almost constant indulgences. The
most acceptable gift which one could bestow on another,
was a large and handsome shaped gold drinking cup or
goblet, which was supposed to be greatly enriched in proportion as it was covered with accounts of the drinking
exploits of its owner. Almost every event in life, from
birth to death, was celebrated by drinking, and bargains of
whatever kind were concluded over the cup, a stipulated
amount for the supply of which, formed part of the most
trifling contract. The Pope's representative at the Court of
Frederick III., wrote home to his master, —"Living here is
naught but drinking." Wine was so cheap that it became
a proverb :

> "In fifteen hundred and thirty-nine,
> The casks were valued at more than the wine."

" At the beginning of this century," says Keysler, "Germany
saw three empty wine casks, from the construction of which no
great honor could redound to our country among foreigners.
The first is that of Tubingen; the second that of Heidelberg; and
the third, at Gruningen, near Halberstadt; and their dimen-
sions are not greatly different : the Tubingen cask is in length
24, in depth 16 feet; that of Heidelberg, 31 feet in length, and
21 deep; and that of Gruningen, 30 feet long, and 18 deep.
These enormous vessels were sufficient to create in foreigners a
suspicion of our degeneracy ; but to complete the disgrace of
Germany, in the year 1725, a fourth was made at Konigstein,
larger than any of the former." *

Unfortunately, in tracing out the causes or occasions for
this extensive demand for wine, we are forced to place the
responsibility for it, in a large measure, on those who under
Charlemagne established what they thought were the institutions of the Christian church in Germany. The founding
of the monastery was supposed to require the use of wine
in the celebration of Mass, and so necessitated the planting

---

* Travels, vol. I. p. 97.

of vines to supply that demand. Soon the desire for more wine to please the palate, brought a large portion of the grounds connected with the religious houses under cultivation, until at last the long famous vineyards have been those planted by the monks.

Bridgett * quotes from a sermon preached by Rabamus Maurus in the beginning of the ninth century.

"There are some vices, dearest brethren, which, though very great, yet in our days to some appear so small, that they reckon them either the least of evils or no evils at all. They have so spread by the abuse of men that instead of being blamed as crimes and sins, they are praised as if they were virtues. * * * * Among these vices feasting and drunkenness especially reign, since not only the rude and vulgar people, but the noble and powerful of the land, are given up to them. Both sexes and all ages have made intemperance into a custom. * * * And so greatly has this plague spread that it has infected some of our own order in the priesthood, so that not only they do not correct the drunkards, but become drunkards themselves. Oh! what wickedness is this, brethren, what bitter evil is this, which does not leave unhurt even rulers and dignities, until virtues are spoken ill of, and vices extolled ? * * * * Tell me, you who praise feasting and drunkenness, whether it is a good thing or an evil, to extinguish the light of the mind by excess, to disturb the reason, to obscure the sight, to lose speech and the use of the limbs, and to become like a madman or one possessed? Did God make man thus? Is this the glory of God's image, of which it is written : 'God made man in His own likeness ?' What an intolerable blasphemy would it be to assert such a thing, when God is supremely good and alone blessed and powerful, the King of kings and Lord of lords. ' He saw all things that He had made, and behold they were very good.' Hence drunkenness was not made by him."

"In the monastery of St. Gall, during the tenth century, each monk received daily five measures of beer, besides occasional allowances of wine, which were consumed at breakfast, dinner, and supper ; and healths were often pledged by the abbots." †

"When the Pope reproved the German priesthood for their

---

* The Discipline of Drink.   By the Rev. T. E. Bridgett, C. S. S. R. pp. 49, 50.

† Eckehard's, Jun., quoted by Samuelson, p. 114.

luxurious habits, they uproariously returned for answer: 'We have no more wine than is needed for the Mass; and not enough to turn our mills with!'"

Good living, as it was erroneously called, was certainly, at one time, an universal observance in Germany, when the sole wish of man was, that he might have short sermons and long puddings. When this wish prevailed, every dining-room had its *faulbett*, or sot's couch, in one corner, for the accommodation of the first couple of guests who might chance to be too drunk to be removed. Indeed, in German village inns, the most drunken guests were, in former days, by far the best off: for, while they had the beds allotted them, as standing in most need of the same, the guests of every degree, whether rich or poor, the perfectly sober, wherever such phenomena were to be found, and those not so intoxicated but they could stagger out of the room, all lodged with the cows among the straw. Probably, no country on the earth presented such scenes, arising from excessive drinking, as were witnessed in Saxony and Bohemia, a few generations back. These scenes were so commonly attended by murder, or followed by death, that it was said to be better for a man to fall among the thickest of his enemies fighting, than among his friends when drinking. There were deadly brawls in taverns, deadly feuds in the family circle, and not less deadly contentions in the streets. * * * This is no overdrawn picture of an ancient German period.

"It is on record that once, on the banks of the Bohemian Sazawa, a party of husbandmen met for the purpose of drinking twelve casks of wine. There were ten of them who addressed themselves to this feat; but one of the ten, attempting to retire from the contest before any of his fellows, the remaining nine seized, bound him, and roasted him alive on a spit. The murderers were subsequently carried to the palace for judgment; but the Duke's funeral was taking place as they entered the hall, and the Princes who administered justice were all so intoxicated, that they looked upon the matter in the light of a joke that might be compensated for by a slight fine. There was a joyous revelry at that time in every direction. A father would not receive a man for a son-in-law who could not drink ; and in universities the conferring of a degree was always followed by a carouse, the length of which was fixed, by College rules, as not to exceed eight hours' duration." *

The following, from the same author, given as an extract from a sermon by the Bishop of Triers or Treves, will show how religious teachers became debauched, and so corrupters of their people, by wine :

"Brethren, to whom the high·privilege of repentance and penance has been conceded, you feel the sin of abusing the gifts of Providence. But, *abusum non tollit usum.* It is written, 'Wine maketh glad the heart of man.' It follows, then, that to use wine moderately is our duty. Now there is, doubtless, none of my male hearers who cannot drink his four bottles without affecting his brain. Let him, however,—if by the fifth or sixth bottle he no longer knoweth his own wife,—if he beat and kick his children, and look on his dearest friend as an enemy,—refrain from an excess displeasing to God and man, and which renders him contemptible in the eyes of his fellows. But whoever, after drinking his ten or twelve bottles, retains his senses sufficiently to support his tottering neighbor, or manage his household affairs, or execute the commands of his temporal and spiritual superiors, let him take his share quietly, and be thankful for his talent. Still, let him be cautious how he exceed this : for man is weak, and his powers limited. It is but seldom that our kind Creator extends to any one the grace to be able to drink safely sixteen bottles, of which privilege he hath held me, the meanest of his servants, worthy. And since no man can say of me that I ever broke out in causeless rage, or failed to recognize my household friends or relations, or neglected the performance of my spiritual duties, I may, with thankfulness and a good conscience, use the gift which hath been intrusted to me. And you, my pious hearers, each take modestly your allotted portion ; and to avoid all excess, follow the precept of St. Peter, —'Try all, and stick to the best.'" †

A story is told of a German Bishop, named Defoucris, that, being exceedingly fond of wine, it was his custom in travelling to send his valet forward a post, with instructions

---

* Table Traits, with Something on Them. By Dr. Doran, pp. 263-265.

† Ibid, p. 262.

that he should taste the wine at every place where he stopped, and write under the " bush," (a bunch of evergreens hung up over the entrance of houses in Italy where wine was sold,) the word "est," if it was tolerable, and " est, est," if it was very good ; but where it was indifferent, he should not write anything. The valet arrived at Monte Fiascone, and so much admired the wine that he wrote up " est, est." The bishop soon followed, found the wine so palatable that he got drunk, and repeating the experiment too often, drank himself dead. His valet thereupon wrote his epitaph, as follows :

> " ' Est, est,' propter nimimum ' est,'
> Dominus meus, mortuus ' est.' "

Which may be rendered :

> ' 'Tis, 'tis,' from too much ' 'tis,'
> My master dead ' is.'

A sadly blasphemous custom of mixing sacred things with the most profane prevails to this day in Germany; as Mrs. Trollope mentions having seen over the door of a brewery in the city of Bruges ; a group in alto-relievo, representing the whole process of brewing; several figures are employed in mashing, cooling, and putting the beer into casks, while winged seraphs are seen tasting the liquid, and the Blessed Virgin and her infant are admiringly looking on ! *

From time to time, owing, as we shall see in another chapter, to religious efforts and also to civil enactments, temperance checks have been placed on the downward tendency of such indulgence in wine and beer, but the general drift, is here as everywhere else where wine and beer are used, to the demand for more potent liquors. Morewood, an English Surveyor of Excise, and writing in the interest of the liquor trade in England, and so above suspicion of being prejudiced by any Temperance notions, says: " In Germany,

---

* Belgium and Western Germany, Vol. I. p. 28.

of late years, distilleries have increased, while breweries have decreased in the same ratio." *

Student life in Germany is still beset by drunkenness from beer drinking. "An American Student" who has recently concluded a course of study at Leipzig, writes the following:

" When the student has made his examination, after his hard pull of the last six months, hilarity reigns supreme. He gives drinking feasts first to all his acquaintances, and last to his most intimate friends. Wild is the sport, and no one unless he is a cynic leaves the hall of friendship in a presentable condition. The German student is by no means modest in his beer. The most quiet and sedate speak openly of being slightly intoxicated, as if it were, as it is here, a mere matter of course; but at an entertainment consequent on having made his degree, he is indeed a cold friend who does not complain on the morrow of excessive feline combats in the regions of the brain. It is more than probable that coming from the scene of festivity, a desire to sing on the street occurs. No police is more strict than the German, and semi-wild singing on the street brings down the entire police force. The students are brought to the police station, the college beadle conducts them politely to the college prison, where they remain a few days living on the fat of the land, and seeing their friends whenever they wish; only one however being allowed admission at one time." †

Rev. William F. Warren, D. D., now President of Boston University, testified before a committee of the Legislature of Massachusetts, in 1867, that he spent eight years in Europe, for the most part in Germany.

" The result of my observation was, that there was double the amount of drinking and of drunkenness among the students that there is among the same class in this country. As regards the people, I can only say that during the last five years, drunken people have gone past my house, I suppose, every evening, sometimes boisterously drunk and sometimes reelingly drunk. In a street but a few rods from where I lived, there were brawls almost every Sunday afternoon." ‡

---

* Essay on Inebriating Liquors, p. 459.
† Universalist Quarterly, July, 1878, p. 356.
‡ Report and Testimony, 1867, p. 807.

In an account given by Rev. William Reid,* of the "German Protestant Conference for Inner Missions, at Bremen, Sept. 16, 1852," are quite copious extracts from a Paper by Dr. Wald, on the Progress of Intemperance in Germany, to the effect that the use of Brantivein, (the general name for distilled liquors,) had increased nine-fold in thirty-five years; and in consequence, prisons and lunatic asylums were being overcrowded, ignorance was increasing, and physical deterioration was becoming so general, that on the occasion of a conscription in one district, "out of one hundred and seventy-four young men, only four were declared admissible by the reviewing army surgeons, the rest being physically incapacitated by the use of alcohol."

IX. RUSSIA.—In this great Empire drunkenness is not only unchecked by the government, but the distilleries and liquor stores, yielding more than one-third the entire revenue of the nation, it is encouraged and sometimes enforced. The chief drink is *Vodki*, or corn brandy.

"Until 1752 it was farmed for £540,000; until 1774 for £900,-000; and until 1778 for £1,500,000; in 1779 it was let for four years, at the sum of £1,800,000; since which time it has been gradually increasing. So far back as 1789 the licenses to inns and taverns yielded £1,708,338, and the brandy sold in the cities of Petersburg, Moscow, and the parts adjacent, amounted to 3,330,000 rubles per annum; but this is not remarkable, when, in the city of Moscow alone, there were no fewer than 4,000 kabaks or shops for the retail of brandy. The crown, or rather the chamber of revenue, farms all the kabaks, and the contractor or merchant who supplies them with spirits is prohibited from distilling himself, but is obliged to buy all from the functionaries of government, who either draw the brandy from their own distilleries, or obtain it by contract from those of the privileged provinces." †

In 1847 the brandy monopoly yielded a revenue to the crown, of £9,774,176. In 1854, this was the condition of affairs :

---

* The Temperance Cyclopædia, p. 388.
† Morewood, p. 253.

"In the central provinces the farmer of the duty on spirits buys the assistance of the local authorities, and between them it is arranged that all business shall be carried on at the public-house, glass in hand. In the other provinces, where the farmer of the duty has also an exclusive right of sale in his own district, he makes each commune take a certain quantity per head, or else he forces the peasants to pay a certain sum for permission to buy spirits elsewhere, threatening, in case of refusal, to accuse them of a breach of the revenue laws ; and they know that whether innocent or guilty, if once accused, they are sure to be condemned. The result is, in the words of Haxthausen (a favorable authority), that in the provinces of Central Russia, the peasants are *seduced* into drunkenness, while in the other provinces they are *forced* into it." *

. And William Howitt, in his work on the Revenue of Russia, written before the abolition of serfdom, said : " The nobles of Russia, who own vast numbers of serfs, are rather pleased than otherwise to find them indulging in drink ; it blinds them to their degradation ; and, in their cups they forget that they ought to be free men." On this head, he says:

"Notice the remark of a writer, in a work recently published, viz., ' Take care how you advise a Russian noble to proscribe drunkenness in his domains,' the noble is so enchanted with the *happiness !* it procures for his peasants, that, far from putting any obstacle to it, he encourages it with all his power. The government supports a considerable number of public-houses on the land of the nobles, from whence a large revenue is drawn. Again, let the reader mark an important fact, viz., *The Temperance Societies have never been able to take root in Russia.*"

A writer in the *Pall Mall Gazette*, says that about 1859, as these farmers, having a monopoly of the trade, were charging exorbitant rates for their liquors, some of the peasants,

" Banded themselves into temperance societies, with a view to forcing down the prices. Hereupon the farmers complained to Government, and the teetotal leagues were dissolved, as illegal secret societies ; and summary measures were taken towards forcing the people to contribute to the revenue by their intem-

---

* Gentleman's Magazine, 1854, pp. 481-2.

perance. Policemen and soldiers were sent into the disaffected districts, and the teetotallers were flogged into drinking; some who doggedly held out had liquor poured into their mouths through funnels, and were afterwards hauled off to prison as rebels; at the same time the clergy were ordered to preach in their churches against the new form of sedition, and the press-censorship thenceforth laid its veto upon all publications in which the immorality of the liquor traffic was denounced. In 1865 the people fancied that because they were no longer serfs they could not be treated so unceremoniously as of yore, but they found out their mistake. They were simply dealt with as insurgents, and, though not beaten, were fined, bullied, and preached at till there was no spirit of resistance left in them. However, this new rising led to the abolition of the monopolies. An excise was substituted, the price of vodki fell by competition, and the lower orders of Russia are now drunker than ever. According to the latest returns (Wesselowski's *Annual Register*,) the liquor duties yield the revenue 800,000,000 roubles (£32,-000,000 sterling) a year." Equal to, $160,000,000.

"Before the abolition of the monopolies a land-owner might set up a distillery on his estate, but he was compelled to sell the produce to the vodki farmers, and these speculators might build a public-house on his land against his consent, though he was entitled to fix the spot and to receive a fair rent. At present, the trade being free, licenses to distil and sell are conferred by Government (*i. e.*, virtually bought of the Tschinn,) and almost every land-owner of consequence has one. Prince Wiskoff might get one if he pleased, and has more than once thought of doing so; but he has been deterred for want of capital to compete with his neighbor, Prince Runoff, who has a distillery in full swing, and floods the whole district with its produce.

"The Prince's chief agents are the priests, who in the farming days were allowed a regular percentage on the drink sold in their parishes, but who now receive a lump-sum, nominally as an Easter gift, but on the tacit understanding that they are to push the sale of vodki by every means in their power. These pious men do not go the length of urging their parishioners to get drunk, but they multiply the Church feasts whereon revelry is the custom; they affirm that stimulants are good for the health, because of the cold climate, and they never reprove a peasant whose habitual intemperance is notorious. The Prince's land agent, the tax collectors, the conscription officers, all join in promoting the consumption of vodki by transacting their business at the village dram-shop, with glasses before them;

and even the doctor, who lives by the Prince's patronage, prescribes vodki for every imaginable ailment. The inducements
to drink in the towns are not less than in the country. When
the coachman, Ivan Ivanowitch, goes out for a stroll among
the fine shops of Odessa, he is lured into the tea shops by the
loud music of the barrel-organs, and vodki is served him with
his tea, as a matter of course. If he drives his master to a party,
he has no sooner drawn up his trap under the shed in the host's
yard, than the servants invite him into a lower room and give
him as much spirit as he will drink; if he goes to the corn-
chandler's for oats, to the veterinary surgeon about his horse's
legs, to the harness makers or coachmakers, the preface to all
business is vodki; and when he sets out to visit his kinsman
upon holidays, vodki greets him upon every threshold. It is
the same with the doornick when he ascends to the different flats
of the house to collect rent or carry letters; vodki is offered him
before he has time to state his business; and under these hos-
pitable circumstances the wonder is not that the man should
occasionally exceed sobriety, but that he should so often be
sober. But in Russia a sober servant means—*exceptis excipiendis*
—one who only gets drunk upon the festivals of the Church."
"We have not," says a recent writer from St. Petersburg, "a
single temperance society; and the intemperance cause in Russia is flourishing." *

IX. ENGLAND.—Although English history begins with
the landing of Julius Cæsar on the Island of Britain, in
the year 56 B. C., all ancient writers agree that the inhabitants found there by the Romans, were a tribe of the Gauls,
who at a not very remote period had emigrated from the
neighboring continent.† Doubtless they took with them
the general manners and customs of their Fatherland,
among which was the manufacture and use of beer. According to Macrobius, the Gauls had no knowledge of the
cultivation of the vine till Rome had arrived at a high
state of prosperity. Some Roman wine, given by a Helve-
tian to the Gauls, so delighted them, that they were induced
to attack the Roman capital in order to obtain unlimited

---

* Centennial Temperance Vol., p. 354.
† Hume's History of England, Vol. I. chap. i.

supplies of this beverage.* Subsequently the use of the
vine was taught in Gaul, but no advances were made in its
culture till the arrival and conquests of the Romans. So
late as the sixth century, beer was the common drink in
Paris, a circumstance which drew from Julian, who had
been appointed Cæsar for Gaul, the following epigram:

> "Whence art thou, thou false Bacchus, fierce and hot?
> By the true Bacchus, I do know thee not!
> He smells of nectar ;—thy brain-burning smell
> Is not of flowers of heav'n, but weeds of hell.
> The lack-vine Celts, impoverish'd, breech'd, and rude,
> From prickly barley-spikes thy beverage brew'd:
> Whence I should style thee, to approve thee right,
> Not the rich blood of Bacchus, bounding, bright,
> But the thin ichor of old Ceres' veins,
> Express'd by flames from hungry barley grains,
> Child-born of Vulcan's fire to burn up human brains."

Mead was also held in great esteem by the ancient Britons.

" In the court of the ancient Princes of Wales the mead maker
was held as the eleventh person in point of dignity. By an ancient law of the principality, three things in the principality
were ordered to be communicated to the King, before they were
made known to any other person. First, every sentence of the
judge ; second, every new song; and third, every cask of
mead." †

When the Britons were finally subdued by the Romans,
near the close of the first century, they begun speedily to
imitate the manners of their conquerors, not simply says
Tacitus,‡ learning the Latin language, style of architecture
and modes of dress; but "by degrees they fell even into a
relish of our vices."

On their abandonment by the Romans, early in the fifth
century, they were in danger of being overrun by the Picts
and Scots, when a few Saxons coming to their relief, the
invaders were defeated. In turn, the Saxons joined by the

* See also Livy's History, Book V. sect. 33.
† Bacchus, p. 208.
‡ Agricola, xxi.

Angles, conquered the natives, and established the Anglo-Saxon government. We have already seen from the testimony of Tacitus, that the German people were, early in their history, known to us as drinkers, and Wright, in his history of Domestic Manners and Sentiments in England, says of the Saxons, who were a German tribe, that it is "evident from the Romance of Beoroulf,* that they were drinkers before they settled in Britain." "Their drinking cups," he says, are frequently found in their burrows or graves."† Miller, says that "the Saxons were hard drinkers, mead, wine and ale flowed freely at their feasts."‡ Turner describes their drinks as "wine, mead, ale, pigment, morat and cider. The pigment was a sweet and odoriferous liquor, made of honey, wine and spiceries of various kinds. The morat was made of honey diluted with the juice of mulberries. Three sorts of ale are mentioned, clear, Welsh and mild."§ Feasting was frequent with them, often uproariously jolly, and not unfrequently ending in strife and bloodshed. Wright gives a translation of the legend of Juliana, in which the Evil Spirit describes his influence at the festive board :

> " Some I by wiles have drawn
> To strife prepared,
> That they suddenly
> Old grudges
> Have renewed,
> Drunken with beer;
> I to them poured
> Discord from the cup,
> So that in the social hall,
> Through gripe of sword,
> The soul let forth
> From the body." ‖

---

* See copious extracts from this poem in Taine's History of English Literature. Book I. chap. i.

† Pp. 2, 5.

‡ History of the Anglo-Saxons. By Thomas Miller, p. 359.

§ History of the Anglo-Saxons. By Sharon Turner, Vol. II. pp. 203, 204.

‖ Homes of Other Days, p. 50.

Indulgence in the intoxicating cup was not confined to the secular days, nor to those who made no profession of religious faith and conviction; but was common with the clergy and with their congregations. Wright [*] quotes this record from the Ecclesiastical Institutes : "It is a very bad custom that many men practise, both on Sundays and also other Mass-days ; that is, that straightways at early morn they desire to hear mass, and immediately after the mass, from early morn the whole day over, in drunkenness and feasting they minister to their belly, not to God."

St. Boniface writes in the eighth century to the Archbishop of Canterbury :

"It is reported that in your dioceses the vice of drunkenness is too frequent ; so that not only certain bishops do not hinder it, but they themselves indulge in excess of drink, and force others to drink till they are intoxicated. This is certainly a great crime for a servant of God to do or to have done, since the ancient canons decree that a bishop or a priest given to drink should either resign or be deposed." [†]

As it became necessary to define the extent to which intoxication should go in order to be improper or penal, the following definition was given: "This is drunkenness, when the state of the mind is changed, the tongue stammers, the eyes are disturbed, the head is giddy, the belly is swelled, and pain follows." [‡]

In the Pagan ceremonies of the Saxons, the first day of November was dedicated to the Angel presiding over fruits, seeds, etc., and was therefore called " La Mas Ubhal, *i. e.*, the day of the apple fruit; and being pronounced Lamaswool, the English have corrupted the name to Lamb's-wool." [§] When the Saxons became Christians, their Papal teachers made no attempt to abolish wassailing, as the observance

---

[*] Domestic Manners and Sentiments, p. 77.

[†] Discipline of Drink, p. 77.

[‡] Spelm, Concilia, 286. Quoted by Turner, ii. p. 204 ; attributed by Bridgett, p. 148, to Abp. Egbert.

[§] Vallancey, Collectanea, de Rebus Hibernicos, Vol. iii. p. 464.

of this feast was called, but on the contrary, caused it to assume a kind of religious aspect, conformed in some respect to their new religious views. The wassel bowl was placed, in the great monasteries, on the Abbot's table, at the upper end of the refectory or eating hall, to be circulated among the community at his discretion, whence it received the honorable name of "Poculum Charitas." Still in use among the students of the English Universities, it is called the Grace Cup.* One verse of ancient song thus describes the mixture:

> "Next crown the bowle, full
>   With gentle lambs-wool; †
> Add sugar, nutmeg, and ginger,
>   With store of ale too;
> And thus ye must doe
>   To make the Wassaile a swinger." ‡

Fosbrooke,§ states that both the monasteries and convents of the Anglo-Saxons were nurseries of the worst imaginable vices; that dissolute nobles and other persons of rank and wealth often purchased crown lands on pretence of founding Religious Houses, and that making themselves abbots, they gathered about them dissolute monks who had been expelled from the more strictly managed monasteries, and brought their wives and other women into their monasteries. Some of the nunneries were dissolute, especially at Coldingham, where the nuns are said to have spent their time in feasting, drinking and gossipping; to have "employed themselves in working fine clothes, dressing themselves like brides, and acquiring the favor of strange men."

The Church authorities, as we shall see, fought against these evils, but even Bridgett concedes that it was not always with much success.

"A monastery was sometimes a village or town, with many

---

* Milner, Archæologic, Vol. xi. p. 420.
† Roasted apples.
‡ Herrick's Hesperides, p. 376.
§ British Monachism, vol. i. pp. 16, 17.

hundred inmates. Most of these were laymen. They were re-
cruited from all classes of society, and great criminals, no less
than those who had been always pious and innocent, thronged
into them. It would have been strange had they not brought
with them some of their old bad habits. Again, long fasting
united with hard manual labor, was their daily discipline. No
wonder that when the refreshment hour came, the beer got into
the heads of some." *

The various accounts of the customs in both high and
low life, show that excessive drinking was common with all
classes.

" We have an account of Ethelstan's dining with his relative
Ethelfleda. The royal providers, it says, knowing that the
king had promised her the visit, came the day before to see if
every preparation was ready and suitable. Having inspected
all, they told her: ' You have plenty of everything, provided
your mead holds out.' The king came with a great number of
attendants, at the appointed time, and after hearing mass,
entered joyfully in the dinner apartment ; but unfortunately in
the first salutation, their copious draughts exhausted the mead
vessels. Dunstan's sagacity had foreseen the event and provided against it ; and though ' the cup-bearers, as is the custom at royal feasts, were all the day serving it up in cut horns,
and other vessels of various sizes,' the liquor was not found to
be deficient. This, of course, very much delighted his majesty
and his companions, and as Dunstan chose to give it a miracu-
lous appearance, it procured him infinite credit." †

Wright quotes the Chronicler Wallingford, as saying of
an early Saxon dinner party, that " after dinner they went to
their cups, to which the English were very much accustomed." He also shows, from the story of Dunstan and
king Eadeny, that it was considered a mark of disrespect to
the guests, even in a king, to leave the drinking early after
dinner. " In the latter part of the day they were accustomed to sitting in their halls and drinking. At such times
they rehearsed their adventures, sung songs, and made proof
of their powers in hard-drinking." † From an Anglo-Saxon

* Discipline of Drink, p. 136.
† Turner, Vol. ii. p. 202.
† History of Domestic Manners and Sentiments, pp. 30, 168.

poem entitled Judith, an extract is given, showing that the idea of a more ancient feast is borrowed from scenes in the time of the writer:

> "Then was Holofernes
> Enchanted with the wine of men:
> In the hall of the guests
> He laughed and shouted,
> He roared and dinned,
> That the children of men might hear **afar,**
> How the sturdy one
> Stormed and clamored,
> Animated and elated with wine.
> He admonished amply
> Those sitting on the bench
> That they should bear it well.
> So was the wicked one all day,
> The lord and his men,
> Drunk with wine;
> The stern dispenser of wealth;
> Till that they swimming lay
> Over drunk,
> All his nobility
> As they were death slain,
> Their property poured about.
> So commanded the lord of men
> To fill to those sitting at the feast,
> Till the dark night
> Approached the children of men." *

The condition of holding and occupying the crown lands was often made to be the supplying of the King, at stated times, with liquors. Thus, in the reign of King Edward, A.D. 901: "One William de Insula held one carucate of land, with the appurtenances, in West Hundred, by the Serjeanty of buying ale for the use of our Lord the King, and it is worth by the year one hundred shillings."† A carucate of land is defined as being no certain quantity, " but as much as a plow can, by course of

---

* Turner, vol. ii. p. 204.

† Blount's Ancient Tenures of Land, and jocular Customs of some Manors, pp. 63,131, 133.

husbandry, plow in a year; and may contain a messuage, wood, meadow and pasture."

In the same reign, "Bartholomew Peytenyn holds two carucates of land at Stoney Aston in the County of Somerset, of our Lord the King in Capite, by the service of one Sextary of Clove wine [about a pint and a half of spiced wine] to be paid to the King yearly, at Christmas. And the said land is worth ten pounds a year." In the reign of Edward the Confessor, A.D. 1042, "John de Roches holds the Manor of Winterslew in the County of Wilts, by the service, that when our Lord the King should abide at Clarendon, he should come to the Palace of the King there, and go into the Buttery, and draw out of any vessel he should find in said Buttery, at his choice, as much wine as should be needful for making a pitcher of Claret, which he should make at the King's Charge; and that he should serve the King with a cup, and should have the vessel from whence he took the wine, with all the remainder of the wine left in the vessel, together with the cup from whence the King should drink that Claret."

In the reign of King John, A.D. 1199, "Walter de Burgh and his Partners, hold sixteen Pounds land [as much as would pay a yearly rent of an English Pound of twenty shillings, ordinarily fifty-two acres,] in Rakey in the County of Norfolk, by the Serjeanty of paying two Muids [two hogsheads] of Red Wine and two Hundred Pears called Permeines, to be paid at the Feast of St. Michael yearly, at the King's Exchequer." In the same reign, "Walter de Hevene held the Manor of Runham in the County of Norfolk, in Capite of our Lord the King, by the Serjeanty of two Muids of Wine made of Permains to be paid to the King at his Exchequer, yearly, at the feast of St. Michael."

Not far from the year 1000, King Edgar endeavored to check the vice of drinking, and to put an end to the disputes and violence arising from the practice of handing round to the guests on every social occasion, a common drinking vessel of large size, which they were expected to attempt to rival each other in draining. He ordered that all such vessels should be made with knobs or pegs of brass, at certain intervals, so that no one should be compelled to drink more at a draught than from one peg to another. Before long it became customary to insist that the full space between the pegs should be drained, and thence to see how many could, undetected, exceed the allowance, till at last it was customary to say of one who became inebriated sooner than the others, "he has got a peg too low." *

"The two gallon measure had eight pegs: and the half pint, from peg to peg, was deemed a fitting draught for an honest man; but as the statute, or custom, did not define how often the toper might be permitted to indulge in this measure, people of thirsty propensities got rather more inebriated than they had dared to be previously. As the half-pint was roughly set down as the maximum of their draught, it was a point of honor with them never to drink less,—and to drink to that extent as often as opportunity offered."†

The Danes, who at this time had made settlements in various parts of England, were notoriously hard drinkers, their soldiers setting no bounds to their debaucheries. Their habits in this respect caused them frequent surprises in their camps. The visit of King Alfred, in the disguise of a minstrel, to the camp of Gunthrum, where he found the soldiers steeped in drunkenness, is one of many incidents of a similar nature.

The Ramsey History tells a story of a Saxon bishop, who invited a Dane to his house in order to obtain some land from him, and that he might drive a better bargain, he determined to make his guest drunk. He therefore pressed him to prolong his stay, and when they had all eaten enough at dinner, "the tables were taken away, and they passed the rest of the day, till late in the evening,

---

* Club Life of London, vol. ii. p. 111.
† Doran's Table Traits, p. 298.

drinking. He who held the office of cup-bearer managed that the Dane's turn at the cup came round oftener than the others, as the bishop had directed him."*

The Danes, it is said, so tyrannized over the conquered Saxons as not to allow them to drink in their presence without first asking permission, under penalty of death; a regulation which so terrified the Saxons that they dared not even take advantage of the privilege to drink unless a pledge was given that they should not be harmed in consequence of it. Hence the custom of pledging in drinks.

Towards the close of the eleventh century, the Normans conquered England. For a time there was a marked contrast in their habits and those of the conquered Saxons. But soon they learned the vices of the people they had subdued by arms. William of Malmsbury is quoted as saying of the Saxons at this period : " They passed entire days and nights in drinking." Rioting in gluttony and drunkenness, they were " accustomed to eat till they became surfeited, and to drink till they were sick. These latter qualities they imparted to their conquerors : whose manners in other respects they adopted. " †

In the twelfth century drunkenness was on the increase, and more attention to the description and praise of intoxicants was given by writers. One thus enumerates the qualities of good wine :

"It should be as clear as the tears of a penitent, so that one may see distinctly to the bottom of the glass. Its color should represent the greenness of the buffalo's horn ; when drunk, it should descend impetuously like thunder, sweet-tasted as an almond, creeping like a squirrel, leaping like a roebuck, strong like the building of a Cistercian monastery, glittering like a spark of fire, subtle as the logic of the schools of Paris, delicate as fine silk, and colder than crystal." ‡

---

* Wright, p. 30.
† Ibid, p. 81. Also Taine, Book I. chapter ii.
‡ Wright, p. 90.

The monks were so gluttonous and dissipated that even contemporary ecclesiastical writers of this century upbraided them in severe terms. Giraldus Gambrinus, one of these writers, complains with great indignation of the table kept by the monks of Canterbury; and he relates this incident to show that the clergy were more extravagant in this respect than even the highest among the laity.

"One day, when Henry II. paid a visit to Winchester, the prior and monks of St. Swithin met him, and fell on their knees before him to complain of the tyranny of their bishop. When the king asked what was their grievance, they said that their table had been curtailed of three dishes. The king, somewhat surprised at this complaint, and imagining, no doubt, that the bishop had not left them enough to eat, inquired how many dishes he had left them. They replied, ten; at which the king, in a fit of indignation, told them that he himself had no more than three dishes at his table, and uttered an imprecation against the bishop, unless he reduced them to the same number." *

Bridgett bears testimony to the drunkenness of the clergy in the following century, in recording that when the advocate of the Bishop of Worcester appeared before the Pope to argue against the exemption of Evesham Abbey, he said : " Holy Father, we have learned in the schools, and this is the opinion of our masters, that there is no prescription against the rights of bishops : " the Pope replied :

"Certainly, both you and your masters had drunk too much English beer when you learnt this." †

He also quotes an Archdeacon as adding, after extolling the zeal of the Irish clergy, of this period :

"Among so many thousands you will not find one who, after all his vigorous observance of fasts and prayer, will not make up at night for the labors of the day, by drinking wine and other liquors beyond all bounds of decorum." ‡

It was at the latter part of the twelfth century, or early in the thirteenth, that duties were levied on imported wines, and

---

* Ibid, p. 248.　† Discipline of Drink, p. 79.　‡ Ibid, pp. 79, 80.

"A small license of four pence a year was paid by brewers. The publican sold the liquor he brewed himself, and was forbidden to convey it to another burgh for sale. Outside a burgh no one could have a brew-house unless he had in the place furcam and fossam,—'gallows and pit.' [The gallows for hanging men, the pit for drowning women.] No one could sell ale unless it had been brewed for sale and previously tasted. The provost and other public officials of the burgh were altogether forbidden to brew ale or bake bread for sale; no doubt lest they should be bribed indirectly in the administration of justice, or lest they should draw customers by intimidation. Public tasters were appointed, who had to make oath to taste and lawfully apprise the ale, according to the price of malt, and in so doing to spare or favor no one. The brewing and selling of the ale seems to have been an exclusively female occupation. One law runs as follows: 'What woman that will brew ale to sell shall brew all the year through, after the custom of the town. And if she does not, she shall be suspended of her office by the space of a year and a day. And she shall make good ale and approvable as the time asks. And if she makes evil ale, and does against the custom of the town, and be convicted of it, she shall give to her amercement eight shillings, or be put on the cuckstool, and the ale shall be given to the poor folk, the two parts, and the third part sent to the brethren of the hospital. And right so doom shall be done of mead as of ale. And each brewer shall put her ale-wand outside her house at her window or above her door, that it may be visible to all men. And if she do not she shall pay 4d fine." *

"The same writer quotes Burton's Annals to the effect, that "in A.D. 1200, prices were fixed for the different kinds of wines, both wholesale and retail. The retail price, however, was found to be impracticable, as it allowed no profit, and was immediately changed, and so the land was filled with drink and drinkers." The original law in regard to duties paid by importers was: "that the King seized one tun before and one behind the mast." In the time of Henry III. the duty was changed to "one penny on a tun."

It was in his reign also, A.D. 1266, that the price of beer or ale was established by law, an act being passed

---

* Ibid, pp. 120-122.

which provided "that when a quarter of wheat is sold for 3s. or 3s. 4d., and a quarter of barley for 1s. 8d., and a quarter of oats for 1s. 4d., then brewers in cities ought, and may well afford to sell two gallons of beer or ale for a penny; and out of cities to sell three or four gallons for a penny." This was supplemented the same year by a "statute enacting penalties against brewers and venders who charged too much." Subsequently the justices in each shire, and the mayor and sheriffs of the cities fixed the price, and "every beer and ale brewer was forbidden to take more than such prices and rates as should be thought sufficient" by these authorities. *

Wickliffe, denouncing the clergy, in the middle of the fourteenth century, says

"That they haunt taverns out of measure, and stir up laymen to drunkenness, idleness, and cursed swearing, chiding and fighting. For they will not follow earnestly in their spiritual office, after Christ and his apostles, therefore they resort to plays at tables, chess and hazard, and roar in the streets, and sit at the taverns till they have lost their wits, and then chide and strive, and fight sometimes. And sometimes they have neither eye, nor tongue, nor hand, nor foot, to help themselves for drunkenness. By this example the ignorant people suppose that drunkenness is no sin; but he that wasteth most of poor men's goods at taverns, making himself and other men drunken, is most praised for nobleness, courtesy, freeness and worthiness." †

Jeaffreson says that in the fourteenth century, " Sunday was the day of the whole week for revels in the tavern and feasts at the Squire's table."‡

On account of "drinking and buffooneries," it became necessary for a council held in London in 1342, to abolish wakes over the dead. § Twenty-five years later the Abp. of York "complains that in vigils men come together in

---

* Ibid, pp. 124-128.
† A Book about the Clergy, by J. C. Jeaffreson, Vol. I. p. 47.
‡ Ibid, p. 149.
§ Discipline of Drink, p. 177.

the churches and at funerals, as if to pray ; and then turning to a reprobate sense, they indulge in games and van-
ities, and even worse, by which they greatly offend God
and the saints, whom they pretend to venerate ; and they
make the house of mourning at funerals a house of laughter and excess, to the great ruin of their souls." *

Singular enough, the dying made provisions in their
wills for " solace " and " recreation " for the mourners at
their funerals, to be obtained in eating and drinking :

"Katharine Cooke, widow of John Cooke, sometime Mayor
of Cambridge, dying in 1496, left fifteen pence in money 'to the
mayor, bailiffs, and such of their brethren there being present
at the said dirge, at the calling of the said mayor and bailiffs
to the tavern for a solace there among them to be had.' John
Keynsham, alderman of Cambridge in 1502, appointed by his
will an obit, at which the mayor, bailiffs, etc., shall assist, and
that immediately after the dirge, 'a recreation, otherwise called
a pinkett or banquet, to be had within the Abbey of Barnewell,
at cost and charge of the treasurers, at which to be spent six
shillings and eighteen pence in bread, cheese, a hogget of good
ale and another of hostel ale ;' for which he leaves foundation." †

In " A Relation of the Island of England about the
year 1500," ‡ supposed to be the work of a Venetian nobleman who accompanied an Ambassador from Venice to England in 1497, it is stated that, " Few people keep wine in
their own houses, but buy it, for the most part, at a tavern ;
and when they mean to drink a great deal, they go to the
tavern ; and this is done not only by the men, but by ladies of distinction." Wright says : §

"The tavern was also the resort of women of the middle or
lower orders, who assembled there to drink and gossip. In
the Mysteries, or Religious Plays, Noah was represented as finding his wife drinking with her gossips at the tavern when
he wanted to take her into the ark. The meetings of gossips in

---

* Ibid, p. 177.
† Ibid, p. 110.
‡ Printed by the Camden Society, in 1847.
§ Domestic Manners and Sentiments, pp. 437-439.

taverns form the subjects of many of the popular songs of the fifteenth and sixteenth centuries, both in England and France. It appears that these meetings of gossips in taverns were the first examples of what we now call a picnic, for each woman took with her some provisions, and with these the whole party made a feast in common. One of the songs of the fifteenth century gives a picturesque description of one of these gossip-meetings. The women, having met accidentally, the question is put where the best wine was to be had, and one of them replies that she knows where could be procured the best drink in the town, but that she did not want her husband to be acquainted with it—

> 'I know a draught of mery-go-downe,
> The best it is in all this towne;
> But yet wold I not, for my gowne,
> My husband it wyst, ye may me trust.'

The place of meeting having thus been fixed, they are represented as proceeding thither two and two, not to attract observation, lest their husbands might hear of their meeting. 'God might send me a stripe of two,' said one, 'if my husband should see me here.' 'Nay,' said Alice, another, 'she that is afraid had better go home; I dread no man.' Each was to carry with her some goose, or pork, or the wing of a capon, or a pigeon pie, or some similar article—

> 'And ech off them wyll sumwhat bryng,
> Gosse, pygee, or capon's wyng,
> Past'es off pigeons or sum other thyng.'

Accordingly, on arriving at the tavern, they call for wine 'of the best,' and then—

> 'Ech of them brought forth ther dysch;
> Sum brought flesh, and sume fysh.'

Their conversation runs first on the goodness of the wines, and next on the behavior of their husbands, with whom they are all dissatisfied. When they pay their reckoning, they find, in one copy of the song, that it amounts to three pence each, and rejoice that it is so little; while in another, they find that each had to pay sixpence, and are alarmed at the greatness of the amount. They agree to separate, and go home by different streets, and they are represented as telling their husbands that they had been to church."

During this period the singular charities and fairs known as "Ales," greatly flourished. When an unfortunate tradesman failed in business, his neighbors sent him a purse of money, which he was expected to convert into materials for a feast or " ale," to which the donors were invited, each paying a stipulated price for what he might eat and drink, and with the sum thus obtained, the bankrupt was started in business again. This was called, a " Bid Ale." A similar device for increasing the parish-clerk's meagre salary was designated the " Clerk Ale." * Stubbs, in his " Anatomie of Abuses," has the following, on

"The Maner of Church Ales in England : In certaine townes where dronken Bacchus beares swaie, against Christmas, and Easter, Whitsondaie, or some other tyme, the church-wardens of every parishe, with the consent of the whole parishe, provide halfe a score of twenty quarters of mault, whereof some they buy of the churche stocke, and some is given them of the parishioners themselves, every one conferring somewhat, according to his abilitie : whiche mault being made into very strong ale or bere, is sette to sale, either in the church or some other place assigned to that purpose. Then when this is set abroche, well is he that can gete the soonest to it, and spend the most at it. In this kind of practice they continue sixe weekes, a quarter of a year, yea, a halfe a year together."

William Kethe, in a sermon in 1570, complains that these Church-Ales are kept on the Sabbath day, " which holy day the multitude call their revelyng day, which day is spent in bullbeatings, bearebeatings, bowlings, dicying, cardyng, daunsynges, drunkenness, and whoredome, in so much, as men could not keepe their servauntes from lyinge out of theyr owne houses the same Sabbath-day at night."

The satires written by the clergy themselves, give us a lively picture of the times, the general dissoluteness, and yet the fact that the various ranks and degrees in so-called Holy Orders, experienced different treatment and different fare. In Forbrooke's " British Monachism," a number of these satires are preserved. We give extracts from two of different dates :

---

* Jeaffreson, Vol. i. p. 351.

" The abbot and prior of Gloucester and suite,
Were lately invited to share a good treat:
The first seat took the abbot, the prior hard by;
With the rag, tag, and bobtail below was **poor I.**
For wine for the abbot and friar the call;
To us poor devils nothing, but to the rich all.
The blustering abbot drinks health to the prior;
Give wine to my lordship, who am of rank higher;
If people below us but wisely behave,
They are sure from so doing advantage to have;
We'll have all, and leave naught for our brothers to take,
For which shocking complaints in the chapter they'll **make.**
Says the prior, 'My lord, let's be jogging away,
And to keep up appearances, now go and pray.'
'You're a man of good habits, and give good advice,'
The abbot replies:—they returned in a twice,
And then without flinching stuck to it amain,
Till out of their eyes ran the liquor again."

This is from the other, of a later date :

" One law for our rulers, another for us,
To us wretches the smell ev'n of wine is unknown;
The vinegar 's ours,—the wine all their own;
Not a peg from the cloister must we dare to roam,
While the lords of a dwelling withdraw to their home,
To a smoking good fire, then set themselves down,
And with nectar of heaven their best moments crown." *

According to the same authority, the nuns were no better than the monks. † One of the questions to be asked was : "Item, whether any of the sisters be commonly drunke?" "They were accused of avarice, voluptuousness and sloth ; and one of them, the Prioress of Rumsey, was a notorious drunkard."

"The abominable reputation of these religious houses in 1523, led to a visitation by Wolsey ; the immediate effect of which was, the suppression of from twenty to forty convents—authorities being somewhat at variance as to the number—and the conversion of their estates to the founding of Christ Church College, at Oxford. Subsequently a number of abbots, through fear of the visitation, voluntarily surrendered their property to the king ;

* Quoted by Samuelson, p. 143.　　† Ibid, p. 145.

and parliament, at the next session, suppressed three hundred and seventy-six, and vested their estates in the crown. The result shows that the common reputation of these establishments was no exaggeration. The preamble to the Act giving these estates to the king, recites that 'manifest sin, vicious, carnal, and abominable living' characterized these 'religious houses of monks, canons, and nuns.'" *

Hallam, while thinking that it is not to be doubted that in these visitations, "many things were done in an arbitrary manner, and much was unfairly represented," adds : †

"Yet the reports of these visitors are so minute and specific, that it is rather a preposterous degree of incredulity to reject their testimony, whenever it bears hard on the regulars. It is always to be remembered that the vices to which they bear witness, are not only probable from the nature of such foundations, but are imputed to them by the most respectable writers of preceding ages. Nor do I find that the reports of this visitation were impeached for general falsehood in that age, whatever exaggeration there might be in particular cases. And surely the commendation bestowed on some religious houses as pure and unexceptionable, may afford a presumption that the censure of others was not an indiscriminate prejudging of their merits."

The sixteenth century seems to have been characterized by as general dissipation as its predecessor had been. Strong beer was a penny a gallon; table beer less than a half-penny; Spanish and Portuguese wines a shilling; French and German wines, eightpence. For a penny, then, the laborer, whose wages were from threepence to six pence a day, according to his skill, could buy as much as the laborer of to-day can for a shilling. ‡ Erasmus, who visited England in the early part of this century, gives a curious description of an English interior of the better class. The furniture was rough, the walls unplastered, but sometimes wainscotted or hung with tapestry; and the

---

* Constitutional History of England. By Henry Hallam, Vol. i. p. 72.
† Ibid, p. 71.
‡ Fronde's History of England, Vol. I. chap. i.

floors covered with rushes, which were not changed for months. The dogs and cats had free access to the eating-rooms, and fragments of meat and bones were thrown to them, which they devoured among the rushes, leaving what they could not eat to rot there, with the drainings of beer vessels and all manner of unmentionable abominations. Of the moral and intellectual condition of the people he exclaims:

"Oh, strange vicissitudes of human things! Heretofore the heart of learning was among such as professed religion. Now, while they for the most part give themselves up, *ventri luxui pecuniæque*, the love of learning is gone from them to secular princes, the court and the nobility. May we not justly be ashamed of ourselves? The feasts of priests and divines are drowned in wine, are filled with scurrilous jests, abound with intemperate noise and tumult, flow with spiteful slanders and defamation of others; while at princes' tables modest disputations are held concerning things which make for learning and piety." *

Alexander Barclay, who wrote early in this century, says in his poem, the "Ship of Fools:"

"The holy day we fill with eche unlefull thing,
  As late feastes and bankettes saused with gluttony,
  And that from morn to night continually.
   *   *   *   *   *   *
The tavern is open before the church be;
The pots are ronge as bels of droukennesse,
Before the church bels with great solemnitie.
There have these wretches their mattins and their masse.
Who listeth to take heede shall often see dowtless,
The stalles of the tavern stuffed with eche one,
When in the church stalles he shall see few or none." †

Another poet is quoted as saying of taverns:

"They are become places of waste and excess,
An harbor for such men as live in idleness.
And lyghtly on the contry they be placed so,
That they stand in men's way when they should to church go.

---

* Ibid, p. 48. See also Hume, Vol. III. p. 448.
† Jeaffreson, vol. II. p. 123.

And such as love not to hear theyr faults told,
By the minister that readeth the New Testament and Old,
Do turn into the alehouse and let the church go ;
And men accompted wise and honest do so." *

The early taverns made no provision for supplying their
visitors with food, beyond a crust to relish the wine ; and
those who wished to dine before they drink, must go to the
cooks.†

In 1551 the following song appeared, and at once became
popular.   It has more poetic merit than any drinking song
that had been sung before.   Sixteen years later it appeared
in a play entitled " Gammer Guston's Needle," written by
Bishop Still.‡   The presumption is that he was the author
of the song.   The song commences wtth the chorus, which
in singing, is repeated at the close of each verse :

"Back and side go bare, go bare,
    Both foot and hand go cold:
But belly, God send thee good ale enough,
    Whether it be new or old.

        I cannot eat but little meat,
        My stomach is not good ;
        But sure I think, that I can drink
        With him that wears a hood.
        Though I go bare, take ye no care,
        I am nothing a cold ;
        I stuff my skin so full within
        Of jolly good ale and old.

I love no roast, but a nut-brown toast,
    And a crab laid in the fire ;
A little bread shall do me stead,
    Much bread I do not desire.
No frost nor snow, no wind, I trow,
    Can hurt me if I wold,
I am so wrapt, and throughly lapt
    Of jolly good ale and old.

* Ibid, p. 125.
† Club Life of London, vol. II. p. 113.
‡ The Works of the British Dramatists.   By John S. Keltie,
p. xxxviii.

And Tyb my wife, that as her life
Loveth well good ale to seek,
Full oft drinks she, till ye may see
The tears run down her cheek;
Then doth she trowl to me the bowl,
Even as a malt worm should;
And saith, Sweet heart, I took my part
Of this jolly good ale and old.

Now let them drink, till they nod and wink,
Even as good fellows should do,
They shall not miss to have the bliss
Good ale doth bring men to :
And all poor souls, that have scorned bowls,
Or have them lustily trold,
God save the lives of them and their wives,
Whether they be young or old."

In the reign of Elizabeth, so great was the drunkenness
on the Lord's day, and so general the acquiescence in it, that
parliament failed to pass a law, "That no victualler have
his shop open before the service be done in his parish where
he dwelleth." In a homily on the "Place and Time of Prayer," the people are represented as resting "not in holiness,
as God commandeth ; but they rest in ungodliness and
filthiness, * * * in excess and superfluity, in gluttony
and drunkenness, like rats and swine; * * * so that
doth too evidently appear that God is more dishonored,
and the devil better served on the Sunday, than upon all
the days in the week beside." *   A great variety of names
now begin to be attached to the ales, as single beer, or
small ale, which is represented as being very mild : double
beer, which was recommended as containing a double quantity of malt and hops ; double double beer, twice as strong
as the last ; dagger ale, a particularly sharp and dangerous
drink ; and a special favorite, the chief article of vulgar debauch, was commonly called Huffcap, but was also termed
by frequenters of ale houses, Mad Dog, Angel's Food, and
Dragon's Milk. "And never," says Harrison, " did Romu-

* Jeaffreson, Vol. II. p. 129.

lus and Remus seek their she-wolf with such eager and sharp devotion as these men hale at Huffcap, till they be as red as cocks, and little wiser than their combs."

The wealthy brewed a generous liquor for their own consumption, which was not brought to the table till it was two years old. This was called March ale, from the month in which it was made. The poorer classes and the servants had to content themselves with a simpler beverage, which was seldom more than a month old.* Drunken feasting seems to have characterized both ecclesiastical and secular occasions. An instance of the form(˙˙found in the account that when the Archbishop of Canterbury was enthroned, a fish banquet was given, at which the following drink provision was made: "Six tuns of red wine, four of claret wine, one of choice white wine, one of white wine for the kitchen, one butt of Malmsey, one pipe of wine of Osey, two tierces of Rhenish wine, four tuns of London ale, six of Kentish ale, and twenty of English beer." † A striking instance of the latter is found in the fact of the general statement of Harrison, that Queen Elizabeth's visits to her nobility were a great oppression to them by reason of the cost of her luxurious entertainment; and by the special instance cited by Hume, of her visit to the Earl of Leicester, whereat, "among other particulars, we are told that three hundred and sixty-five hogsheads of beer were drunk." ‡

Health drinking, as it was called, was observed with a great deal of formality, the toasts beir_ given, not to any person present at the feast, but to some one for whom the drinker had great partiality. Wright§ quotes from a little book published in 1623, the following description of it:

"He that begins the health, first uncovering his head, takes a full cup in his hand, and setting his countenance with a grave aspect, he craves for audience. Silence being once obtained, he

* Wine and Wine Countries. By Charles Tovey, p. 46.
† Samuelson, p. 137.
‡ History of England, Vol. IV. p. 372.
§ Domestic Manners, pp. 467, 468.

begins to breathe out the name, peradventure of some honorable personage, whose health is drunk to, and he that pledges must likewise off with his cap, kiss his fingers, and bow himself in sign of a reverent acceptance. When the leader sees his follower thus prepared, he sups up his broth, turns the bottom of the cup upward, and, in ostentation, gives the cup a phillip to make it cry twango. And thus the first scene is enacted. The cup being newly replenished to the breadth of a hair, he that is the pledger must now begin his part, and thus it goes round throughout the whole company. In order to ascertain that each person has fairly drunk off his cup, in turning it up he was to pour all that remained in it on his thumb nail, and if there was too much to remain as a drop on the nail without running off, he was made to drink his cup full again."

Max Müller has brought to light a volume of the Travels of Paul Hentzner in England, in 1598,* in which the German traveller, after remarking on the " clever, perfidious, and thievish " character of the English, and saying that " they are very fond of noises that fill the ears," adds this curious statement: " In London, persons who have got drunk are wont to mount a church tower, for the sake of exercise, and to ring the bells for several hours." The statement of a popular historian † is to the effect that: " Excess in the use of wine and intoxicating liquors was now the common charge against the English, and it seems to be borne out, not only by the quantity consumed, but by the extent to which taverns had multiplied by the end of Elizabeth's reign."

To this state of thing Shakespeare alludes in his Hamlet :

"Is it a custom ?
Aye marry is't :
But to my mind, though I am native here,
And, to the manner born, it is a custom
More honored in the breach than the observance.
This heavy-headed revel, east and west
Makes us traduced and taxed of other nations ;
They class us drunkards, and with swinish phrase
Soil our addition: and, indeed, it takes

* Chips from a German Workshop, Vol. III. pp. 232-237.
† Knight's Pictorial History, Vol. II. p. 884.

From our achievements, though performed at height,
The pith and marrow of our attribute."

In consequence, there was appalling insecurity of life, great increase of crimes against person and property, and although arrests were numerous and penalties severe, it is the uniform testimony of history that not more than a fifth part of these offences were ever punished by the civil law. Hume says that

"There were at least three or four hundred able-bodied vagabonds in every county, who lived by theft and rapine; and who sometimes met in troops to the number of sixty, and committed spoil on the inhabitants: * * * * and that the magistrates themselves were intimidated from executing the laws upon them; and there were instances of justices of the peace who, after giving sentence against rogues, had interposed to stop the execution of their own sentence, on account of the danger which hung over them from the confederates of these felons." *

The seventeenth century opens with the reign of James I., under whose administration drunkenness did not decrease. He was known to be an habitual drunkard, and in his court men and women of high rank, copying the royal manners, rolled intoxicated at his feet.† Involved in difficulties with his parliament, and stinted by them in his allowance of money, he contributed greatly to the spread of intemperance by licensing an immense number of tippling houses, in order to increase his revenue. The visit of the Danish king and his courtiers, whose example of constant intoxication, the English people readily imitated, led to the remark that the Danes had again conquered England. Cecil gave a feast to the two monarchs, on which occasion both got so drunk that James was carried to bed in the arms of his courtiers, and Christian IV., in his stumbling intoxication, mistook his own chamber and offered the grossest insults to the Countess of Nottingham.

---

* History, Vol. IV. p. 359.
† Green's History of the English People, p. 473.

Nearly the whole company gave proof that they were capable of following these examples.

Sir John Harrington, an eye-witness, says:

"The ladies abandon their sobriety, and are seen to roll about in intoxication. . . . . The lady who did play the Queen's part (in the Masque of the Queen of Sheba) did carry most precious gifts to both their Majesties; but forgetting the steppes arising to the canopy, overset her caskets into his Danish Majesty's lap, and fell at his feet, though I rather think it was in his face. Much was the hurry and confusion; cloths and napkins were at hand, to make all clean. His Majesty then got up and would dance with the Queen of Sheba; but he fell down and humbled himself before her, and was carried to an inner chamber and laid on a bed of state, which was not a little defiled with the presents of the Queen which had been bestowed on his garments; such as wine, cream, jelly, beverage, cakes, spices, and other good matters. The entertainment and show went forward, and most of the presenters went backward, or fell down; wine did so occupy their upper chambers. Now did appear in rich dress, Hope, Faith, and Charity: Hope did assay to speak, but wine rendered her endeavors so feeble that she withdrew, and hoped the king would excuse her brevity: Faith . . . . left the court in a staggering condition. . . . . They were both sick and spewing in the lower hall. Next came Victory, who . . . . by a strange medley of versification . . . . and after much lamentable utterance, was led away like a silly captive, and laid to sleep in the outer steps of the anti-chamber. As for Peace, she most rudely made war with her olive branch, and laid on the pates of those who did oppose her coming. I ne'er did see such lack of good order, discretion, and sobriety in our Queen's days." *

Scotland and Ireland becoming parts of the same empire with England at the commencement of the seventeenth century, there was naturally an interchangeable influence from one portion of the nation to the other. Heretofore the drunkenness of England had been almost wholly caused by fermented drinks; now distilled liquors came in, and by the middle of the century the use had become common. Distillation had been practiced in Ireland for

---

* Taine, Vol. I. Book II. chap. i.

nearly, if not quite a century,—the precise date is unknown,—and as early as 1558 such a drain had thus been
made on the country's supply of corn as to call for legislative interference with the manufacture of ardent spirits,
in order to avert a famine. The Irish called their new
drink *usique ratha*, or usquebaugh, and also *bulcaan*, the
latter word derived from *buile*, madness, and *caan*, the
head, was descriptive of the fiery properties of the liquor;
and from *usique*, or usque, is derived whiskey. Moryson's
History of Ireland is quoted by several writers as authority for the statement that in 1600, men and women never
go to Dublin for the purpose of disposing of any article in
the market, but they stay till they have spent the price received in usquebaugh, and have outslept two or three
days' drunkenness.

Sir John Parrot, Lord Deputy of Ireland, 1584, addressing the mayor and corporation of Galway, said: "The
aqua vitæ that is sold in towns, ought rather to be called
aqua mortis, to poysen the people, than to comfort them in
any good sorte." *

The dissolute habits of the English seem to have kept
pace with the rapid political changes which characterized
the seventeenth century. Tavern life grew more frequent
with the people, and more debauched and dangerous.
Bishop Earle, writing in 1650, thus describes a tavern :

" A tavern is a degree, or (if you will) a pair of stairs above
an alehouse, where men are drunk with more credit and apolo-
gy. If the vinter's nose be at the door, it is a sign sufficient.
* * * Men come here to make merry, but indeed make a
noise, and this music above is answered with a clinking below.
* * * A melancholy man would find here matter to work upon, to see heads as brittle as glass, and often broken. * * *
A house of sin you may call it, but not a house of darkness, for
the candles are never out ; and it is like those countries far in
the north, where it is as clear at midnight as at mid-day." †

---

* War of Four Thousand Years, p. 138.
† Club Life of London, Vol. II. pp. 118–119

All the taverns had a bad reputation, although some were more uniformly the scene of violence than others. Among these was the Rose Tavern, in Covent Garden, which "was constantly a scene of drunken broils, midnight orgies, and murderous assaults by men of fashion, who were designated 'Hectors,' and whose chief pleasure lay in frequenting it for the running through of some fuddled toper, whom wine had made valiant." * This was in the days of the Commonwealth, when the high pitch of dissoluteness gave to England the name of the "land of Drunkards." But bad as the condition of the people then was, it was aggravated beyond all computation when Charles II. succeeded the Protector. Beyond all question his was the most dissolute court, and his subjects the most drunken people known to the history of the English-speaking people. Pepys says of the Court: "Things are in a very ill condition, there being so much emulation, poverty, and the vices of drinking, swearing, and loose amours, that I know not what will be the end of it, but confusion."

The clergy, blind to the debaucheries around them, were attributing the misfortunes and miseries of the times to the judgment of God on the people for not putting to death the murderers of Charles I.† In a short time "the profligacy of the Court begun to show itself in more daring outrages than the indecencies and riots which rivalled the orgies of the lowest of mankind. The jolly blades racing, dancing, feasting and revelling, more resembling a luxurious and abandoned rout than a Christian court."‡ The Parliament which met at Edinburgh on the 1st of January, 1661, to accommodate the people and the laws to their changed condition on the accession of Charles, " has been honored," says Knight, " with the name of the ' drunken Parliament.' " He quotes Burnett as saying, "It was a mad,

---

* Ibid, p. 192.
† Knight's History of England, Vol. IV. p. 259.
‡ Ibid, p. 312.

roaring time, full of extravagance ; and no wonder it was
so, when the men of affairs were almost perpetually drunk."
The historian adds : " The violence of the drunken Parliament was finally shown in the wanton absurdity of what
was called the 'Act Rescissory,' by which every law that
had been passed in the Scottish parliament during twenty-
eight years was wholly annulled." *

A writer in Addison's Spectator, relates an incident, of
which he says he was an eye-witness, of the king's dining
with the Lord Mayor of London, when the latter, over-
weighted with wine, grew more familiar with the king than
was seemly at such a place ; whereupon the king withdrew
to his coach, but was pursued by the mayor, who with an
oath insisted, " Sir, you shall stay and take t'other bottle ! "
The king, far from resenting it, looked kindly at the
mayor, and repeating a line of an old song, " He that is
drunk is as great as a King," immediately turned back and
complied with the demand. † Of the general debauchery
and absence of moral sense characterizing this period, Ma-
caulay closes a satirical description by saying : " It is an un-
questionable and most instructive fact, that the years during which the political power of the Anglican hierarchy
was in the zenith were precisely the years during which
national virtue was at its lowest point." ‡

Unfortunately, this condition of things characterized the
most of the century, making religion disreputable through
the example of the dignities of the Church, and defeating
all political fairness and honesty. Lecky, thus describes
the manner in which disputed elections were decided in
Parliament, in 1672 :

"It is impossible to conceive a more grotesque travesty of a
judicial proceeding than was habitually exhibited on these occasions, when private friends of each candidate and the members of the rival parties mustered their forces to vote entirely

---

* Ibid, p. 258.
† The Spectator, No. 462.
‡ History of England, Vol. I. p. 169.

irrespective of the merits of the case; when, the farce of hearing evidence having been gone through with in an empty House, the members, who had been waiting without, streamed in, half intoxicated, to the division, and when the plainest and most incontestable testimony was set aside without scruple, if it clashed with the party interests of the majority." *

Clubs with outrageous names, and addicted to still more outrageous acts, were organized in this century, and continued their existence and depredations far into the next. Such were the " Thieves," who gloried in stealing and destroying property ; the " Lying Club," any member of which telling the truth between the hours of six and ten in the evening, paid a fine of a gallon of wine ; the " Bold Bucks," whose members all denounced the claims of God, and who, after disturbing divine service by parading back and forth before the churches with bands of music and boisterous shouts, sat down to dine on dishes named in blasphemous derision of sacred things, prominent among which was " Holy Ghost Pie," after which they rushed into the streets, and shouting their motto : " Blind and Bold Love," committed the most horrible and disgusting atrocities ; and the " Sword Clubs," whose members, after getting roaring drunk at their suppers, took possession of the town, rushing violently about with sword in hand, demanding of all passers to defend themselves or suffer.

In 1688, when the population did not much exceed 5,000,000, there were 12,400,000 barrels of ale brewed in England, † about a third part of the arable land of the country being devoted to barley, raised for this purpose. About the same time the restrictions placed on distillations were removed, and the manufacture rose from 527,000 gallons of gin in 1684, to 2,000,000 gallons in 1714, to 3,601,000 in 1727, and to 5,394,000 gallons in 1735.

Early in the eighteenth century,—as soon as 1724,—

---

* England in the Eighteenth Century, Vol. I. p. 477.
† Ibid, p. 518.

the passion for gin drinking, spreading like an epidemic, infected the masses of the population. Says Lecky :

" Small as is the place which this fact occupies in English history, it was probably, if we consider all the consequences that have flowed from it, the most momentous in that of the eighteenth century—incomparably more so than any event in the purely political or military annals of the country." *

As the evil progressed, as indicated above in the enumeration of the supply of the poison, the clergy, the medical profession and the county grand juries brought to bear all the arguments derived from increased immorality, disease and crime, to induce the law-makers to arrest the evil. An attempt was made in 1736, by the passage of a bill intended to have the force of a prohibitory law,—a measure which will be more fully noticed in another place. The state of morals at that time must have been appalling. Smollett thus gives us a glimpse of the desperate condition :

" The populace of London were sunk into the most brutal degeneracy, by drinking to excess the pernicious spirit called gin, which was sold so cheap that the lowest class of the people could afford to indulge themselves in one continued state of intoxication, to the destruction of all morals, industry and order. Such a shameful degree of profligacy prevailed, that the retailers of this poisonous compound set up painted boards in public, inviting people to be drunk for the small expense of one penny; assuring them they might be dead drunk for two-pence, and have straw for nothing. They accordingly provided cellars and places strewed with straw, to which they conveyed those wretches who were overwhelmed with intoxication. In these dismal caverns they lay until they recovered some use of their faculties, and then they had recourse to the same mischievous potion; thus consuming their health, and ruining their families, in hideous receptacles of the most filthy vice, resounding with riot, execration and blasphemy. Such beastly practices too plainly denoted a total want of all police and civil regulations, and would have reflected disgrace upon the most barbarous community." †

---

* Ibid, p. 519.

† History of England, by T. Smollett, M. D. Vol. III. chap. vii. p. 36.

Hogarth's picture gives a vivid and frightful view of the physical degradation of the frequenters of "Gin Lane;" and Lecky thus sums up the insecurity and immorality caused by the general dissipation :

"A club of young men of the higher classes, who assumed the name of Mohocks, were accustomed nightly to sally out drunk into the streets to hunt the passers-by, and to subject them in mere wantonness to the most atrocious outrages. One of their favorite amusements, called 'tipping the lion,' was to squeeze the nose of their victim flat upon his face and to bore out his eyes with their fingers. Among them were the 'sweaters,' who formed a circle round their prisoner and pricked him with their swords till he sank exhausted to the ground ; the 'dancing masters,' so called, from their skill in making men caper by thrusting swords into their legs, the 'tumblers,' whose favorite amusement was to set women on their heads and commit various indecencies and barbarities on the limbs that were exposed. Maid servants as they opened their master's doors were waylaid, beaten, and their faces cut. Matrons enclosed in barrels were rolled down the steep and stony incline of Snow Hill. Watchmen were unmercifully beaten and their noses slit. Country gentlemen went to the theatre as if in time of war, accompanied by their armed retainers. . . . . Long after the Revolution, the policy of the Government was to rely mainly upon informers for the repression of crime, but the large rewards that were offered were in a great degree neutralized by the popular feeling against the class. The watchmen or constables were as a rule utterly inefficient, were to be found much more frequently in beer-shops than in the streets, and were often themselves a serious danger to the community. Fielding, who knew them well, has left a graphic description of one class. 'They were chosen out of those poor, decrepit people who are, from their want of bodily strength, incapable of getting a livelihood by work. These men, armed only with a pole, which some of them are scarcely able to lift, are to secure the persons and houses of his Majesty's subjects from the attacks of gangs of young, bold, desperate, and well-armed villains. If the poor old fellows should run away, no one, I think, can wonder, unless it be that they were able to make their escape.' Of others an opinion may be formed from an incident related by Horace Walpole in 1742. 'A parcel of drunken constables took it into their heads to put the laws in execution against disorderly persons, and so took up every woman they met, till they had collected five or six and twenty, all of whom they thrust into St. Martin's roundhouse, where they kept them all night, with doors and windows closed. The poor creatures, who could not stir or breathe, screamed as long as they had any breath left, begging at least for water, but in vain. In the morning four were found stifled to death, two died soon after, and a dozen more are in a shocking way. Several of them were beggars, who from having no lodging, were necessarily found in the street, and others honest, laboring women. One of the dead was a poor washerwoman, big with child, who was retiring home late from washing. One of the constables is taken, and others absconded; but I question if any of them will suffer death, though the greatest criminals in this town are the officers of justice; there is no tyranny they do not exercise, no villany of which they do not partake.'

"The magistrates were in many cases not only notoriously ignorant and inefficient, but also what was called ' trading justices,' men of whom Fielding said, ' they were never indifferent in a cause, but when they could get nothing on either side.' The daring and number of robbers increased till London hardly resembled a civilized town. ' Thieves and robbers,' said Smollett, speaking of 1730, ' were now become more desperate and savage than they had ever appeared since mankind were civilized.'

" The Mayor and Aldermen of London in 1744 drew up an address to the king, in which they stated that ' divers confederacies of great numbers of evil-disposed persons, armed with bludgeons, pistols, cutlasses, and other dangerous weapons, infest not only the private lanes and passages, but likewise the public streets and places of usual concourse, and commit most daring outrages upon the persons of your Majesty's good subjects whose affairs oblige them to pass through the streets, by robbing and wounding them, and these acts are frequently perpetrated at such times as were heretofore deemed hours of security.' The same complaints were echoed in the same year in the ' Proposals of the Justices of the Peace for Suppressing Street Robberies,' and the magistrates who drew them up specially noticed, and ascribed to the use of spirituous liquors ' the cruelties which are now exercised on the persons robbed, which before the excessive use of these liquors were unknown in this nation.' . . . . ' One is forced to travel,' wrote Horace Walpole in 1751, ' even at noon, as if one were going to battle! . . . . The more experienced robbers for a time completely overawed the authorities.' ' Officers of justice,' wrote Fielding, ' have owned to me that they have passed by such, with warrants in their pockets against them, without daring to apprehend them; and, indeed, they could not be blamed for not exposing themselves to sure destruction ; for it is a melancholy truth that at this very day a rogue no sooner gives the alarm within certain purlieus than twenty to thirty armed villains are found ready to come to his assistance.'

"When the eighteenth century had far advanced, robbers for whose apprehension large rewards were offered, have been known to ride publicly and unmolested, before dusk, in the streets of London, surrounded by their armed adherents, through the midst of a half-terrified, half curious crowd. . .

. . . . A multitude of clergymen, usually prisoners for debt and almost always men of notoriously infamous lives, made it their business to celebrate clandestine marriages in or near the Fleet. They performed the ceremony without license or question, sometimes without even knowing the names of the persons they united, in public-houses, brothels, or garrets. They acknowledged no ecclesiastical superior. Almost every tavern or brandy shop in the neighborhood had a Fleet parson in its pay. Notices were placed in the windows, and agents went out in every direction to solicit the passers-by.

"A more pretentious, and perhaps more popular establishment was the Chapel in Curzon street, where the Rev. Alexander Keith officiated. He was said to have made a 'very bishopric of revenue' by clandestine marriages, and the expression can hardly be exaggerated if it be true, as was asserted in Parliament, that he had married on an average 6,000 couples every year. He himself stated that he had married many thousand, the great majority of whom had not known each other more than a week, and many only a day or half a day. Young and inexperienced heirs, fresh from college, or even from school, were thus continually entrapped. A passing frolic, the excitement of drink, an almost momentary passion, the deception or intimidation of a few unprincipled confederates, were often sufficient to drive or inveigle them into sudden marriages, which blasted all the prospects of their lives. In some cases, when men slept off a drunken fit, they heard to their astonishment that, during its continuance, they had gone through the ceremony. When a fleet came in and the sailors flocked on shore to spend their pay in drink and among prostitutes, they were speedily beleaguered, and 200 or 300 marriages constantly took place within a week. . . . . In many cases in the Fleet registers, names were suppressed or falsified, and marriages fraudulently antedated, and many households, after years of peace, were convulsed by some alleged pre-contract or clandestine tie. It was proved before Parliament that on one occasion there had been 2,954 Fleet marriages in four months, and it appeared from the memorandum books of Fleet parsons, that one of them made £57 in marriage fees in a single month; that another had married 173 couples in a single day." *

In the last half of the century there was visible improvement, but as late as 1780, a candidate for public honors :

"If not defeated at the polls, by riots and open violence,—or defrauded of his votes by the partiality of the returning officer, or the factious manœuvres of his opponents—was ruined by the extravagant costs of his victory. The poll was liable to be kept open for forty days, entailing an enormous expense upon the candidates, and prolific of bribery, treating and riots. During this period, the public houses were thrown open; and drunkenness and disorder prevailed in the streets, and at the hustings. Bands of hired ruffians, armed with bludgeons, and inflated by drink,—paraded the public thoroughfares, intimidating voters, and resisting their access to the polling places. Candidates assailed with offensive, and often dangerous missiles, braved the penalties of the pillory; while their supporters were exposed to the fury of a drunken mob." †

To an alarming extent, notwithstanding the general improvement of the English people, these evils extended into and characterized the first third of the present century. In 1830, the Temperance movement gained a foothold throughout the kingdom of Great Britain. But in spite of the great work accomplished through the various agencies put into operation by this movement, intemperance continues to be the great political, social and moral scourge of the land of our fathers. For the first twenty years of the nineteenth century the spirits distilled in England, averaged about 4,000,000 gallons per year, while the additional importations were about 3,400,000 gallons per year. ‡

In 1822, 7,584,807 barrels of beer were brewed in England, 2,000,875 barrels of which were brewed in London. § In 1875 a careful and exact writer published the following :

---

* England in the Eighteenth Century, Vol. I. pp. 522–532.
† May's Constitutional History of England, Vol. I. p. 280.
‡ Morewood, pp. 293, 294.
§ Ibid, p. 279.

"The British people annually expend on intoxicating liquors a sum of above a hundred and thirty millions sterling, the great bulk of it coming from the pockets of men and women who would be seriously affronted if any doubt were cast upon their religious sincerity. This sum is sixty millions in excess of the National Revenue. It is one-sixth of the National Debt. It is one-fifth the value of all the railway property of the United Kingdom. It is equal to one-fourth of the income of the wage-receiving class, and one-eighth of the income of all classes united. It is equal to a yearly expenditure of £4 per head, and of £22 per family, in the United Kingdom. Bulky figures are seldom realized unless by illustrations drawn from familiar objects. One ingenious means of impressing the mind with a total so stupendous as that just named is the following: There are in the Old and New Testaments together 66 books, 1189 chapters, 31,173 verses, 773,746 words, and 3,566,480 letters. Now if these £130,000,000 sterling were distributed over each of these respectively there would lie on each letter £36, 10s.; or on each word, £168; or on each verse £4170; or on each chapter £110,775; or on each book £1,969,696. Put edge to edge 130 million sovereigns would form a golden belt (reckoning 41 to a yard) 1800 miles in length; or a golden column (reckoning 15 to an inch) 140 miles in height. And this, be it remembered, is the drink money of the British people *for one year only,* and year by year." *

To what end this great amount is expended, take the testimony of Charles Buxton, Esq., M. P., a London brewer, given in 1855:

"Startling as it may appear, it is the truth, that the destruction of human life, and the waste of national wealth, which must arise from this tremendous Russian war, are outrun every year by the devastation caused by national drunkenness. Nay, add together all the miseries generated in our times by war, famine, and pestilence, the three great scourges of mankind, and they do not exceed those that spring from this one calamity. This assertion will not be readily believed by those who have not reflected on the subject. But the fact is, that hundreds of thousands of our countrymen are daily sinking themselves into deeper misery; destroying their health, peace of mind, domestic comfort, and usefulness; and ruining every fac-

* Christendom and the Drink Curse. By Rev. Dawson Burns, A. M. P. 39.

ulty of mind and body, from indulgence in this propensity. And then what multitudes do these suicides drag down along with them! It would not be too much to say, that there are at this moment *half a million homes* in the United Kingdom, where home happiness is never felt, owing to this cause alone; where the wives are broken-hearted, and the children are brought up in misery. . . . .

"Then the sober part of a community pays a heavy penalty for the vices of the drunkard. Drink is the great parent of crime. One of the witnesses before the Committee of the House of Commons, states that he went through the New Prison at Manchester (it contained 550 criminals) with Thomas Wright, the prison philanthropist. 'I spent an entire day,' he says, 'in speaking with the prisoners, and in every case, without exception, *drinking was the cause of their crime.*' One of the Judges stated, some time ago at the Circuit Court in Glasgow, that 'every evil seemed to begin and end in whiskey.' Judge Erskine in the same way declared at the Salisbury Assizes, in 1844, that ninety-nine cases out of every hundred arose from strong drink. . . . . Not only does this vice produce all kinds of positive mischief, but it also has a negative effect of great importance. It is the mightiest of all the forces that clog the progress of good. It is in vain that every engine is set to work that philanthropy can devise when those whom we seek to benefit are habitually tampering with their faculties of reason and will,—soaking their brains with beer, or inflaming them with ardent spirits. The struggle of the school, and the library, and the church all united, against the beer-house and gin-palace, is but one development of the war between heaven and hell. Well may we say with Shakespeare, 'O that men should put an enemy in their mouths to steal away their brains! that we should, with joy, pleasance, revel, and applause, transform ourselves into beasts!'

. . . .

"Looking, then, at the manifold and frightful evils that spring from drunkenness, we think we are justified in saying that it is the most dreadful of all the ills that afflict the British isles. We are convinced, that if a statesman who heartily wished to do the utmost possible good to his country, were thoughtfully to inquire which of the topics of the day deserved the most intense force of his attention, the true reply—the reply which would be exacted by a full deliberation,—would be, that he should study the means by which this worst of plagues can be stayed. The intellectual, the moral, and the religious welfare of our people, their material comforts, their domestic happiness, are all involved. The question is, whether millions of our countrymen shall be helped to become happier and wiser,—whether pauperism, lunacy, disease and crime, shall be diminished,—whether multitudes of men, women, and children, shall be aided to escape from utter ruin of body and soul. Surely such a question as this, enclosing within its limits consequences so momentous, ought to be weighed with earnest thought by all our patriots."*

Still more recent testimonies are to the same effect:

"The amount expended on intoxicating drinks in this country is larger than ever, and this increased drinking has been mainly induced by the greatly increased wages of the working-classes during the last few years. It has, however, produced such a fearful amount of social and moral evil that public attention has been aroused to the question with a more earnest desire to do something to mitigate or prevent this great national vice."

"The crime and misery that are daily chronicled in our public papers almost invariably can be traced to drink." "I deeply regret to say that Great Britain is more than ever cursed by intemperance. The people will have it, and a foolish and wicked government will pander to them, so that our country is becoming one *universal grog shop.* London has about 2,000 churches open on the Sabbath, in order to raise the masses to God and to heaven, and about 11,000 public-houses to drag them down to hell." †

In 1879, the Bishop of Manchester said: "I do not know what is to become of this country if the terrible drinking habits are to be persevered in by the great mass of the people, high and low, rich and poor, for I am afraid the curse is spreading like the leprosy everywhere. And when we say we hope God will give England back its days of prosperity, I am not quite sure that the days of prosperity will come back till England has become a sober and industrious land."

THE UNITED STATES.—Just at this point we propose to set forth a few of the facts corroborative of the statement made near the beginning of this chapter, that the aborigines of North America knew nothing of any kind of intoxicating drinks till the arrival of Europeans among them.

* How to Stop Drunkenness, pp. 8–13.
† Letters of Samuel Bowly, Esq., Lord Claud Hamilton, and Rev. John Jones, D.D., in Centennial Temperance Volume, pp. 346, 347, 355.

In September, 1609, Henry Hudson, the British naviga-
tor, sailed into New York bay. Some Indians who were
fishing caught sight of his vessel, and in their wonder at so
strange a sight hurried to the shore to inform their countrymen, who soon assembled, set their conjurers to work to
determine what it might be and mean, and how they ought
to receive the strange people whom they could now see on
its deck. They concluded that the chief man of the group,
distinguished by his red coat and glittering gold lace,
must be the Manitou, the Great or Supreme Being, come
to bring them some kind of game, such as they had not been
favored with before, and so prepared an abundance of meat
for a sacrifice and feast. At last the house, or as some
say, the large canoe, stops, and a canoe of smaller size
comes to the shore, bearing among others, the Manitou
himself. The chiefs and wise men form a circle and receive their visitors, who salute them with a friendly countenance. Then one of the strangers produces a large bottle
from which he pours an unknown substance into a small
glass, and hands it to the supposed Manitou.

"He drinks, has the glass filled again, and hands it to the
chief standing next to him. The chief receives it, but only
smells the contents and passes it on to the next chief, who does
the same. The glass or cup passes through the circle, without
the liquor being tasted by any one, and is upon the point of being returned to the red clothed Manitou, when one of the Indians, a brave man and a great warrior, suddenly jumps up and
harangues the assembly on the impropriety of returning the cup
with its contents. It was handed to them, says he, by the Manitou, that they should drink out of it, as he himself had done.
To follow his example would be pleasing to him; but to return
what he had given them might provoke his wrath, and bring
destruction on them. And since the orator believed it for the
good of the nation that the contents offered them should be
drunk, and as no one else would do it, he would drink it himself, let the consequence be what it might; it was better for one
man to die, than that a whole nation should be destroyed. He
then took the glass, and bidding the assembly a solemn fare-
well, at once drank up its whole contents. Every eye was fixed
on the resolute chief, to see what effect the unknown liquor

would produce. He so~~ ~~began ~~ ~~ ~~~~ ~ ~ ~~ - -   . .
trate on the ground. His companions now bemoan his fate, he
falls into a sound sleep, and they think he has expired. He
wakes again, jumps up and declares that he has enjoyed the
most delicious sensations, and that he never before felt himself
so happy as after he had drunk the cup. He asks for more, his
wish is granted; the whole assembly then imitate him, and all
become intoxicated."

"I have no doubt," says Heckewelder, from whom we have
quoted the foregoing, "that this tradition is substantially
founded on fact. Indeed, it is strongly corroborated by the
name which, in consequence of this adventure, those people
gave at the time, to that island, and which it has retained to
this day. They called it *Manahachta-nienk,* which in the Delaware language means '*the island where we all became intoxicated.*'
We have corrupted this name into *Manhattan,* but not so as to
conceal its meaning, or conceal its origin. The last syllable,
which we have left out, is only a termination, implying locality,
and in this word signifies as much as *where we.* There are few
Indian traditions so well supported as this." *

The following year, the Dutch made settlements on the
island. As is almost uniformly the case, dissolute and dishonest men were among the early settlers, and true to
their base instincts, these liberally supplied the Indians
with intoxicants, that they might more easily overreach
and rob them. Angry and bloody quarrels were the consequence, and at times the settlements were wholly depop-
ulated by the maddened natives.

"But," says Bancroft, "the traders did not learn humanity,
nor the savage forget revenge; and the son of a chief, stung by
the conviction of having been defrauded and robbed, aimed an
unerring arrow at the first Hollander exposed to his fury. A
deputation of the river chieftains hastened to express their sorrow, and deplore the alternate, never-ending libations of blood.
* * * 'You yourselves,' they said, 'are the cause of this evil.
You ought not to craze the young Indians with brandy. Your
own people, when drunk, fight with knives, and do foolish

* History, Manners, and Customs of the Native Indians, pp.
71–74; 262.

things ; and you cannot prevent mischief, till you cease to sell strong drink to the Indians.'" *

In the Dutch settlements on the Delaware the same disgraceful and dangerous traffic was carried on, and in 1660, "the greatest chief of the Minquas," complains of D'Hinoyossa, the Director of the colony, that the outrageous conduct of the Indians arises from his not restricting the sale of liquors. Beckman, the commissary, charges the same negligence on his superior, and says that allowing drink to be sold to the savages, they behave shamefully." † Two years later, D'Hinoyossa, yielding to the solicitations of the Indians, prohibits the sale of liquors to the Indians, under penalty of 300 guilders, and authorizes the savages to rob those who bring them strong liquors.‡

In 1668, the English being then in possession, a messenger conveys the request of the Indians that there shall be an absolute prohibition on the whole river, of selling strong liquors to their people. In 1671, Deputy Governor Lovelace "leaves to the discretion of the military officers the selling of liquor to the Indians :" and in 1675, there is a special order of the Court: " Strong liquors not to be sold to the Indians less than two gallons, under penalty of five shillings sterling." The chiefs finding it impossible to obtain a general prohibition, unite in a petition to the Governor and Council of Pennsylvania, in 1681, asking that the local prohibition may be removed, for reasons which they thus set forth :

"Whereas the selling of strong liquors was prohibited in Pennsylvania, and not at New Castle, we find it a greater ill-convenience than before, our Indians going down to New Castle, and there buying rum, and making them more debauched than before, in spite of the prohibition. Therefore, we, whose names are here under written, do desire that the prohibition may

---

* History of the United States, by George Bancroft. Vol. II. p. 289.
† Annals of Pennsylvania, by Samuel Hazard, pp. 314, 316.
‡ Ibid, p. 333.

be taken off, and rum and strong liquors may be sold (in the foresaid province) as formerly, until it be prohibited in New Castle, and in the government of Delaware." *

The same year, William Penn writes from London, to the Indians, and sends by the hands of his commissioners, as follows:

"I am very sensible of the unkindness and injustice that hath been too much exercised toward you by the people of these parts of the world, who sought themselves, and to make great advantages by you, rather than to be examples of justice and goodness unto you, which I hear hath been matter of trouble to you, and caused great grudgings and animosities, sometimes to the shedding of blood, which hath made the great God angry; but I am not such a man, as is well known in my own country; I have great love and regard towards you, and I desire to win and gain your love and friendship, by a kind, just and peaceful life, and the people I send are of the same mind, and shall in all things behave themselves accordingly." †

"In this spirit Penn and his religious associates, the Friends, conducted all their dealings with the Indians, and were especially zealous to keep the intoxicating bowl away from them. Large sums were offered him for the monopoly of the Indian trade, but he sternly refused, being resolved, he said, "not to act unworthy of God's providence and so defile what came to me clean." "To have sold that monopoly," says Janney, "would have frustrated the efforts made by him and his friends to prevent the sale of rum to the Indians, and to promote their civilization." ‡

Perhaps the only instance of departure from this policy, on the part of the Friends, was on the occasion of the purchase of lands in New Jersey, by the Colony led by John Fenwick, in 1675, when among the articles paid to the Indians, "there were included more than 300 gallons of rum. The colonists themselves, not having yet seen

* Ibid, pp. 372, 387, 418, 532.
† Ibid, p. 533.
‡ History of the Religious Society of Friends. By Samuel M. Janney, Vol. II. p. 388.

the propriety of abstaining from intoxicating drinks as a beverage, were probably not aware of the fearful scourge they were introducing among the simple children of nature. It was but a few years after this, when the Friends settled in New Jersey adopted measures to prevent the sale of rum to the Indians." *

In 1685, the Yearly Meeting for Pennsylvania and New Jersey adopted the following minute:

"This meeting doth unanimously agree and give as their judgment, that it is not consistent with the honor of Truth, for any that make profession thereof, to sell rum or any strong liquors to the Indians, because they use them not to moderation, but to excess and drunkenness."

The same was reaffirmed in 1687, the following clause being added in the latter year:

"And for the more effectually preventing this evil practice, we advise that this our testimony may be entered in every monthly-meeting book, and every Friend belonging to the said meeting to subscribe the same." †

Thomas Campanius Holm, in his "History of the Province of New Sweden, now Pennsylvania," published in 1702, says of the Indians:

"As to their manners and customs, they have greatly changed since the Swedes first came among them. It has been observed and been a subject of regret, as Sir William Penn and others relate, that they have learned many vices by their intercourse with the Christians; particularly drunkenness, which was before unknown to them, as they drank nothing but pure water."‡

Francis Daniel Pastorius, who came to this country in 1683, and settled Germantown, near Philadelphia, said of the earlier settlers:

"These never had the proper motives in settling here, for in⸳ ⸳₋ad of instructing the poor Indians in the Christian vir-

---

* Ibid, p. 368
† Ibid, Vol. III. pp. 501, 502.
‡ Du Ponceau's translation, Memoirs of The Historical Society of Pennsylvania, Vol. III. Part I. p. 118.

tues, their only desire was gain, without ever scrupling about
the means employed in obtaining it. * * * * These wicked people make it a custom to pay the savages in rum and other
liquors for the furs they bring to them, so that these poor delu-
ded Indians have become very intemperate, and sometimes
drink to such excess that they can neither walk nor stand." *

In 1753, Rev. Timothy Woodbridge attempted to do
some missionary work among the Indians in New York,
and mentioning to them that one great impediment to successful work was "their intemperate use of Strong
Lyquors," they desired him to communicate this reply to
Sir William Johnson :

"My Brother, my dear Brother, pity us. Your Batoe is often
here at our place, and brings us rum, and that has undone us.
Sometimes on Sunday our people drink and can't attend to their
duty, which makes it extreamly difficult. But now we have cut
it off, we have put a stop to it. You must not think, one man,
or a few men, have done it ; we all of us, both old and young,
have done it. It is done by the whole. My Brother, I would
have you tell the great men at Albany, Skenectctee, and Sko-
harry not to bring us any more rum. I would have you bring
powder, lead and clothing, what we want, and other things
what you please, only don't bring us any Strong Lyquors . . . .
You live nearer your brother than I do, and you are more intimate together ; I would have you tell him to bring no more rum
to my place. He has sent a great deal of it there, and we die
many of us only by strong drink. I would have you take care
that no more is brought to us. Now my Brother pity us ; rum
is not good, we have had enough of it. This is the third time
that I have sent to you that I would have no more rum brought
here." †

At a Council held by Col. Johnson with the Indians, a
few months later, one of the chiefs said : "We Return you
a great many thanks for stopping the Rum coming to the
Six Nations, and would be very glad the same Prohibition would have effect at Oswego." ‡

---

* Memoirs of the Historical Society of Pennsylvania, Vol. IV.
Part II. p. 98.

† Documentary History of New York, Vol II. p. 627, 628.

‡ Ibid, p. 640.

The following year the Mohawks made a similar request.* But the trouble then was the same as it has always been in the intercourse of the baser sort of settlers and adventurers with the Indians, greed overbears all other considerations, and the prospect of immediate gain blinds them to all future consequences. Sir William Johnson complained, in 1770, that

" Many traders carry little or nothing except Rum, because their profits upon it are so considerable. Again, whenever Indians are assembled on public affairs, there are always traders secreted in the neighborhood, and some publicly, who not only make them intoxicated during the time intended for Public business, but afterwards get back great part of their presents in exchange for Spirituous Liquors." †

" The traders and hunters of this period," says the biographer of David Zeisberger, the Western Picneer and Apostle of the Indians, " formed a class of their own ; bold, courageous, and with a sagacity almost equal to that of the Indians, but unscrupulous and dishonest, of degraded morals, intent upon their own advantage, and indifferent to the rights of the natives." ‡

The authority whom we have several times quoted,—and there is no more reliable or competent one on this subject, —says :

" It is a common saying with those white traders who find it their interest to make the Indians drunk, in order to obtain their peltry at a cheaper rate, that they *will* have strong liquors, and will not enter upon a bargain unless they are sure of getting it. I acknowledge that I have seen some such cases ; but I could also state many from my own knowledge, where the Indians not only refused liquor, but resisted during several days all the attempts that were made to induce them even to taste it, being well aware, as well as those who offered it to them, that if they should once put it to their lips, such was their weakness on that score, that intoxication would inevitably follow. * * * * The Indians are very sensible of the state of degradation to which they have been brought by the abuse of strong liquors ;

* Ibid, 592.
† Ibid, p. 976.
‡ Life of Zeisberger, by Edmund De Schwenitz, p. 255.

and whenever they speak of it, never fail to reproach the whites, for having enticed them into that vicious habit. I could easily prove how guilty the whites are in this respect, if I were to relate a number of anecdotes, which I rather wish to consign to oblivion." *

The Indians found in New England had also no knowledge of intoxicants before their intercourse with the whites. Roger Williams † testifies that their only drink was water. Robertson says : " They were not acquainted with any intoxicating drink." ‡   Gov. Hutchinson says:

"They had nothing that would intoxicate them. As soon as they had a taste of the English sack, and strong waters, they were bewitched with them, and by this means more have been destroyed than have fallen by the sword." §

Palfrey in his more recent work, says , " Water was their only drink, except when they could flavor it with the sweet juice for which in Spring they tapped the rock maple trees." ‖

In 1629, the Massachusetts Company in London, sent among their instructions to Gov. Endicott, the following :

" We pray you endeavor, though there be much strong water for sale, yet so to order it as that the savages may not, for our lucre sake, be induced to the excessive use, or rather abuse of it, and at any time take care our people give no ill example : and if any shall exceed in that inordinate kind of drinking as to become drunk, we hope you will take care his punishment be made exemplary for all others." ¶

Whereupon we find that one of the earliest laws of the Colony was the following, passed in 1633 : " No man shall sell or (being in a course of trade) give any strong water to any Indian." **

---

* Heckewelder, pp. 266, 267.
† Key to the Indian Language.
‡ History of America, Vol. I. p. 218.
§ History of Massachusetts, Vol. I. chap. vi.
‖ History of New England, Vol. I. p. 32.
¶ Young's Chronicles of the First Planters of Massachusetts Bay, p. 190.
** Massachusetts Colony Records, Vol. I. p. 16.

In 1644 there was a temporary letting down of this prohibition by the following order :

" The court, apprehending that it is not fit to deprive the Indians of any lawful comfort which God alloweth to all men by the use of wine, orders that it shall be lawful for all who are licensed to retail wines to sell also to Indians." * Four years later, mischief having already come from this opening of the door, it was attempted to partially close it by ordering that " only one person in Boston be allowed to sell wine to the Indians." †

In 1657, confession is made of inability to confine drinking by the Indians to moderation, and so, " all persons are wholly prohibited to sell, truck, barter, or give any strong liquors to any Indian, directly or indirectly, whether known by the name of rum, strong waters, wine, strong beer, brandy, cider, or perry, or any other strong liquors going under any other name whatsoever." ‡ I do not find that the Plymouth Colony ever permitted intoxicants to be sold to the Indians ; but it imposed a fine on all who should engage in such traffic with the natives. §

The melancholy and disgraceful fact stares us in the face, therefore, that the white people are responsible for the intemperance of the aborigines of North America ; and if responsible for the cause of their demoralization, and cruel barbarities, on whom but on themselves can they place the responsibility for the consequences of such degradation, shame and violence ? While the policy of our General Government has been to protect the Indians, by prohibiting the traffic of intoxicants among them, the agents of the government have not been ignorant of the presence among the Indians, and even at the frequent councils held with the various tribes, of a set of lawless, unprincipled adventurers, whose sole object has been to make them drunk in

---

* Ibid, Vol. II. p. 85.
† Ibid, p. 258.
‡ Ibid, Vol. III. p. 425.
§ Plymouth Colony Records, Vol. II. pp. 150, 207.

order that they may obtain for themselves the money and numerous stores paid by the government to its dependent wards; and not unfrequently these very agents have resorted to this unlawful and rascally course, for their own more sudden enrichment. Our broken treaties with the Indians are also, in the majority of cases, due to the influence of rum. Reckless and abandoned men, the scum of our great cities, fugitives from justice, utterly unprincipled and debauched, hang on the borders of civilization, making constant aggression on the lands reserved for the Indians, until at last, dissipation leads to violence, and in order to prevent bloodshed, the Indians are driven still farther into the wilderness, to be again in like manner dispossessed of their homes there. The problem is confessedly a difficult one to solve, since so extended is our frontier line that it seems utterly impossible to guard it from such wanton intrusion; but nobody can doubt that the just and wise thing to do would be by the severest measures to make an example of the dissolute intruders, rather than by breaking faith with the Indians, invite a repetition of such disgrace and wrong.

Intemperance among the white people of the United States has a sad history, and is to the present an extensive evil. The early adventurers in the New World, were, whether Spanish, French, or English, impelled thereto by the prospect of gain; some sought it through the legitimate channels of trade, and many others were drawn here by the expectation of finding gold in such immense quantities that they had only to land, fill their vessels and return home rich. As is always the case in such expeditions, the worst characters engaged in them; not only the broken down in business, the indolent and improvident, but criminals, also, men of the worst passions and of the basest habits. Even the prisons were opened that their inmates might brave the dangers of exploration and the subjugation of the savage inhabitants of America, and so prepare the way for the less reckless to follow with greater safety; and the most abandoned portions of the European cities furnished recruits whose chief characteristics were dissipation and licentiousness.*

As late as the beginning of the seventeenth century, when religious motives, the desire to escape persecution for opinion's sake, and to enjoy liberty of conscience, led to the attempt to plant permanent colonies that should be established in the fear of God, a chief hindrance, as is shown in all the early colonial records, was the intrusion of reckless and dissipated emigrants, who frustrated the efforts of the well-disposed in gaining the confidence and securing the peaceable attitude of the natives. A notable instance of this kind was the settlement made at Mt. Wallaston in 1625, and the change of the name of the place to Mare, or Merrymount, by Thomas Morton, the leader of the jovial crew, who, partly for trade and partly for pleasure, made their abode there. Well furnished with "strong beer and a liberal supply of bottles containing yet stronger fluid," he instituted Bacchanalian riotings, in which the Indians, both men and women, joined him, to the so great scandal of the settlers at Plymouth, that they took him captive, exiled him to England, and utterly broke up his settlement.*

As early as 1633 legislation against intemperance commenced in the New England colonies, as will be more fully seen in a subsequent chapter; but a great difficulty then in the way was the fact that there were few, if any, total abstainers among those who made and executed the laws, and therefore it was impossible to prevent a recruiting of the army of inebriates. Rev. Thomas Mayhew wrote from Martha's Vineyard, in 1678, to the Commissioners of the United Colonies:

"Drunkenness is severely punished in every place. It is

---

*Bancroft's History of the United States, Vol. I. chap. I. II.; Vol. II. Chap. XIV.

† See the story as graphically told by Charles Francis Adams, Jr., with full references to the sources of information, in the Atlantic Monthly Magazine for May and June, 1877.

strange to see how readyly they stripp themselves to receive
punishment for this sin of which our nation is much guylty.
All vessels that com hither and passe through the Sound, Roade
Islanders and some of our Inhabitants, doe supply them, and it's
very hard to take them. I am not out of hope that the gene-
rallity will be convinced of their folly and gyve it quite over,
that is the use of rum." *

Gov. Winthrop, in 1630, " upon consideration of the inconveniences which had grown in England by drinking
healths one to another, restrained it at his own table, and
wished others to do the like, so as it grew, by little and little, to disuse." †

" In 1639, the General Court attempted by an order " to abolish that vain custom of drinking one to another, and that upon
these and other grounds : 1. It was a thing of no good use : 2.
It was an inducement to drunkenness, and occasion of quarrelling and bloodshed. 3. It occasioned much waste of wine and
beer. 4. It was very troublesome to many, especially the masters
and mistresses of the feast, who were forced thereby to drink
more oft than they would, &c. Yet divers (even godly persons)
were very loath to part with this idle ceremony, though (when
disputation was tendered) they had no life, nor indeed could
find any arguments, to maintain it. Such power hath custom
&c." ‡

There is good reason to believe, although the mention of
dealing with and punishing intemperate persons, occurs
several times both in the Massachusetts and the Plymouth
Records, that drunkenness was not common during the early
period of the New England Colonies. Increase Mather,
D.D., preached and published "Two Sermons Testifying
Against the Sin of Drunkenness," in 1673. In the " Note
to the Reader" prefixed to the second edition, 1712, he
said : " There was a time when a man might live seven
years in New England, and not see a drunken man." And
in one of the sermons:

---

* The New England Historical and Genealogical Register, Vol.
IV p. 17
† Winthrop's History of New England, Vol. I. p. 37.
‡ Ibid, p. 324.

" Time was when there was no need for Ministers to preach much against this sin in New England. Oh! that it was so now. . . . . It is sad that ever this serpent should creep over into this wilderness, where threescore years ago he never had any footing. . . . Some there are amongst us (who they are the Lord knoweth), out of covetousness have sold intoxicating liquors to the poor Indians." Pp. 33–35.

In "A Serious Address to those who unnecessarily Frequent the Tavern," written by Cotton Mather, D.D., and published by himself and twenty-two other ministers, in 1726, there is the following allusion to the customs of their fathers:

" The practice we are now reproving, isn't it what your pious forefathers were very much strangers to? Yourselves know how ye ought to be followers of them. . . . Did they (think you) so frequent Drinking-Houses, and customarily trifle away their evenings there as many now-a-days do? . . . . May you not well blush to think how their example reproaches you, and what a different figure (in this regard) you make from them?"

In this address eight reasons are offered and enlarged upon why the people should be dissuaded from frequenting such places:

"It is a very faulty mispence of time; no good account can be given of the money that goes to support the expense; it occasions much vain conversation; they are put in the way of temptation and exposed to many dangers; it ill-affects their spiritual and best interests; obstructs family order and religion; the example is of ill-influence, and hurtful to others; it is a great grief to your ministers, who watch for your souls."

To show that even at that time Church Members incurred discipline for frequenting taverns, there is appended to the Address, a Letter of Dr. Increase Mather, then recently deceased, in answer to the following question: " Whether it be lawful for a Church-Member among us to be frequently in Taverns?" Four answers are given, and reasons adduced for each. The answers were:

" It is not lawful for a Church-Member to be in Taverns oftener than necessity calleth for it. For Church-Members to transact their Civil Affairs in the Tavern, when they might as well do it in their own houses, is an Evil in the Sight of God. It is not unlawful for a Church Member to go into a Tavern, when the business of his Civil calling does necessarily call him there. If a Church Member be necessitated to be in a Tavern, he ought to carry himself circumspectly."

In those days ale or beer, and wine, were the chief intoxicants known to the colonies. There were, however, imported and domestic distilled liquors in use, both in the Plymouth and in the Massachusetts colony. In the latter, the General Court in December, 1661, passed a law that "No person shall practice this craft of stilling strong water, nor shall sell or retail any by lesser quantity than a quarter caske." And in 1688, the Treasurer was authorized "to rent, set, or farme let the Impost of wine, brandy and rhum, and the rates upon beere, cider, ale and mum." * In Plymouth Colony a law was made in 1662, that " All persons that doe or shall still any strong waters, shall give account of their disposal of them, both of the quantity and the persons to whom sold." †

As early as 1650, an attempt to land rum in Connecticut, was made and resisted. The Swedish settlers on the Delaware were great brewers and drinkers of beer; and the Dutch, on Long Island, in 1644, were so addicted to the use of malt liquor, that James, Duke of York and Albany, issued an ordinance against its manufacture and use. And as early as this, the Dutch on the Delaware, manufactured distilled liquors, with such evil results that on the arrival of D'Hinoyossa, at Altona, in 1663, he " prohibits distilling and brewing in the colony, even for domestic use; he means to extend it to the Swedes." ‡ Just previous to this prohibition, and probably the chief occasion for it, some of the soldiers of the colony had been grossly intoxicated, and committed great outrages both on the whites and the In-

---

* Massachusetts Colony Records, Vol. IV. Part II. pp. 37, 366.
† Plymouth Colony Records, Vol. XI. p. 136.
‡ Hazard's Annals of Pennsylvania, p. 356.

dians.* Only the year before, the Director, himself, had taken away the palisades of the fort, and burned them in his brewery.† In Gov. Dougan's Report to the Committee of Trade on the Province of New York, in 1687, mention is made that rum, brandy and other distilled liquors were imported, and yielded a revenue.‡ And in 1691 flour is sent from the same Province "to the West Indies, and there is brought in returne from thence, amongst other things, a liquor called Rumm, the duty whereof considerably increaseth your Majesties revenue."§

About this time emigrants were flocking to our shores in large numbers, and drinking spread rapidly. The description which Dr. John Watson gives of the early settlement of Buckingham and Solebury, in Pennsylvania, is probably applicable to other localities.

"It is probable that the first settlers used spirits principally to prevent the bad effects of drinking water, to which they had not been accustomed in Europe. They imagined the air and water of this hot climate to be unwholesome. The immediate bad effect of cold water, when heated with exercise in summer, and the fevers and agues which seized many in the autumn, confirmed them in this opinion; and not having conveniences to make beer that would keep in hot weather, they at once adopted the practise of the laboring people in the West Indies, and drank rum. This being countenanced by general opinion, and brought into general practice as far as their limited ability would admit, bottles of rum were handed about at vendues, and mixed and stewed spirits were repeatedly given to those who attended funerals. . . . . At births many good women were collected; wine or cordial waters were esteemed suitable to the occasion for the guests; but besides these, rum, either buttered or made into hot-tiff, was believed to be essentially necessary for the lying-in woman. The tender infant must be straightly rolled round the waist with a linen swathe, and loaded with clothes until he could scarcely breathe: and when unwell or fretful, was dosed with spirit and water stewed with

---

* Ibid, p. 301.
† Ibid, p. 335.
‡ Documentary History of New York, Vol. I. pp. 147–189.
§ Ibid, p. 407.

spicery. . . . . As money was scarce, and laborers few, and business often to be done that required many hands, friends and neighbors were commonly invited to raisings of houses and barns, grubbing, chopping, and rolling logs, that required to be done in haste to get in the crop in season. Rum and a dinner were provided on these occasions. . . . . Rum was drunk in proportion to the hurry of business, and long intervals of rest employed in merry and sometimes angry conversation. . . . A considerable degree of roughness and rusticity of mind and manners prevailed, and for some time increased in the generations that succeeded the first settlers. For this I shall call to view several reasons; . . . . but more than all, the free use of rum at vendues, at frolics, and in hay time and harvest." *

The first thirty years of the eighteenth century witnessed constantly increasing drunkenness. West India rum was the principal intoxicant, which every year came in more plentifully, as flour, lumber, and general produce could be furnished in exchange, commodities which the West Indians must have, and which they came to rely on the American Colonies to furnish. But quite early in the century, distilleries were established in various parts of North America, and so numerously in Massachusetts,—the famous Medford Rum being made as early as 1735,—that in 1748 complaint was made by the planters of West India that the distillers of Massachusetts were carrying on a direct trade with France and other European countries, to their own immediate advantage, and against the interests of the mother country. To the commissioners appointed by Parliament to investigate this charge, the authorities of Massachusetts made a lengthy reply, in the course of which, they said:

"Rum is the chief manufacture in Massachusetts; there being upward of 15,000 hogsheads of rum manufactured in the Province annually. This, with what they get from the English Islands, is the grand support of all their trades and fishery; without which they can no longer subsist. Rum is a standing article in the Indian trade, and the common drink of all the laborers, timber men, mast men, loggers and fishermen in the

---

* Memoirs of the Historical Society of Pennsylvania, Vol. I. pp. 296–299.

Province. . . . . Rum is the merchandise principally made use of to procure corn and pork, for their fishermen and other navigation. The best and cheapest provision in this way of life. . . . . The rum carried from Massachusetts Bay, and the other northern colonies to the coast of Guinea, is exchanged for gold and slaves." *

In consequence of this great flood of liquors, drinking places increased, and from 1702 on, for many years, licenses to sell allowed the trade to be carried on both within doors and without.† In a Fast Sermon, preached April, 1753, by Rev. Andrew Eliot, " 'Tis surprising," he said, " what prodigious sums are expended for spirituous liquors in this one poor Province. If things are not greatly exaggerated, more than a million of our old currency in a year."‡

Nor was it in Massachusetts alone, that this evil grew. Of Huntington County, Pennsylvania, settled in 1754, it is said that:

" The deadly practice of drinking whiskey prevailed among our whole community, among Judges of the Courts, members of the bar, ministers of the gospel, physicians and patients, farmers and mechanics, servants and laborers. It was used when we were born, when we were buried; when we rose in the morning, when we went to bed at night; before dinner and after dinner; when we were full and when we were hungry; when we were sick and when we were well; when we were cold and when we were hot. It was the universal panacea." §

In 1744 the Grand Jury of Philadelphia, of which Benjamin Franklin was a member, made the following presentment:

" The Grand Jury do therefore still think it their Duty to complain of the enormous increase of Publick Houses in Philadelphia, especially since it now appears by the Constable's Returns that there are upwards of one hundred that have Licenses which, with the Retailers, make the Houses that sell

---

* Minot's Continuation of the History of the Province of Massachusetts Bay, Vol. I. pp. 155, 157.
† Drake's History of Boston, p. 525.
‡ Ibid, p. 6:5.
§ Collections of the Historical Society of Pennsylvania, p. 409.

strong drink by our Computation near a tenth part of the City,
a Proportion that appears to us much too great, since by their
number they impoverish one another, as well as the neighborhood they live in; and for the want of better customers, may
thro' necessity, be under greater temptations to entertain Apprentices, Servants and even Negroes. The Jury therefore are
glad to hear from the Bench that the Magistrates are become
sensible of this evil, and purpose to apply a remedy; for which
they will deserve the Thanks of all good Citizens. . . . . The
Jury observ'd with concern in the Course of Evidence, that a
neighborhood in which some of these disorderly houses are, is
so generally thought to be vitiated, as to obtain among the common people the shocking name of *Hell Town.*" *

Acrelius, in his History of New Sweden, published in 1759,
enumerates forty-eight drinks then in use in North America,
forty-three of which were intoxicating.†

The war with the French Colonies, which lasted nearly
ten years, closing in 1759, contributed no little, as all wars
do, to the demoralization and drunkenness of the people.
The colonies furnished large numbers of troops, to whom the
English government, as was then its custom, dealt out rum
as part of the regular rations. In addition to this supply,
rumsellers followed the army wherever it moved, establishing themselves in proximity to the camps, to the serious
detriment of the service. Measures of extreme severity
were resorted to for the suppression of this traffic, even to
the inflicting of twenty lashes per day on the soldiers who
should become intoxicated on the contraband liquor, for the
purpose of forcing a disclosure of the name of the person
who furnished it; but in many cases with no desired results.
The allowed supply created a thirst for more, which not
only braved all penalty, but also formed habits which the
soldiers did not throw off on their returning to civil life.
With what ruin to the morals of the people, and to the political interests of the colonies in the critical times then ra-

* Ibid, p. 268.
† Memoirs of The Historical Society of Pennsylvania, Vol. XI.
pp. 160, 164.

pidly hastening on, these habits were formed, we may learn
from many sources. The Diary of John Adams, under date
of February 29, 1760, gives a striking instance :

" At the present day, licensed houses are becoming the eternal haunt of loose, disorderly people of the same town, which
renders them offensive, and unfit for the entertainment of a
traveller of the least delicacy ; and it seems that poverty
and distressed circumstances are become the strongest arguments to procure an approbation ; and for these assigned reasons, such multitudes have been late licensed that none can
afford to make provisions for any but the tippling, nasty, vicious crew that most frequent them. The consequences of these
abuses are obvious. Young people are tempted to waste their
time and money, and to acquire habits of intemperance and
idleness, that we often see reduce many to beggary and vice,
and lead some of them, at last, to prison and the gallows.
The reputation of our country is ruined among strangers, who
are apt to infer the character of a place from that of the taverns
and the people they see there. But the worst effect of all, and
which ought to make every man who has the least sense of his
privileges tremble, these houses are become, in many places, the
nurseries of our legislators. An artful man, who has neither
sense nor sentiment, may, by gaining a little sway among the
rabble of a town, multiply taverns and dram-shops, and thereby secure the votes of taverner, and retailer, and of all ; and the
multiplication of taverns will make many, who may be induced
by flip and rum, to vote for any man whatever. I dare not presume to point out any method to suppress or restrain these increasing evils, but I think, for these reasons, it would be well
worth the attention of our Legislature to confine the number
of, and retrieve the character of, licensed houses, lest that im-
piety, and profaneness, that abandoned intemperance and prodigality, that impudence and brawling temper, which these
abominable nurseries daily propagate, should arise at length to
a degree of strength that even the Legislature will not be able
to control." *

But he made an attempt at reform, though he soon gave
up in despair. " I applied," he says, " to the Court of Sessions, procured a Committee of Inspection and Inquiry, reduced the number of licensed houses, etc.; but I only

acquired the reputation of a hypocrite and an ambitious demagogue by it. The number of licensed houses was soon reinstated; drams, grog, and sotting were not diminished."*

Rev. S. Kirkland, thus describes the manner of keeping Christmas on the Mohawk, N. Y., in 1769 :

"They generally assemble for read'g prayers, or Divine service—but after they eat, drink and make merry. They allow of no work or servile labor on ys day and ye follow'g—their servants are free—but drink'g, swear'g, fight'g and frolic'g are not only allowed, but seem to be essential to ye joy of ye day." †

In those days, events which were to culminate in the Independence of the Colonies, were hurrying on, some of the incidents relating thereto being greatly aggravated by the intemperance of those whose presence as an armed guard over the people was daily widening the breach between the Government and its subjects. The King's troops were in Boston, and were often intoxicated.

"Some outrage was complained of every day, and the nights were rendered hideous by drunken brawls and revels. The regular Town-watch were insulted during their rounds, and invaded in their watch-houses in the night. Distilled spirits were so cheap that the soldiers could easily command them; and hence scenes of drunkenness and debauchery were constantly exhibited before the people, vastly to the prejudice of the morals of the young. As a remedy for such conduct, the equally demoralizing exhibition of whippings was put in practice." ‡

The Revolutionary struggle which soon followed led to new supplies and extended use of intoxicants. Before that event, except in New England, and by a few small private stills established in other localities, the country was wholly dependent on, as it was deluged with, West India rum. The war cut off all foreign supply, and at once distilleries arose in all directions. The waste of grain being enormous, the prospect of famine in the army created general

---

* Ibid, Vol. IX. p. 657.
† Documentary History of New York, Vol. IV. p. 1059,
‡ Drake's History of Boston, p. 725.

alarm. Washington denounced in severest terms the general dissipation which followed, and many of the clergy, taking up the alarm, spoke from the pulpit against the fearful waste of grain, and the moral curse to the country, as well as the danger of want, that was imminent.

Congress, in session in Philadelphia, in February, 1777, unanimously :

> "*Resolved,* That it be recommended to the several legislatures in the United States immediately to pass laws the most effective for putting an immediate stop to the pernicious practice of distilling grain, by which the most extensive evils are likely to be derived if not quickly prevented."

Pennsylvania took vigorous measures to crush the evil, in 1779, by laying an embargo on the exportation of wheat and flour,. and prohibiting the distillation of all kinds of grain or meal ; but before long the gates of destruction were again thrown wide open by an exception being made in favor of barley and rye. The paper money created to meet immediately pressing demands, so rapidly depreciated that the existence of the army was jeopardized, and Congress, in 1780, called on the states to provide in some way to make up the deficiency. New Jersey and Pennsylvania attempted it by placing an excise duty on the stills. In the former state the attempt to apply the law met with such uniform opposition that it was wholly defeated. In Pennsylvania, where such an excise had been for some time established, and had yielded not far from $16,000 per annum, Robert Morris, the great financier of the times, offered to farm it at $300,000.

After the close of the war various efforts were made to diminish the evil of drinking :

> " ' Upwards of two hundred of the most respectable farmers of the County of Litchfield, Connecticut, formed (in 1789) an association to discourage the use of spirituous liquors, and determined not to use any kind of distilled liquors in doing their farming work the ensuing season.' The following year, 1790, a volume of sermons, supposed to have been written by Dr. Rush, was published in Philadelphia, which awakened such an interest among the medical men of that city, that on December

29, 1790 they sent a memorial to Congress, in which they said: 'They rejoice to find, among the powers that belong to this Government, that of restraining by certain duties the consumption of distilled spirits in our country. It belongs more peculiarly to men of other professions to enumerate the pernicious effects of these liquors upon morals and manners. Your memorialists will only remark, that a great portion of the most obstinate, painful, and mortal disorders which afflict the human body, are produced by distilled spirits; and they are not only destructive to health and life, but they impair the faculties of the mind, and thereby tend equally to dishonor our character as a nation and degrade our species as intelligent beings.

'Your memorialists have no doubt that the rumor of a plague or other pestilential disorder, which might sweep away thousands of their fellow-citizens, would produce the most vigorous and effective measures in our Government to prevent or subdue it.

'Your memorialists can see no just cause why the more certain and extensive ravages of distilled spirits upon life should not be guarded against with corresponding vigilance and exertion by the present rulers of the United States. . . . Your memorialists have beheld with regret the feeble influence of reason and religion in restraining the evils which they have enumerated. . . . . They thus publicly entreat the Congress, by their obligations to protect the lives of their constituents, and by their regard to the character of our nation and to the rank of our species in the scale of being, to impose such heavy duties upon all distilled spirits as shall be effectual to restrain their intemperate use in our country.' " *

In March,1791, Congress passed a general excise law, from which Alexander Hamilton, then Secretary of the Treasury, expected a revenue of $826,000. Western Pennsylvania, largely settled by emigrants from the north of Ireland,—famous whiskey drinkers,—was more extensively engaged in distilling than any other part of the country. " Upon a fair calculation, every sixth man became a distiller, but all equally bound to resist the excise law, which would fall heavily upon every farmer, as the money which they would procure in the east from the sale of their liquor

---

* History of the Temperance Movement, by Rev. J. B. Dunn, D.D. Centennial Volume, pp. 423, 424.

would, on their return, be demanded by the excise officer.*

A public convention of the people, at Pittsburg, in Sept. 1791, denounced the principle of excise as unjustly discriminating, and secret meetings in various parts of the disaffected counties determined on resistance by force of arms. Washington issued a Proclamation of warning against all who might be concerned in such resistance, but it produced no good result, the Marshal, while in the execution of his duty, being resisted by an armed force, and the Inspector, with the force collected for his defence, was attacked and captured. The President then placed 14,000 troops under General Henry Lee, and the rebellion was at once quelled, so large a force overcoming all opposition, without the firing of a gun.

From the 1st of July, 1791, to the 30th of September, 1792, the aggregate amount of intoxicants distilled in the country was reported as 5,171,564 gallons. " The returns aforesaid were incomplete and below the truth. From some states they were made only for a part of the year."† The aggregate amount of the Internal Revenue from 1792 to 1798, was, from all sources, $4,308,383.59. Of which $3,201,150.58 was " On domestic distilled spirits, and on stills," and on "Licenses granted to retailers, $286,286.95."‡

Nearly everybody drank, and the chief items in the expenses of town officials, religious conventions or associations, ordinations of ministers, raising the frames of church edifices, or dedicating the completed churches, were generally for liquors furnished and consumed. " Two barrels of New England Rum" were among the articles which the Parish Committee of North Carver, Mass., were ordered to procure for the use of the visitors invited to assist in raising the frame of their new Meeting House. " Eight barrels of

---

* History of Washington County. By Alfred Creigh, LL.D. Appendix, p. C1.

† Statistical Annals of the United States, by Adam Seybert, M. D., p. 460.

‡ Ibid, p. 477.

Rum," are among the items of a bill in the writer's possession for extensive alterations, repairs and enlargement of a Church edifice in Boston, in 1792; and the following items are in the expenses of the auditing committee who examined the accounts of a Church treasurer, at the close of a long term of service:

```
" 1794. Oct. 14.  3 Bowls Punch,........   12s. 0
                  2 Bottles Wine,.......    8  0
           19.  5 Bowls Punch,......£1. 0  0
                  2 Bottles Wine, .....     8  0
           24.  3 Bowls Punch, .......     12  0
                  2 Bottles Wine,.....      8  0
                       Brandy, ...........  2  2d.
```

In Gloucester, the " expense for the Selectmen and Licker at the house of Mr. James Stevens," was £3. 18s. 2d. In a description of funeral customs in New York city in 1790, given by the wife of the Rev. John Murray, occurs the following item: " Every person who attends the funeral, both within and without doors, is, previous to the interment, plentifully supplied with wine. A waiter is appointed to every room, and they are very attentive. Large quantities are often swallowed. Ten gallons of prime Madeira was lately expended at a funeral."

The Mendon Association of Ministers, presided over by Rev. Dr. Nathaniel Emmons, regaled themselves with liquors as regularly as with food, until their meeting in October, 1826, when they:

" Voted, that it be the rule of this association that no ardent spirits be presented at their meetings." The origin of this vote was given in the following incident:—The host of the association, the Rev. James O. Barney, then of Seekonk, went into Providence on the day preceding the meeting, to procure the due assortment of spirits, which immemorial usage had made an important part of his preparation.

" He accomplished his object, and at sunset commenced his return with a choice variety of liquors. Driving rapidly out of the city in his haste to reach home, he was startled from his reverie by the loud laughter of some men upon the staging round a new house in the outskirts of the city. Instantly thinking of his freight, he looked behind him, when lo! fragments of jugs, demijohns, and bottles, were dancing in and out the basket, and a ruby stream of wines, brandies, and cordials, was allaying the excited dust of the street. What was to be done? Should he go back and replenish, or take it as a providential hint, and go on? The lateness of the hour decided him to proceed, and to state the calamity to the venerable body when they should assemble. He did so, and they *took the hint*, and promptly banished the sideboard from their meetings. The Rev. Mr. Barney, from whom, in his advanced age, these facts were received, thus writes in closing his narration:

' I have lived to see and watch the rise, progress, and blessed fruits of the temperance cause; and what I once regarded as a calamity to me, in the loss of my liquor, God overruled to be one of the greatest favors He has conferred upon the clergy, the church, and the world.' " *

The following account, from an article in the Encyclopædia Americana, shows the custom of the people of the southern part of the United States, after the close of the war:

"A fashion, at the south, was to take a glass of whiskey, flavored with mint, soon after waking; and so conducive to health was this nostrum esteemed, that no sex, and scarcely any age, was deemed exempt from its application. At eleven o'clock, while mixtures under various peculiar names—sling, toddy, flip, etc.,—solicited the appetite at the bar of the common tippling shop, the offices of professional men, and the counting-room, dismissed their occupants for a half hour, to regale themselves at a neighbor's, or at a coffee-house, with punch, hot or cold, according to the season; and females, or valetudinarians, courted an appetite with medicated rum, disguised under the chaste name of "Huxham's Tincture," or "Stoughton's Elixir!" The dinner hour arrived, according to the different customs of different districts of the country, whiskey and water, curiously flavored with apples, or brandy and water, introduced the feast; whiskey or brandy, with water, helped it through, and whiskey or brandy, without water, often secured its safe digestion, not again to be used in any more formal manner than for the relief of occasional thirst, or for the entertainment of a friend, until the last appeal

---

* Life of E. N. Kirk, D.D. By Rev. David O. Mears, pp. 80, 81.

should be made to them to secure a sound night's sleep. Rum, seasoned with cherries, protected against the cold; rum, made astringent with peach-meats, concluded the repast at the confectioner's; rum made nutritious with milk, prepared for the maternal office; and, under the Greek name of paregoric, rum, doubly poisoned with opium, quieted the infant's cries. No doubt there were numbers that did not use ardent spirits; but it was not because they were not perpetually in their way. They were an established article of diet, almost as much as bread; and, with very many, they were in much more frequent use.

" The friend who did not testify his welcome, and the master who did not provide bountifully of them for his servants, was held niggardly; and there was no special meeting, not even of the most formal or sacred kind, where it was considered indecorous, scarcely any where it was not thought necessary to produce them. The consequence was, that what the great majority indulged in without scruple, large numbers indulged in without restraint. Sots were common of both sexes, various ages, and all conditions; and though no statistics of the vice were yet embodied, it was quite plain that it was constantly making large numbers bankrupt in character, property, and prospects, and inflicting on the community a vast amount of physical and mental ill in their worst forms." This picture of the south has no local coloring. "Everybody furnished and everybody drank intoxicating drinks, without shame, fear, or remorse. Every man, woman, and child viewed them as a luxury, an essential part of daily diet, the first expression of hospitality, the necessary accompaniment of labor, the best refreshment for the wearied traveller, the preventive of disease, the panacea of all ills, the joy of youth, and the support and comfort of old age. True, the loss of property in forty years, by the consumption of ardent spirits, had amounted to a greater sum than the value of all the houses and lands in the United States. True, scarcely a family was to be found in the land entirely disconnected with some miserable inebriate; each town and village had its score of drunken husbands and fathers; poorhouses groaned under their heavy burdens; church and state often saw their brightest ornaments fallen, degraded, the sport of idle boys; and year by year, from twenty to thirty thousand lost beings were hurried to the grave; but this was only the incidental and unavoidable accompaniment." *

" In 1810, the marshals returned 14,191 distilleries within the

---

* Half Century Tribute to the Cause of Temperance, by Rev. John Marsh, p. 4.

United States, and 22,977,167 gallons of spirits distilled during that year, from fruits and grains, beside 2,827,625 gallons distilled from molasses, making an annual product of 25,704,892 gallons, valued at $15,558,010. In the same year only 133,853 gallons of domestic distilled spirits from grain, and 474,900 gallons from molasses, making an aggregate of 608,843 gallons were exported from the United States, leaving of that distilled during the year, 25,006,094 gallons for consumption. On the average of the ten years from 1803 to 1812 inclusive, 7,512,415 gallons of foreign distilled spirits were annually imported into the United States, of which there was annually re-exported on the same average, only 679,322 gallons; it thence appears, that 31,929,142 gallons of spirits remained within the United States in 1810, which, if consumed in the year, was equal to four and one quarter gallons for each inhabitant." * To produce that which was distilled in this country, between five and six millions of bushels of rye and corn must have been made into spirits." †

As the Temperance Reformation,—started about 1810— progressed and gained attention, statistics began to be gathered from other sources, which disclosed the fearful extent and havoc of our drinking customs. Hartley showed that for fifteen years from the last war with Great Britain, the people of the United States had consumed on an average, every year more than 80,500,000 gallons of distilled spirits, at an annual cost of not less than $35,500,000. Barbour, in his Statistics of Intemperance in Churches, showed that out of 1634 cases of discipline whose history had been traced, more than 800 were for intemperance and more than 400 for immoralities occasioned by the use of intoxicants. Rev. Leonard Woods, D.D., stated in 1836 :

" That, at a period prior to the Temperance Reformation, he was able to count up nearly forty ministers, none of whom resided at a great distance, who were either drunkards, or so far addicted to intemperate drinking, that their reputation and usefulness were injured, if not utterly ruined. He mentions also, an ordination that took place about twenty years ago, at

---

* Saybert's Statistical Annals, p. 463.
† Pitkin's Statistical View of the Commerce of the United States, p. 122.

which he was ashamed and grieved to see two aged ministers literally drunk ; and a third indecently excited by strong drink."

Thomas Jefferson, near the close of his life, gave emphatic testimony to the mischievous effects of drinking on political affairs, and is reported to have said :

"During my administration I had more trouble from men who used ardent spirits than from all others whatever; and were I to go through my administration again, the first question I would ask of every candiate for office should be, Does he use ardent spirits ?"

The Secretary of War stated, that " during 1830, nearly 1000 men deserted from the army, and that nearly all the desertions were caused by drink ; and that from 1823 to 1829, nearly 800 deserted annually, or one-seventh of the whole army, from the same cause."

The Attorney-General of the United States published statistics in 1832 showing that the cost of spirit-drinking in the United States was $100,000,000 per year. More accurate estimates followed, making the total $150,000,000, which, with our population at 13,000,000, made the cost per capita, $11.50. Mr. Hopkins gathered up the statistics of crime chargeable to ardent spirits, at a cost of $6,525,000.*

The Temperance reformers having devoted their energies during the first twenty years of their organized work, wholly to efforts for the disuse of distilled liquors, becoming apologists for, and,—as we shall see in the next chapter,—sometimes advocating the use of fermented drinks, took no note of the intemperance caused by fermented beverages. They thus crippled their own efforts, and in many instances, as they afterwards saw, increased, instead of diminishing intemperance. When at last, they turned their attention to total abstinence from all that intoxicates, a field almost entirely new was open to them ; a field whose fruitfulness

* Dr. Barton's Discourse on Temperance, at New Orleans, 1837, pp. 11, 12.

in mischief we are in no danger of over-estimating, and
which, in spite of earnest, constant and greatly varied effort,
still lies before us with its sad harvests of incalculable mischief and misery. Except in a few localities, where the
liquor traffic is put and kept under the ban of the law,
drunkenness prevails in our land and thoroughly demoralizes
both our politics and our religion.

The yearly influx of thousands of foreigners with whiskey
and beer-loving proclivities; the opening of new lands by
our own adventurers; the lust for gain; the mad ambition
for party supremacy in politics; have made us too willing
to allow, then to foster, and at last to risk all in perpetuating this gigantic crime of crimes. We have made even
our Christian enterprises the occasion for the triumph of
anti-Christ, and brought down on our heads the curses of
the heathen, to whom we have sent,—and in the same ships
with the Gospel preachers, who were to save them,—the
vilest and most ruinous intoxicants that were ever made.
And that we may secure a partizan triumph, having in most
cases, only inconsequential ends in view by its supremacy,
we cater to the worst elements in society, the vilest traffic
that man can engage in, and suffer our Government officials,
unrebuked, to counsel with distillers and brewers as to how
legislation shall be framed in order to give the greatest
facility for the continuance and growth of "the gigantic
crime of the age."

As a consequence, we are fostering intemperance, and
that is but another way of saying, we are debauching the
nation: impoverishing the many, for the enrichment of a
few; increasing crime; destroying morality, and putting
our religion to an open shame. Our national liquor bill is
simply enormous. For thirteen years, from 1860 to 1872
inclusive, we have paid the sum of $6,780,161,805, for
2,762,926,006 gallons of liquors on which the General Government has received a revenue. It is not too much to
say,—in view of the whiskey frauds, and the brewers' frauds
on the Treasury,—we have consumed as much more that the

government received nothing for. Internal Revenue Commissioner Wells, said on the floor of the Brewers' Convention, in October, 1865: "By statistical reports it has been proven that six millions of barrels of beer are brewed annually, whilst only two and one-half millions had paid tax."

In 1878, the total revenue receipts of the National Government from distilled and fermented liquors, was $60,357,867.58. Annually we are consuming 20,000,000 bushels of corn, and nearly 39,000,000 bushels of barley in our distilled and fermented drinks. One brewery reports an annual production of 4,225,000 gallons of beer,— more than four-fifths of all the manufactured intoxicants of the United States, reported to the Internal Revenue Bureau for fifteen months from July, 1791, to September, 1792.

A Milwaukee paper states that the retail price of beer and whiskey manufactured in that city during the year ending July 1, 1878, was $21,336,000,—nearly five times the sum received by the United States Internal Revenue from all sources from 1792 to 1798. This being so, we need not wonder at the report which comes to us from Chicago, one of the chief markets for Milwaukee intoxicants, that prominent police officers stated at a public meeting in that city, not long since, that nine-tenths of all criminal cases there, including those of juveniles, grow out of the use of liquor; and that many of the drinking saloons could not exist if it were not for the boys and girls that patronize them.

The New York *Evening Post*, commenting on a recent article in the London *Gentleman's Magazine*, written by a distinguished physician, in regard to the growing disposition of many English women of every class to "habitually overstimulate," reminds its readers that "the habits of New York and of London are in this wise very similar;" and adds: "There is certainly more drinking in society — among both sexes—than before our Civil War. The statistics show a formidable increase in the consumption of spirituous liquors in the United States, and this consumption is not confined to saloons, hotels, and the jugs of the working classes." This is too true ; the sources of supply, and the numerous classes indulging in the use of intoxicants being alarmingly on the increase. Druggists are becoming liquor sellers on a large scale; a careful writer in the *Western Christian Advocate* estimating that " not less than 5,400,000 gallons of whiskey are sold annually by the druggists of the United States," and in addition to this, the drug-store traffic in " brandies, cordials, wines, bitters, and other forms of intoxicating drinks, sold in the name of medicine, will amount to 10,000,000 gallons more." That any considerable portion of this is desired as medicine, or supposed to be sought as medicine, is not entitled to belief. Our condition with reference to the use of intoxicants, and to the ease with which they can be obtained, never was more critical than it is at the present time.

CHAPTER III.

The Annual Cost of Intoxicants to the Leading Nations, and to the World—The Connection of Intemperance with Crime, Prostitution, Pauperism, Physical Decay, Mental Disease and Heredity.

COST OF INTOXICANTS.—The exact cost of intoxicating drinks in the United States, and in other parts of the world, through a series of years, it is not possible to arrive at ; but an approximation can be made. Rev. T. F. Parker has carefully compiled statistics from the best authorities, and presents this result, which is as nearly correct as figures setting forth this matter can be:

Liquors consumed in the United States :

Spirituous liquors, . . .	. 69,572,062	gallons annually.
Beer, ...	279,746,044	" "
Imported wines, . . .	10,700,000	" "

Liquors consumed in Great Britain :

Spirituous liquors, . . .	. 33,090,377	gallons annually.
Beer and ale, . . .	906,340,399	" "
Foreign and British wines, .	. 17,144,539	" "

Liquors consumed in Germany :

Beer, ...	146,000,000	gallons annually.
Wine, ...	121,000,000	" "

Liquors consumed in France :

Spirituous liquors, . . .	. 27,000,000	gallons annually.
Beer, ...	. 51,800,000	" "
Wine, ...	600,000,000	" "

We estimate that the world consumes twice as much as these four nations:

Spirituous liquors, . . . 314,031,882 gallons annually.
Beer, ... 2,797,291,632 " "
Wine, ... 1,482,239,914 " "

Cost of liquors in the world in ten years, $64,405,042,234, or twice the value of the United States of America. Allowing the average value of the world, per square mile, to equal the United States, and every one hundred and twenty years the actual cash value of the world is consumed in these drinks.

The materials used in the manufacture are annually as follows:

	Bushels of Grain.	Bushels of Grapes.	Value.
United States, . .	39,349,520	2,364,312	$42,895,984
Great Britain and Ireland, . . .	63,929,550	3,784,246	69,605,920
Germany, . . .	9,125,000	34,714,285	61,196,428
France, . . .	9,237,500	171,428,571	366,380,357
The World, . . .	242,971,145	432,634,261	891,922,536

The cost in France and Germany would be modified by the cost of grapes, which are much cheaper there.

The land, buildings, machinery, labor, etc., invested in the traffic is about as follows:

	Acres.	Buildings and Machinery.	Labor.
United States, . .	903,414	$74,041,044	$9,405,104
Great Britain and Ireland, . . .	1,629,773	92,116,883	15,271,432
Germany, . . .	517,410	46,120,535	6,304,892
France, ...	1,576,017	190,967,633	27,929,283
The World, . . .	9,253,223	746,488,070	117,821,020

	Value of Land.	Total Investment.
United States, ...	$45,170,500	128,616,848
Great Britain and Ireland, . .	81,488,650	188,876,965
Germany, ...	25,870,000	78,395,427
France, ...	78,800,850	297,697,766
The World, ...	462,660,400	1,326,909,493

Cost of alcoholic drinks in the United States annually:

Direct outlay for drink, ... $725,407,028
Seven per cent. on the $10,000,000,000 which the nation should possess, but has been destroyed by the traffic, ... 700,000,000

14

Direct loss of wages, ...	7,903,844
Ten per cent. on capital employed in the manufacture, ...	25,848,081
Ten per cent. on capital employed in saloons, .	36,254,700
Charity bestowed on the poor, ...	14,000,000
Loss by sea and land, ...	50,000,000
Court, police, hospital expenses, charity, litigation, insurance, ...	207,266,550

Total, . $1,866,642,203

"In return for this," says Mr. Parker, "the nation receives : 500 murders, 5 J0 suicides, 100,000 criminals, 200,000 paupers, 60,-000 deaths from drunkenness, 600,000 besotted drunkards, 600,000 moderate drinkers, who will be sots ten years hence, 500,000 homes destroyed,1,000,000 children worse than orphaned. And if the country should be searched from centre to circumference, it would be impossible to find *any* good resulting from the traffic, or a single reason why it should exist longer."

In *The Western Brewer*, for October, 1880, is the following : " Prof. Thausing has compiled the following statistics of beer production for the year 1879, in hectolitres :

Countries.	Quantities brewed.
German Empire................38,946,510 hect.
Great Britain..................36,597,550 "
United States..................15,400,000 "
Austro-Hungary ....11,184,681 "
France...... .... 8,721,000 "
Belgium.......... .... 7,854,000 "
Russia .... 2,300,000 "
Holland .... 1,600,000 "
Denmark...................... 1,160,000 "
Sweden................... .. ... 960,000 "
Italy.......................... 870,000 "
Switzerland...... .... 724,000 "
Norway................. ..... 615,000 "

In all, 120,842,741 hectolitres (2,660,000,000 imp. galls.), among 332,000,000 people. The average consumption was largest in Belgium, 147 litres per head; and smallest in Russia, 3 litres per head."

In the November number, it adds : "Official statistics show that in 1879 there were brewed in Bavaria 11,925,345 hect. of beer, of which 651,431 hect. were exported. The imports amounted to 16,104 hect. The consumption during the year was at the rate of 225 litres per head of the population." A litre being .946 of a quart, the consumption per head in Belgium is 35 gallons ; and in Bavaria, 53 gallons per head.

These figures, representing quantity and cost, are appalling. Consider then, the fact that the Temperance cause has done not a little to diminish the consumption of intoxicants in all these countries, and think what a vast amount of worse than worthless beverages have been made and consumed in the last one hundred years, and at what an expense of values, and the mind is confounded in its effort to grasp and realize either. We had prepared several tables of statistics, gleaned from various sources, but none of them can begin to set forth the truth in its awfulness. Turn back to the sketch already ·given of the History of Intemperance through long series of years in so many countries, and then grasp, if you can, on the basis of the figures just given, an idea of the cost, quantity, and waste, involved in this fearful traffic.

It may help in our effort to approximate the facts in this case, to take a very careful statement made by Hon. Henry W. Blair, of New Hampshire, in the House of Representatives, Washington, December 27th, 1876. Much of what he says is true of all the nations using intoxicants,—bearing in mind, of course, their relative population and the statistics before given in regard to their annual consumption of distilled and fermented liquors. Mr. Blair said :

"I now desire to present in the best manner I can a statement of facts bearing upon the effect of the manufacture and use of intoxicating liquors on the wealth, industries, and productive powers of the nation ; also upon its ignorance, pauperism, and crime. I have endeavored to authenticate every statement by careful inquiry. The information is drawn from the census returns, from records of the Departments of Government, reports of State authorities, declarations from prominent statisticians and responsible gentlemen in different parts of the country.

Much of it is to be found, with a great deal more of similar matter, in a very valuable book published the present year. The author is William Hargreaves, M. D., of Philadelphia. No one who has not fought with figures, like Paul of old with the beasts at Ephesus, knows how it taxes the utmost powers of man to classify, condense, and present intelligibly to the mind the mathematical or statistical demonstration of these tremendous social and economic facts. The truths they teach involve the fate of modern civilization.

"In 1870 the tax collected by the Internal Revenue Department was upon 72,425,353 gallons of proof spirits and 6,081,520 barrels of fermented liquors. Commissioner Delano estimates the consumption of distilled spirits in 1869 at 80,000,000 gallons. By the census returns, June 1, 1860, there were *produced* in the United States 90,412,581 gallons of domestic spirits—and of course this was consumed, with large amounts imported besides —but there are very large items which escape the official enumeration. These have been carefully estimated as follows:

	Gallons.
Domestic liquors evading tax and imported smuggled, at least	5,000,000
Domestic wines	10,000,000
Domestic wines made on farms	3,092,330
Domestic wines made and used in private families	1,000,000
Dilutions of liquors paying tax by dealers	7,500,000
	26,592,330

"This amount added to the total produced in 1860, would be 107,004,911; added to amount on which was collected tax in 1870, would be 99,017,683.

"It is well known that the great mass of alcoholic liquor is consumed as a beverage, and it will fall below the fact to place the amount paid for it at retail by the American drinker at 75,000,000 gallons yearly. But take the very modest estimate of Dr. Young, Chief of the Bureau of Statistics, who makes the following estimate of the sales of liquors in the fiscal year ending June 1, 1871:

Whiskey, (alone)	60,000,000 gallons at $6, at retail	$360,000,000
Imported spirits	2,500,000 gallons at $10, at retail	25,000,000
Imported wine	10,700,000 gallons at $5, at retail	53,500,000

Ale, beer, and porter ..... 6,500,000 gallons at $20
a bbl. at retail........ 130,000,000
Native wines, brandies, cordials, estimated........ 31,500,000

Total................................ 600,000,000

" I am satisfied that this is much below the real amount, but it is enough.

"This is one-seventh the value of all our manufactures for that year, more than one-fourth that of farm productions, betterments, and stock, as shown by the census."

"Dr. Hargreaves estimates the retail liquor bill of 1871 at $680,036,042. In 1872, as shown by the internal revenue returns, there was a total of domestic and foreign liquors shown in the hands of the American people of 337,288,066 gallons, the retail cost of which at the estimated prices of Dr. Young is $735,720,-048. The total of liquors paying tax from 1860 to 1872—thirteen years—was 2,762,926,066 gallons, costing the consumer $6,780,161,805. During several of these years, the Government was largely swindled out of the tax, so that no mortal knows how far the truth lies beyond these startling aggregates.

"Dr. Young estimates the cost of liquors in 1876 at the same as in 1871—$600,000,000—and exclaims: "It would pay for 100,000,000 barrels of flour, averaging two and one-half barrels of flour to every man, woman, and child in the country.

"Such facts might well transform the mathematician into an exclamation point. Dr. Hargreaves, who goes into all the *minutiæ* of the demonstrations, dealing, however, only with bureau returns, declares that the annual consumption of distilled spirits in the United States is not less than 100,000,000 gallons annually, and this makes a very small allowance for "crooked whiskey." Take now Dr. Young's moderate estimate of $600,-000,000 annually, and relying upon the official records of the country, and in sixteen years we have destroyed in drink $9,600,000,000—more than four times the amount of the national debt, and once and a half times the whole cost of the war of the rebellion to all sections of the country, while the loss of life, health, spiritual force, and moral power to the people was beyond comparison greater. The lowest estimate I have seen of the annual loss of life *directly* from the use of intoxicating liquor is 60,000, or 960,000 during the period above mentioned; more than three times the whole loss of the North by battle and disease in the war, as shown by the official returns.

"The assessed value of all the real estate in the United States census of 1870 is $9,914,780,825 ; of personal, $4,264,205,-

907. In twenty-five years we drink ourselves out of the value of our country, personal property and all.

"The census shows that in 1870 the State of New York spent for liquors, $106,590,000; more than two-fifths of the value of the products of agriculture and nearly one-seventh the value of all the manufactures, and nearly two-thirds of the wages paid for both agriculture and manufactures, the liquor bill being little less than twice the receipts of her railroads. The liquor bill of Pennsylvania in 1870 was $65,075,000, of Illinois, $42,825,000; Ohio, $58,845,000; Massachusetts, $25,195,000; New Hampshire, $5,800,000; Maine, where the prohibitory law is better enforced than anywhere else, $4,215,000, although Maine has twice the population of New Hampshire.

"Dr. Hargreaves says that there was expended for intoxicating drinks in—

1869	$693,999,509
1870	619,425,110
1871	680,036,042
1872	735,726,048
Total	2,729,186,709
Annual average	682,296,677

"And he says the average is larger since 1872, exceeding $700,000,000.

"Each family by the census averages 5.09 persons, and we spend for liquor at the rate of $81.74 yearly for each. The loss to the nation in perverted labor is very great. In 1872 there were 7,276 licensed wholesale liquor establishments, and 161,144 persons licensed to sell at retail. It is said that there are as many more unlicensed retail liquor shops. All these places of traffic must employ at least half a million of men. There were then 3,132 distilleries, which would employ certainly five men each—say 15,660. The brewers' congress in 1874 said that there were employed in their business 11,698. There would be miscellaneously employed about breweries and distilleries 10,000; in selling say 500,000. In all, say 550,000 able-bodied men, who, so far as distilled liquors are concerned at least, constitute a standing army constantly destroying the American people. They create more havoc than an opposing nation which should maintain a hostile force of half a million armed men constantly making war against us upon our own soil. The temple of this Janus is always open. Why should we thus persevere in self-destruction?

"There are 600,000 habitual drunkards in the United States. If they lose half their time it would be a loss of $150,000,000 to the nation in productive power, and in wages and wealth to both the nation and themselves every year.

" Dr. Hargreaves has constructed the following table :

The yearly loss of time and industry of 545,624 men
    employed in liquor-making and selling.........$272,812,000
Loss of time and industry of 600,000 drunkards 150,000,000
Loss of time of 1,404,323 male tipplers......... 146,849,592

        Total........................... ....$569,661,592

" And he adds that investigation will show this large aggregate is far below the true loss.

" By this same process 40,000,000 bushels of nutritious grain are annually destroyed, equal to 600,000,000 four pound loaves : about 80 loaves for each family in the country.

"Dr. Hitchcock, president of the Michigan State Board of Health, estimates the annual loss of productive life by reason of premature deaths produced by alcohol at 1,127,000 years, and that there are constantly sick or disabled from its use 98,000 persons in this country.

Assuming the annual producing power of an
    able-bodied person to be $500 value, and
    this annual loss of life would otherwise be
    producing, the national loss is the im-
    mense sum of $612,510,000 00
Add to this the losses by the misdirected in-
    dustry of those engaged in the manufac-
    ture and sale ; loss of one-half the time of
    the 600,000 drunkards and of the tipplers,
    as their number is estimated by Dr. Har-
    greaves..................................   568,861,592 00

And we have.... ....$1,181,371,592 00
The grain, etc., destroyed....................   36,000,000 00

        $1,217,371,592 00
Dr. Hitchcock estimates the number of insane,
    made so annually, at 9,338, or loss in effect-
    ive life of 98,250 years, at $500 per year ..   49,129,500 00
Number of idiots from same cause, an annual
    loss of 319,908 years......................   159,954,000 00

        $1,426,455,092 00

Deduct receipts of internal reve-
    nue tax (year 1875)...... ....$C1,225,995 53
Receipts from about 500,000
    State licenses, at $100....... 50,000,000 00

                                           111,225,995 53

Annual loss to the nation of reduction........$1,315,229,096 47
Annual value of all labor in the United States,
    as per census of 1870..................... 1,263,984,003 (0

Losses from alcohol in excess of wages of labor
    yearly....... .. .. $51,245,(93 47

"This calculation includes nothing for interest upon capital invested, for care of the sick, insane, idiotic—it allows alcohol credit for revenue paid on all which is used for legitimate purposes. In England the capital invested in liquor business is $585,000,000, or £117,000,000. It was proved by the liquor dealers before the committee of the Massachusetts Legislature in 1867 that the capital invested in the business in Boston was at least $100,000,000, and in the whole country it can not be less than $1,000,000,000, or ten times the amount invested in Boston. The annual value of *imported* liquors is about $80,000,000. It may be that the above estimate of losses yearly to the nation is too high. Perhaps $500 is more than the average gross earnings of an able-bodied man, and there may be other errors of less consequence. But any gentleman is at liberty to divide and subdivide the dreadful aggregate as often and as long as he pleases, and *then* I would ask him what good reason has he to give why the nation should lose *anything* from these causes."

INTEMPERANCE AND CRIME.—Nothing is more fully proven than that Intemperance is one of the most prolific sources of crime. We present here a mere handful of our gleanings from authorities on this subject. At an International Congress for the Prevention and Repression of Crime," held in London, in July 1872, the question was asked by the United States Commissioner : " What, in your opinion, are the principal causes of crime in your country ? " The following are extracts from the official answers of the several Governments :

Austria; " The desire for luxuries and license."

Belgium, "Drunkenness, libertinism, thoughtlessness, dis-
taste of work, and idleness."

Denmark, "Most frequently idleness, desire for unlawful or
lawful pleasures, and habits of drinking."

France, "The insufficiency of moral education, the general
defect of intellectual culture, and the want of any industrial
calling not opposing to the appetites and instincts a barrier
sufficiently strong, leave an open road to crimes and misde-
meanors."

Bavaria, "In some parts of Bavaria it is still the custom of
the peasants to carry long stiletto-like knives when visiting
public-houses and dancing places, and thus on Sundays and
holidays the smallest cause often leads them to inflict on each
other severe injuries."

Prussia, "Drunkenness, or, rather, a lust after immoderate
and ruinous luxury and debauchery."

Mexico, "Abuse of intoxicating drinks."

Netherlands, "Drunkenness."

Norway, "Laziness, drunkenness, and bad company, into
which these vices will lead."

Russia, "The cause of crimes arises from a certain Oriental
fatalism, which is in the foundation of the character of the people . . . . It results in a kind of slothfulness, which is frequently overcome by the temptations of drunkenness and its
consequences."

Sweden, "An ever constant desire for spirits."

Switzerland, "Drunkenness, often accompanied by other excesses. . . . That which is worst in the vice of drunkenness is
not the criminal act which it has directly or indirectly caused,
but much more the moral waste which the drunkard gradually
suffers, and which causes him to lose all perception of the most
elementary laws of morality "

United States, "Intemperance is the proximate cause of much
crime here."*

To these testimonies add the statements of eminent
Judges in the criminal courts.

Chief-Justice Sir Matthew Hale, of England, said, as
long ago as 1670 :

"The places of judicature I have long held in this kingdom
have given me an opportunity to observe the original cause of
most of the enormities that have been committed for the space

* Report of Proceedings, various pages from p. 20 to 278.

of nearly twenty years; and, by due observation, I have found that if the murders and manslaughters, the burglaries and robberies, the riots and tumults, the adulteries, fornications, rapes and other enormities that have happened in that time, were divided into *five* parts, *four* of them have been the issues and product of excessive drinking—of tavern or ale-house drinking."

Lord Chief-Baron Kelly, perhaps the oldest judge now on the English bench, says in a letter to the Arch-deacon of Canterbury : "Two-thirds of the crimes which come before the courts of law of this country are occasioned chiefly by intemperance."

In 1832, Lord Gillies, "directed the attention of the sheriff and magistrates of Glasgow to the fact, that there were not less than 1,300 licensed public-houses in the Royalty; and large as the city was, and numerous as were its inhabitants, he could not but regard that number as bearing a very extraordinary proportion to the population. They could not but be sensible of the fact that the facilities thus afforded to the indulgence of intemperate habit were the principal cause of the crime that prevailed."

The same year, at the conclusion of the Perth assizes, the Lord Justice Clark, addressing the Sheriff, said :

"He regretted to say, that he could not congratulate him on the decrease of crime in the district ; and he could not help adverting to the numerous instances of assault ; and as these evidently originated in the excitement arising from the immoderate use of spirituous liquors, he was naturally led to condemn the facilities which are too often amply afforded to the thoughtless, the profligate, or the quarrelsome, for the obtaining of ardent spirits : he would, therefore, most earnestly counsel the magistrates and others with whom it lay to grant licenses, not to allow any notion of public economy, however specious, for *increasing the revenue* of the country, to tend to the deterioration of the public morals."

Judge Patteson said, in an address to the Grand Jury : "If it were not for this drinking, you and I would have nothing to do."

Baron Alderson, addressing a Grand Jury, said : "A great proportion of the crime to be brought forward for your consideration, arises from the vice of drunkenness alone;

indeed, if you take away from the calendar all those cases with which drunkenness had any connection, you would make the large calendar a very small one."

Judge Erskine, in passing sentence on a criminal for an offence committed by him while intoxicated, said : " Ninety-nine cases out of every hundred arise from the same cause."

Justice Coleridge said : " Three-fourths of the crimes committed in the country are committed under the influence of liquor. I verily believe that nothing would tend more to make the people of this country moral, and to make the courts of justice empty sinecures, than abstaining from excessive drinking."

Chief-Justice Coleridge says :
" I can keep no terms with a vice that fills our gaols,—that destroys the comfort of homes and the peace of families, and debases and brutalizes the people of these islands."

And quite recently, in a charge to the Grand Jury, at Bristol, he said :

" Persons sitting in his position must by this time be almost tired of saying what was the veriest truism in the world, and what he supposed, because it was so true, nobody paid the slightest attention to—viz., that drunkenness was the vice which filled the jails of England, and that *if they could make England sober they could shut up nine-tenths of her prisons.* It was not only those particular cases to which he had been directing their attention, but other cases ; and indeed, *a large majority of the cases which a judge and jury had to deal with began or ended, or were connected, with the vice of drunkenness.*"

Mr. Justice Keating: " After a long experience I can state that nineteen-twentieths of the acts of violence committed throughout England originate in the public-house." " Drunkenness again ! It is almost the case with every one that is brought before me."

Mr. Justice Lush: " It is my anxiety, and I hope it will be the jurymen's also, to use all possible means to put a stop in a great degree to the drunkenness that prevails. More than half, nay full three-fourths of the cases that have been before me at these Sessions, have their origin, either directly or indirectly, in drunkenness."

Mr. Justice Denman, said at the Leeds Assize : " I may mention as illustrating the connection between excessive drinking and manslaughter, that I found at a Liverpool Assize that of thirteen offences of violence for trial, there was not one which was not directly attributable to excessive drinking. It is so here."

Lord Chief-Justice Whiteside, speaks of intemperance as : " This disgraceful vice, the parent of crime."

A return ordered by the House of Commons in 1852, in reference to the prison of Edinburgh, showed that out of 569 prisoners, 408 assigned strong drink as the cause of their being there. A governor of another prison, states: " As the result of close attention to the state and condition of over 50,000 prisoners, male and female, extending over a period of 19 years, over 75 per cent. are certainly due to intemperance."

So Mr. Justice Lawson, at the Armagh assizes, in 1869, says : " All the crimes we meet with on circuit are more or less directly or indirectly caused by drunkenness."

And Mr. Justice Deasy, at the same assize, in 1871, said : " Drunkenness is the parent of all the crimes committed in Ireland."

The King of France, some years ago, called Mr. Delavan's attention to the fact that the use of wine was the cause of intoxication and its fearful consequences, in that country ; and recently—

"According to a New York paper, the consumption of beer, wine and spirits has materially increased in France, especially within a few years, some persons accounting for it, in part, by the national disappointment and mortification at the result of the German war. The annual quantity of wine drank is declared to be equal to thirty gallons to each inhabitant of the country, while in 1838 it was not more than fifteen gallons. The consumption of beer in the last twenty years has increased three-fold, and of liquor fully fifty per cent. France is no longer a wine-drinking country merely. In many of the northern departments, particularly among the workingmen, cheap and very bad brandy has come into common use, as it has also in Paris and other large cities. The close connection between alcohol and health and vice is shown by the increase of accidental and violent deaths, of mortality generally, and likewise of crime. In the districts where alcohol is freely drank there are five times as many arrests as in the districts in which the inhabitants confine themselves to wine. A number of cases of insanity, directly traceable to alcohol, have declared themselves in different parts of the country, and these, until recently, were almost unknown. The remark, once so frequent, 'You never see a drunken man in France,' can no longer be made with truth. Drunken men, though still very rare compared with Great Britain and the United States, are now quite common; so common, indeed, as to attract no attention. Americans who have been there within three or four years have noticed this, and, if they have been abroad before, have been struck by the difference between what is and what has been."

Not long ago "The Federal Council of Switzerland, having been petitioned for a restoration of the death penalty, on the ground that crime had increased during the four years since its abolition, instituted an investigation of the subject which has occupied a period of six months. By comparison of the statistics of Switzerland with those of other countries, the council find that crime has also increased in other countries during the same period where the death penalty has been retained. Very significant is their finding as to the *cause* of this general increase of crime in their own wine-growing country as well as elsewhere. They say: 'There were five times as many executions in Great Britain in 1877 as in 1871, and nearly twice as many in Belgium, while in Denmark, Holland, Austria, Germany, France, and Italy, murder has been greatly augmented in the same time, the *cause* being, in view of the Swiss Council, the increase in misery, *intemperance* and *licentiousness*, in connection with the greater poverty and wretchedness of the populations.'"

In the United States, the same facts are apparent. The United States Commissioner of Education, in his Report for 1871, says:

"That from 80 to 90 per cent. of our criminals connect their courses of crime with intemperance. Of the 14,315 inmates of the Massachusetts prisons, 12,396 are reported to have been intemperate, or 84 per cent. In the New Hampshire prison sixty-five out of ninety-one admit themselves to have been intemperate. Reports from every State, county, and municipal prison in Connecticut, made in 1871, show that more than 90 per cent. had been in the habit of drinking, by their own admission."

The warden of the Rhode Island State prison estimates 90 per cent. of his prisoners as drinkers."

These relate to those who have been guilty of the most serious offences, not mere every-day arrests for drunkenness and disorderly conduct.

Elisha Harris, M. D., a Prison Inspector, in a Paper on "The Relations of Drunkenness to Crime," presented to the National Temperance Convention, in 1873, says:

"More than half of all the convicts in the State prisons and penitentiaries voluntarily confess the fact that they were intemperate and frequently drunk previous to the crimes for which they are imprisoned, and that such intemperance had an essential influence in preparing them for the acts of crime. About 82 per cent. of the convicts in the United States privately confess their frequent indulgence in intoxicating drinks. The Superintendent of the Detroit House of Correction found that only 18 per cent. of the convicts in fifteen State prisons and a large number of county jails ever claimed to be temperate. This may be taken as a fair statement of percentages of the temperate and intemperate in the prisons and jails of the United States and Great Britain.

"As a physician, familiar with the morbid consequences of alcoholic indulgence in thousands of sufferers from it ; as a student of physiology, interested in the remarkable phenomena and results of inebriation ; and as a close observer of social and moral wants, it was easy for the writer to believe that not less than half of all the crime and pauperism in the State depends upon alcoholic inebriety.

"But after two years of careful inquiry into the history and condition of the criminal population of the State, he finds that the conclusion is inevitable, that, taken in all its relations, alcoholic drinks may justly be charged with far more than half of the crimes that are brought to conviction in the State of New York, and that fully eighty-five per cent. of all convicts give evidence of having in some large degree been prepared or enticed to do criminal acts because of the physical and distracting effects produced upon the human organism by alcohol, and as they indulged in the use of alcoholic drink."

The Board of State Charities of Pennsylvania say in their Report for 1871 : "The most prolific source of disease, poverty and crime, observing men will acknowledge is intemperance."

William J. Mullen, Philadelphia Prison Agent, says in his Report for 1870 :

" An evidence of the bad effects of this unholy business may be seen in the fact that there have been thirty-four murders within this city (Philadelphia) during the last year alone, each one of which was traceable to intemperance, and one hundred and twenty-one assaults for murder proceeding from the same cause. Of over 38,000 arrests in our city within the year, 75 per cent. were caused by intemperance. Of 18,305 persons committed to our prison within the year, more than two-thirds were the consequence of intemperance."

A sketch of the History of the Albany, N. Y. Penitentiary, under the Superintendents Pillsbury, father and son, states that during the period of ten years ending 1876, there have been committed 13,413 criminals. " Of that number, 10,214 have admitted that they were of intemperate habits, while 3,199 claimed to be temperate."

The Indian troubles in America, are greatly aggravated by intemperance. Says the Helena (Montana) *Independent :* " Whiskey is the cause of all the disturbances between the whites and Indians, and no doubt the primary cause of all thefts and outrages by the Indians."

Another journal, the Walla-Walla *Statesman,* says: " Nearly all the trouble with Indians is occasioned by the action of a few depraved whites selling them whiskey."

Nearly twenty years ago, Rowland Burr, Esq., for many years a magistrate at Toronto, stated to the Canadian Parliament :

" That nine-tenths of the male prisoners, and nineteen-twentieths of the female, are sent to jail by intoxicating liquors. In four years there were 25,000 prisoners in Canadian jails, of whom 22,000 owed their imprisonment to drinking habits."

In 1875, a Report of a Committee made to the Dominion House of Commons, contains the following:

" Your Committee further find, on examining the reports of the prison inspectors for the provinces of Ontario and Quebec, that out of 28,289 commitments to the gaols for the three pre-

vious years, 21,236 were committed either for drunkenness or
for crimes perpetrated under the influence of drink."

The connection between drunkenness and crime is further evident from the well-authenticated fact that crime
increases or diminishes in proportion to the amount of intoxicants consumed.

" Witness stated to the Parliamentary Committee, that in
1805, when there was but little drunkenness in Glasgow, the
number of criminals brought up at the court of justice, then
held twice a year, did not exceed ten or twelve; being at the
most twenty-four in the year; but by the year 1830 crime had
become so prevalent, that it was found necessary to hold three
courts in the year, and instead of ten or twelve prisoners being
tried in each, as formerly, the number had increased to eighty!
and, generally, even more than that. This would make, at the
lowest estimate, two hundred and forty, instead of twenty-four
annually! Now, it is a 'contemporaneous fact,' which will
throw suspicion upon the drinking customs, that in the year
1805, the average consumption of spirits in Scotland was one
million and a half gallons; but from 1826 to 1831, the average
consumption was five million and a half gallons. The increase
of population in Glasgow during the above twenty-five years,
could have had but a comparatively slight effect on the calendar; for, at that period its increase would only be about one
hundred per cent., while crime had increased nine hundred per
cent!

" A still more suspicious case, however, is afforded by the reports of the Dublin police, for the three years 1808-9-10. During the first two years, distillation was stopped by law in consequence of scarcity, and crime diminished in a most extraordinary manner. In the third year, distillation was resumed, and
in consequence, liquor became cheaper, and more was drunk.
Now, it is remarkable that in the same year, the citizens of
Dublin seem to have abandoned the notion, held during the
two preceding years, of becoming a remarkably honest people,
for crime, which had been two years on the decrease, suddenly
turned round, and in 1810 increased to an alarming extent, compared with what it was while distillation was stopped. During
half of the year 1812 and the whole of 1813, the distilleries again
ceased working; liquor was consequently raised in price; and
again crime diminished. When, however, the restriction was a
second time taken off distillation, crime, as before, increased.
The number of commitments in each year was in 1811, when the

distilleries were in full work, 10,737; in 1812, six months of
which they ceased working, 9,908; in 1813, during the whole of
which distillation was stopped, 8,985! and in 1814, when the restriction upon distillation came off, 10,249." *

"Lord Morpeth, when Secretary for Ireland, in an address on
the condition of Ireland, gave these statistics : Of cases of murder, attempts at murder, offences against the person, aggravated
assaults, and cutting and maiming – there were, he says, in 1837,
12,096 ; 1838, 11,058; 1839, 1,097; 1840, 173. Between 1838 and
1840 the consumption of spirits in Ireland had fallen off 5,000,-
000 gallons; the public-houses where liquors were retailed had
lessened by 237 in the city of Dublin alone ; the persons imprisoned, in the Bridewell (the principal city prison), had fallen in
a single year from 136 to 23, and more than 100 cells in the
Bridewell being empty, the Smithfield prison was actually
closed."†

"During the seven years between 1812 and 1818, both inclusive, the annual consumption of British spirits in England and
Wales was 5,000,000 gallons, and the annual average number of
prisoners committed for trial was 11,305. During the seven
years between 1826 and 1832, the annual average consumption
had risen to nearly 9,000,000 gallons, and the annual average
commitments to 21,796, both items almost double; while from
1812 to 1832, the population had increased only about one-third.
The amount of crime, then, is not so much measured by the increase of population, as by the increase in the consumption of
intoxicating liquor.

"During the four years succeeding 1820, the consumption of
spirits in England and Wales amounted to 27,000,000 gallons;
the number of licenses granted was 351,647, and the number of
criminals committed for trial was 61,260. In the four years ending 1828, th consumption had increased to 42,000,000 gallons,
the number of licenses granted being 374,794, and the number
of committals rose to 78,345. In the next four years, ending
1832, the amount of spirits consumed was 48,000,000 gallons, the
number of licenses 468,438, when the number of commitments
increased to 91,366.

"Thus during the eight years from 1824 to 1832, the committals had increased 30,000,000, or 50 per cent., and the consumption of spirits increasing in the same time 77 per cent. with a
very decided increase also in the consumption of beer; while

---

* Documents quoted in the Teetotaller's Companion, pp. 34, 35.
† Judge Davis on Intemperance and Crime, p. 12.

15

during the three periods, the licenses had increased from 351,-647 to 468,438, being an increase of 116,794." *

An article in the *New York Commercial Advertiser,* Dec. 1876, on "Drunkenness in England," contains the following:

"As to the number of apprehensions for drunkenness, it is shown that over a course of ten years till September, 1874, the last year of which we have the statistics, the number of persons in England and Wales proceeded against summarily as drunk, and drunk and disorderly, increased 85 per cent., while the population increased only 9 per cent. In 1863-4 there were 100,067 —that is, 33 per 10,000 of population; in 1873-4 there were 185,-730 cases—that is, 57 per 10,000 of population. While, therefore, the mean increase per head of the population in the amounts of spirits and malt consumed between 1860 and 1874 was only 32.1 per cent., the increase in the number of apprehensions for being drunk, and drunk and disorderly, has been 72.7 per cent. since even 1863."

"In Maine, in 1870, the convictions for crime under prohibition were only 431, or one in every 1,689, while in New York (exclusive of the City of New York,) under license, the convictions were 5,473, or one in every 621 souls. Can it be that the rural population of New York is so much more addicted to crime than the people of Maine?

"But take Connecticut—commonly called ' the land of steady habits.' Under the prohibition law of 1854, crime is shown to have diminished 75 per cent. On the restoration of license, in 1873, crime increased 50 per cent., in a single year, and in two years in Hartford, according to official returns presented by the Rev. Mr. Walker, crime increased in that city 400 per cent. In New London the prison was empty, and the jailer out of business for awhile after prohibition went into effect.†

In Massachusetts, after a year's experience in substituting License for Prohibition, Governor Claflin said, in his Inaugural Address, January, 1869:

"A moral and Christian people cannot remain inactive when they see such results as are following, and are sure to follow, the sale of intoxicating drinks to the extent that now prevails

---

* Powell's Bacchus Dethroned, pp. 38, 39.
† Judge Davis on Intemperance and Crime. Pp. 16, 17.

in our hitherto quiet and orderly State. The increase of drunkenness and crime during the last six months, as compared with
the same period in 1867, is very marked and decisive as to the
operation of the law. The State prison, jails, and houses of
correction are being rapidly filled, and will soon require enlarged accommodations if the commitments continue to increase as
they have since the present law went into force. The increase
of commitments for the eight months previous to the 1st of
October, 1868, over the same time in 1867, is remarkable, and
demands the careful attention of the community. In the eight
months alluded to, in 1867, 65 persons were committed to the
States prison; in the same period in 1868, there were 136 commitments—more than double the number of the previous year.
It may be, perhaps, that all this increase is not due to the ease
and freedom with which intoxicating liquors can be obtained,
but few will deny that much the largest part is chargeable to
this cause."

The Chief Constable of the Commonwealth reported to
the Legislature (January, 1869) :

" This law has opened and legalized, in the various cities and
towns, about two thousand five hundred open bars, and over
one thousand other places where liquors are *presumed not* to be
sold by the glass. Of these three thousand five hundred liquor
establishments, Boston has about two thousand, or about five
hundred more than all the other cities and towns of the Commonwealth. Drunkenness is on the increase to a melancholy.
extent.

The State Board of Charities also testified :

" The increase of intemperance, which the reaction of last year
against the strictness of prohibition has greatly promoted, interferes at once with our industrial interests, fosters pauperism
and disease, and swells the lists of criminals. That intemperance has increased will appear from the prison statistics, soon
to be submitted ; that crime and vice have also increased will
be shown by the same impartial test, as well as confirmed by
the observation of all who have attended to that subject, and
noticed what has been going on in the past year. If it is desired to secure in the best manner the repression of crime and
pauperism, the increase of production, the decrease of taxation,
and a genera prosperity of the community, so far as this question
of intemperance is concerned, it is clearly my judgment that
Massachusetts should return to the policy which prohibits the

sale of intoxicating drinks except for mechanical or medical purposes.

"It will be remembered that the election of November, 1867, virtually abolished the prohibitory law, though it remained nominally in force until April 23, 1868. Bearing these facts in mind, and noticing the corresponding decrease in prosecutions for violating the liquor laws, you will also notice the increase of public drunkenness, such as is punished by imprisonment when the fine imposed cannot at once be paid. For the six months ending April 1, 1867, the number committed to jail for drunkenness was 884; for violating the liquor law, 107. In the corresponding six months, beginning October 1, 1867, and ending April 1, 1868, the number of commitments for drunkenness was 1,035; for violation of the liquor law, 47. In houses of correction during the first-named period, 480 commitments for drunkenness, and 58 for violating the law; in the second period 688 and 24; in the Boston House of Industry, 752 commitments for drunkenness in the first period, and 853 in the second. In the whole State during the first period, there were 2,116 commitments for drunkenness, and 165 for violating the liquor laws. In the second period, there were 2,576 commitments for drunkenness, and only 70 for violation of the liquor laws. The whole number of commitments for all offences was 5,977 in the first period, and 6,428 in the second. If we now compare the last six months of the prison year 1867 (from April 1 to October 1) with the last six months of 1868, the figures are equally suggestive. In the jails during this period, in 1867, there were 988 commitments for drunkenness; in the houses of correction, 609; in the House of Industry, 904: total, 2,501. During the corresponding period in 1868 the number of commitments was—to the jails, 1,090; to the houses of correction, 1,020; to the House of Industry, 1,060—total, 3,170; the whole number of commitments for all offences being 6,303 in this period of 1867, and 7,098 in 1868. During the year past, therefore, it appears that while crime in general has only increased about 10 per cent. drunkenness has increased more than twice as much, or 24 per cent. This fact offers the best possible comment on the condition of the public mind and of the legal repression of intemperance since the State election of 1867."

"The prison registers indicate that more than two-thirds of the criminals in the State are the victims of intemperance; but the proportion of crime traceable to this great vice must be set down, as heretofore, *at not less than four-fifths.* Its effects are unusually apparent in almost every grade of crime. A noticeable illustration appears in the number of commitments to the State Prison, which, during eight months of the present year, in which the sale of intoxicating liquors has been almost wholly unrestrained, was 136, against 65 during the corresponding months of the preceding year. *Similar results appear in nearly all the prisons of the Commonwealth.*"

Judge Davis, in his pamphlet previously cited, says: "My sole purpose is to establish that intemperance is an evil factor in crime, by showing that whatever limits or suppresses the one diminishes the other in a ratio almost mathematically certain. Whether judging from the declared judicial experience of others, or from my own, or from carefully collected statistics running through many series of years, I believe it entirely safe to say that one-half of all the crime of this country and of Great Britain is caused by the intemperate use of intoxicating liquors; and that of the crimes involving personal violence certainly three-fourths are chargeable to the same cause." P. 21.

INTEMPERANCE AND PROSTITUTION.— The tendency of intoxicants to inflame the passions, blunt the moral sense, and so throw a pure soul off its guard, that it may fall an easy victim to temptation, has long been noticed by the observing. St. Ambrose, in his first address to Widows, gives this injunction: "Be first pure, O widow! from wine, that thou mayest be pure from adultery." And Lord Bacon, in his "Wisdom of the Ancients," declares: "Above all things known to mankind, wine is the most powerful and efficient agent in stirring up and inflaming passions of every kind, and is of the nature of a common fuel to sensuous desires."

Addison says, in his *Spectator*, No. 569:

"The sober man, by the strength of reason, may keep under and subdue any vice or folly to which he is most inclined; but wine makes every latent seed sprout up in the soul and show itself; it gives fury to the passions, and force to those objects which are apt to produce them. . . . . Nor does this vice only betray the hidden faults of a man, and show them in the most odious colors, but often discovers faults to which he is not naturally subject. There is more of turn than of truth in a saying of Seneca, that drunkenness does not produce, but discovers faults. Common experience teaches the contrary. Wine throws a man out of himself and infuses qualities into the mind which she is a stranger to in her sober moments."

How this is done, Dr. Richardson fully explains in his Lecture on " The Effects of Alcohol on Life and Health," where he traces the action of Alcohol in the system, until—

" The cerebral or brain centres become influenced, reduced in power, and the controlling powers of will and of judgment are lost. As these centres are overbalanced and thrown into chaos, the rational part of the nature of the man gives way before the emotional, passional, or more organic part. The reason is now off duty, and all the mere animal instincts and sentiments are laid more bare—the coward shows more craven, the braggart more boastful, the cruel more savage, the untruthful more false. The reason, the emotions, the instincts, are all in a state of carnival—in chaotic, imbecile disorder."

" There is no question," says Dr. Anstic, "that the great tendency of drinking, in proportion to the frequency with which it is indulged, is to obliterate moral conscience."

Dr. Tait, in his work, " Magdalenism ; being an Inquiry into the Causes and Consequences of Prostitution," says :

" That its ranks are supplied in some measure from those who have been trained from infancy to drinking—who imbibed with their mother's milk the desire for intoxicating drinks, and unconsciously formed a habit which their riper years only confirmed and rendered more inveterate; and others who first formed the habit of intemperance, and subsequently resorted to a life of prostitution in order to procure the means of satiating their desires for alcoholic liquors. Some have recourse to strong liquors to drown remorse and shame, and expel from their minds all uneasy feelings regarding their awful situation. The mental agony which many of them experience in their sober moments, is so afflicting and intolerable that they are glad to intoxicate themselves to gain a moment's ease. The remedy of intoxication is prescribed by their companions in misfortune and associates in wickedness as the only cure for low spirits. The first month of their wicked life of prostitution is thus spent in continuous drunkenness, and the habit of dissipation is formed before they arrive at a sense of their miserable situation. No sacrifice is counted too great so that they may obtain spirituous liquors. Their clamor for drink is incessant, and every artifice is had recourse to in order to obtain it. . . . . There are few causes of prostitution more prevalent and more powerful than inebriety."

Similar testimony is borne by Dr. Vintras, in his work on Prostitution.

So Mr. Logan, in his "Moral Statistics of Glasgow," cites several instances in which he had been told by fallen women in the most explicit terms: "that drink had not only been the cause of their seduction, but it was also partaken of daily to enable them to persevere in their course of wickedness. 'Drink, drink,' said they, 'and nothing but drink, has brought us to this state of shame and degradation.' "

" The brothel," says Judge Pitman, "requires the dram-shop to stimulate the passions and to narcotize the conscience."

The statistics in regard to prostitution are not easily collected, as so many influences combine to purchase or enforce immunity from detection. But in 1839, according to the Report of the Constabulary Force Commissioners, there were about 8,000 in London, all of whom were known to the police. The entire population being then about 1,600,000, the proportion of prostitutes to virtuous adult women in London was just one in forty. In Bath, on the same basis of computation, the proportion was one in twenty-four; in Newcastle-upon-Tyne, one in nineteen; in Bristol, one in thirteen: while in Liverpool it reached the enormous proportion of one professed prostitute to every eight virtuous women! The population has increased in each of these places—in London it has more than doubled, and in the other places more than one-third—in the last forty years, and this evil has more than kept pace with it. If, then, it was claimed, and with ground for belief in the accuracy of the statement, that in 1839 there were 228.000 of these degraded women in the united kingdom, what must be the showing of this enormity now!

"It has been computed," says Mr. Powell, "that the average duration of life of this class is from 4 to 10 years; half-way between the two extremes gives an average of 7 years. Now the social evil has not diminished; in fact, it has grown with our growth, or rather with the growth of the liquor traffic. At the present time there are plying their deadly trade in all our large centres of population, about 10,000 prostitutes. And the bulk of these pass away in seven years—and how? Some perish by their own rash hands; others perish forlorn and forsaken, a mass of loathsome disease; and yet their number is not diminished; other 90,000 are found to have taken their place, to pass through the same brief and blighted career, and in their turn to meet the same sad end. As we gaze upon this diseased and degraded sisterhood—many of them still lovely amid their ruin, we are led to inquire, Whence come they? And the answer is too clear to be mistaken. They are, for the most part, the product of our Ruinous Drink System." *

Europe, generally, is as badly demoralized in this respect as Great Britain is; prostitution being licensed in some countries, and not severely dealt with in any. This significant fact, however, confronts us: "According to exact statistics 700,000 illegitimate children are annually born in Christian Europe, or one illegitimate to every 13.5 legitimate." †

In 1858, William W. Sanger, M. D., published a "History of Prostitution," etc., in which he gives the number of frail women in New York City as 7,860. At that time the number of dancing-saloons, liquor or lager beer stores, where prostitutes assembled, was 151. Now, with 7,000 licensed grog-shops in New York, and more than 2,000 unlicensed; and with the population increased from 800,000 to 1,208,471, what a field of debauchery is furnished in our great metropolis. Consider that about the same proportions are found in all our cities and large towns, and that the result of Dr. Sanger's investigations, is true everywhere: "that not one per cent. of the prostitutes in New York practice their calling without partaking of intoxicating drinks,"—and what a picture we have of the connection of drinking and moral ruin!

---

* Bacchus Dethroned, p. 42.

† Deterioration and Race Education. By Samuel Royce. P. 446.

INTEMPERANCE AND PAUPERISM.—As with crime, so with poverty, intemperance is the most prolific cause.

"It is a curious and important fact," says Dr. Colquhoun, writing in regard to London, "that, during the period when distilleries were stopped in 1796 and 1797, although bread and every necessary of life was considerably higher than during the preceding year, the poor, in that quarter of the town where the chief part resided, were apparently more comfortable, paid their rents more regularly, and were better fed than at any period for some years before; even although they had not the benefit of the extensive charities which were distributed in 1795. This can only be accounted for by their being denied the indulgence of gin, which had become in a great measure inaccessible, from its very high price. It may fairly be concluded, that the money formerly spent in this imprudent manner, had been applied to the purchase of provisions and other necessaries, to the amount of some hundred thousand pounds." *

In 1843, the Home Secretary reported that there were 2,000,000 paupers in Great Britain, the proportion to the whole population being every tenth man, woman, or child; and of this number 1,500,000 were pauperized by intemperance. In 1845, £7,000,000 was levied for the maintenance of English paupers; and during the preceding thirty years there had been an annual average of £6,500,000 levied for the same purpose, being a total of £200,000,000, of which full three-fourths was demanded by intemperance.

Mr. Chadwick, an experienced Poor Law Commissioner, testified before a Parliamentary Committee: "For some months, as I investigated every new case that came under my knowledge, I found, in nine cases out of ten, the main cause of pauperism was the ungovernable inclination for fermented liquors."

Dr. Chalmers said, in 1846: "The public-house is the most deleterious, and by far the most abundant source of pauperism."

And the Chaplain to the Work-house at Birmingham, declared, about the same time:

---

* Colquhoun on the Police of the Metropolis, p. 326.

"From my own actual experience, I am fully convinced of the accuracy of a statement made by the late governor, that of every hundred persons admitted into the Birmingham Work-house, ninety-nine were reduced to this state of humiliation and dependence, either directly or indirectly, through the prevalent and ruinous drinking usages of our country." *

Later, about 1862, the Committee on Intemperance for the Lower House of Convocation in the Province of Canterbury, after the most searching inquiry, reported:

" It appears indeed that at least 75 per cent. of the occupants of our work-houses, and a large proportion of those receiving out-door pay, have become pensioners on the public directly or indirectly through drunkenness, and the improvidence and absence of self-respect which this pestilent vice is known to engender and perpetuate. The loss of strength and wealth to the country, the increase of taxation, the deterioration of national character thus produced, it is at once humiliating and irritating to contemplate." †

In 1873, in a Paper on Poor Law and its Effects on Thrift, read by Mr. G. C. T. Bartley, Hon. Secretary of the Provident Knowledge Society, it is asserted:

"A tithe of the receipts of the public-houses properly expended would render the Poor Law altogether needless. If every man gave up one glass in ten, no Poor Law would be wanted. In my little book, "One Square Mile in the East of London," I showed that one-sixth of the amount expended in drink in one year in that poorest part of London, would build all the schools which were required, at a cost of some £75,000; and that one-twenty-third would maintain them without any Government grant at all. In a little book, 'The Seven Ages of a Village Pauper,' I go somewhat over the same ground with regard to the drink in a remote agricultural village, and the result is very striking, considering the popular notion as to the poverty of the agricultural laborer. Seven public-houses, taking at least £3000 a year, exist in the parish of 1500 souls. Calculating that half this expenditure is necessary and wholesome —and there are several special reasons which render this an excessive estimate, for nearly all the farmers who employ the villagers brew beer themselves for their men's consumption during

---

* Authorities cited in "The Teetotaller's Companion," pp.149-157.
† Britain's Social State. P. 73.

work—it follows that no less than £1500 is wasted in this small village; a sum which would give a pension of £20 a year, or nearly 8*s.* a week, to every person of the industrial class over sixty years of age."

The official " Statistical Summary " of pauperism and poor rates in England for the year 1873, shows that "829,281 paupers, necessitated a tax of £12,657,943." *

"The total income of the people of the United Kingdom was estimated by Mr. Dudley Baxter in 1870 at £860,000,000, which would be an average of more than £132 to every family, even if we take no account of the very considerable increase of wages, &c., which has taken place since then. If we take this into account, the total must be at least £900,000,000, or nearly £140 to every family. The working classes are reckoned at about 20,000,000 of the population, and their income is reckoned by Mr. D. Baxter at £325,000,000, while Professor Leone Levi makes it £418,000,000; these figures give us an average of fully £80 and £104 a year, or 31*s.* and 40*s.* a week to each workman's family. Now, while we may always expect to have some poor in the land, in consequence of disease, misfortune, and death, it is clear that with such resources, poverty should be very exceptional, and pauperism should be all but unknown. The fact is, however, very different. The following figures give the total number of paupers in the United Kingdom during 1873:

England and Wales................3,116,302
Ireland. 278,771
Scotland...........................111,996

Total.................. 3,507,069

It is probable that a few of this number may be reckoned in two or more parishes as ' Casuals,' but as the ' Vagrants' are not included in the figures for England and Ireland, the balance will be fully restored. We have thus 10 per cent. of the population, or nearly 15 per cent. of the manual labor class paupers." †

Prof. Levi, alluded to in the foregoing, estimates that of the $1,500,000,000 cash annually earned by English workmen, they ought to save $75,000,000, but that as a matter

* Cited in " Christendom and the Drink Curse," p. 124.
† " The Temperance Reformation, and its claims upon the Christian Church," pp. 52, 53.

of fact they save only $20,000,000. The bulk of the missing $55,000,000 is wasted mostly in drink."

A Committee on Pauperism appointed by the General Assembly of the Church of Scotland, reported in 1868 that :

"Intemperance creates more than half of the beggary that exists among us. It is intemperance that renders desolate so many homes, it is intemperance that brings ruin to so many families. It is of no use to enlarge upon this cause; but it would be impossible to exaggerate the influence of intemperance in making misery."

Ex-bailie Blackadder stated at a public meeting held at Edinburgh, in 1867 : " After an experience of between twenty and thirty years in connection with the management of the poor, he felt bound to say that the cases of pauperism were rare and exceptional where he did not discover drink to be directly or indirectly the procuring cause."

Rev. Mr. Miller, Superintendent of the Edinburgh City Mission, fully confirmed this testimony by saying: " that the experience of the city missionaries went to prove that nine out of every ten causes of pauperism coming before his notice were in one way or another associated with drinking.

Another clergyman, in answer to a question before a Parliamentary Committee, said : " I know it for a fact, for I have gone over the roll with the Inspector, and I know it is his opinion as well as my own, and the opinion of all who have gone over the roll minutely, that three-fourths of the cases on the roll are attributable to drink, directly or indirectly."

David Lewis, a Magistrate of Edinburgh, and for many years a member of the Parochial Board, also testified :

"Drinking and drunkenness are, I think, the cause of three-fourths of all the pauperism that we have to do with in Edinburgh. In December, 1864, I was anxious to get at something like reliable information on this point, and I applied for a return, which I got after entering a scrutiny into every individual case. The house was then full, that is to the number of 611, and I found that there were 407 of that number reduced to their impoverished condition through drink." *

It is said of this class of people in Scotland:

" Beer is even a standard of value among the lowest classes of the poor. Such expressions as ' the price of a pint,' ' worth a pot,' ' stood a gallon,' are the usual modes of expressing value among the pauperized poor. Dangerous indeed must be that section of society (and it is a large one) whose standard of value is the pot of beer." †

Similar results are noticeable in Ireland.

" During the reign of Philip and Mary, such was the rage for ' Usquebaugh ' that the inhabitants of Ireland converted their grain into spirit to such an extent as not to leave themselves sufficient for food to sustain life. Famine and privation were the result, and to prevent a recurrence of this state of things, the legislature passed an act to check the practice of free distillation. When famine again desolated that ill-fated land, in 1847-8, and the greatest distress and privation were experienced by the poor, it was distinctly proved that we had an ample supply of grain to meet the necessities of the people; but instead of being brought into the market to be disposed of as food, it was locked up in the granaries of breweries and distilleries to be wantonly destroyed in the manufacture of intoxicating liquor; as a terrible result, half a million of people perished by starvation." ‡

Severe as the recent famine has been in that ill-starred land, and enormous in amount as have been the sums raised in this country for the relief of the sufferers, it is stated that not a distillery in Ireland has been closed in consequence, but that the worse than waste of grain which they occasion, has greatly added to the miseries of the Irish. Who can wonder, then, at what is said to be the old Irish proverb: " If you wish for prosperity in Ireland, pray that God may send us a famine ; but if you wish for destruction, petition the legislature to legalize distillation." §

* Britain's Social State, pp. 74, 75.
† Convocation (Canterbury) Report, p. 82.
‡ Powell, p. 46.
§ Teetotaller's Companion, p. 384.

"France expended in 1861, 108,441,828 francs upon its poor, in 1,557 asylums and hospitals, and yet half the cities of France, and rather more, are unprovided for by public assistance, and according to the best French authorities, misery is hardly relieved, notwithstanding the large sum applied for its alleviation, and they had 337,838 vagrant beggars. Paris expended in 1869, 23,806,027 francs for in-and-out-door relief to 317,742 persons out of a population of 1,799,880." *

The following confessions solve the mystery of such poverty and debasement:

The Paris *Constitutionnel* said in 1872 : "The habit of drunkenness has increased in France year by year since the beginning of this century. The French race is deteriorating daily. In forty years the consumption of alcohol has *tripled* in France." A French magazine writes : "Drunkenness is the beginning and end of life in the great French industrial centres, among women as well as men. Twenty-five out of every one hundred men and twelve out of every one hundred women in Lisle are confirmed drunkards."

"The kingdom of Prussia has over 486,179 paupers, and gives public relief to 4.89 per cent. of its entire population. In 60 of its largest towns, 18.12 per cent. of the population are recipients of public charity ; in 238 towns next in rank, 7.38 per cent.; and in 672 of the smallest towns 4.91 per cent. are relieved.

"Saxony, with 2,337,192 population, has 2,540 poor-houses, and relieves 41,547 poor.

"Bavaria, with 4,730,977 population, relieved 79,863 poor, and swarms with tramps and beggars as hardly any other country does.

"Wurtemberg, with a population of 1,400,000, has 1,842 poor-houses, and 16,734 recipients of public charity.

"Austria, exclusive of Hungary, relieves 171,768 poor.

"Italy, exclusive of Rome and other districts, counts, in a population of 18,599,029, 1,115,126 poor, upon whom large sums are expended.

"Belgium had in 1868, 550,000 poor. Of its 908,000 families, 446,000 are public paupers. It spends on its poor $10,673,792.

"Sweden, with a population of 4,114,141, had in 1865, 140,000 poor, at a cost of $1,100,000.

"Denmark had 1,784,741 population, and 74,324 poor relieved.

---

* Royce, on Deterioration, p. 522.

" Norway, with 1,720,500 population, had 180,000 poor relieved.

" Germany has 900,000 paupers."*

In the United States there is less system and accuracy in collecting the statistics of pauperism than in the older countries ; but enough is known to make evident to all who read and observe, that here, as elsewhere, pauperism exists in proportion to drunkenness. For example : The Secretary of the State of New York, reported to the Legislature, in 1863, that the whole number of paupers relieved during the year 1862, was 257,354. These numbers were at that time in the ratio of one to every fifteen inhabitants of the State. A competent Committee made an examination of the history of these paupers, and reported that "seven-eighths of them were reduced to this low and degraded condition, directly or indirectly, through intemperance."

" The Pauper returns, made annually for a long time to the Secretary of the State of Massachusetts, show an average of about 80 per cent. as due to this cause in the County of Suffolk (mainly the city of Boston). Thus, in 1863, the whole number relieved is stated at 12,248. Of these, the number made dependent by their own intemperance is given as 6,048 ; and the number so made by the intemperance of parents and guardians at 3,837 ; making an aggregate of 9,885.

" The Third Report of the Board of State Charities, page 202, (Jan. 1867), declares intemperance to be ' the chief occasion of pauperism ; ' and the Fifth Report says : ' Overseers of the poor variously estimate the proportion of crime and pauperism attributable to the vice of intemperance from one-third in some localities up to nine-tenths in others. This seems large, but is doubtless correct in regard to some localities, and particularly among the class of persons receiving temporary relief, the greater proportion of whom are of foreign birth or descent.'

" In the Sixth Annual Report of the Board of Health (January, 1875,) page 45, under the head of ' Intemperance as a Cause of Pauperism,' the chairman, Dr. Bowditch, gives the result of answers received from 282 of the towns and cities to the two following questions :

---

* Ibid, pp. 523, 524.

1. "What proportion of the inmates of your almshouses are there in consequence of the deleterious use of intoxicating liquors ?"

2. "What proportion of the children in the house are there in consequence of the drunkenness of parents ?"

"While it appears that in the country towns the proportion is quite variable and less than the general current of statistics would lead one to expect, which is fairly attributable in part, at least, to the extent to which both law and public opinion has restricted the use and traffic in liquors, yet we have from the city of Boston, the headquarters of the traffic, this emphatic testimony from the superintendent of the Deer Island Almshouse and Hospital : ' I would answer the above by saying, to the best of my knowledge and belief, 90 per cent. to both questions. Our register shows that full one-third of the inmates received for the last two years are here through the direct cause of drunkenness. Very few inmates (there are exceptions) in this house but what rum brought them here. Setting aside the sentenced boys (sent here for truancy, petty theft, etc.) nine-tenths of the remainder are here through the influence of the use of intoxicating liquors by the parents. The great and almost the only cause for so much poverty and distress in the city can be traced to the use of intoxicating drink either by the husband or wife, or both.' "

" A startling testimony as to the effect of this cause in producing the allied evil and even nuisance of vagrancy, is given in the answer from the city of Springfield : ' In addition to circular, I would say that we have lodged and fed eight thousand and fifty-two persons that we call ' tramps ; ' and I can seldom find a man among them who was not reduced to that condition by intemperance. It is safe to say nine-tenths are drunkards, though we have not the exact records.' " *

EFFECTS OF INTEMPERANCE ON HEALTH.—The testimony of eminent medical men who have studied into the laws of health, and have had extensive practice in the treatment of disease in its various forms, is uniformly to the effect that alcohol is a destructive poison to the human system, deranging all its functions, and producing death.

Sir Henry Thompson, whose warning against so-called

* Alcohol and the State, pp. 30-32.

moderate drinking, we have elsewhere recorded, says in the same letter to the Archbishop of Canterbury:

" I have no hesitation in attributing a very large proportion of some of the most fearful and dangerous maladies which come under my notice, as well as those which every medical man has to treat, to the ordinary and daily use of fermented drinks, taken in the quantity which is conventionally deemed moderate."

Sir William Carpenter, the eminent physiologist and scientist, endorses the following opinion, signed by more than 2,000 medical men, from court physician to country practitioner :

" That the most perfect health is compatible with total abstinence from all intoxicating beverages, whether in the form of ardent spirits or as wine, beer, ale, porter, cider, etc. That total and universal abstinence from alcoholic beverages of all sorts would greatly contribute to the health . . . of the human race."

And Dr. Cheyne, in his practical Essays on the Regimen of Diet, says :

" For fermented liquors, I know no command, counsel, or example. Certainly wise Nature, who has provided liberally supplies for all wants, has furnished none of it. It is the invention of spurious and luxurious art. It is present death to many, and the natural aversion of all animals who follow pure Nature. It certainly shortens the duration of life to all that use it even with moderation, and it is the alone adequate cause of all the mortal, painful, atrocious distempers. As a medicine, for present relief, and as a bitter chalybeate potion, on occasions and extremities, it might be a tolerable medicine ; but as a common beverage, it is a slow but certain poison."

Another English writer,—pronounced by the late Governor John A. Andrew, in his speech before a Committee of the Massachusetts Legislature, in opposition to a Prohibitory Law, to be: " One of the most able English scientific critics,"—says in the " Cornhill Magazine " of September, 1852 :

" And first, as to the effect of long-continued habits of alcoholic excess upon the general health of the body, these may be summed up in brief by one word—*degeneration*. Degeneration

16

of structure and chemical composition is the inevitable fate of the tissues of the drunkard. Apart from moral influences, all that we see of physical misery, of weakened intellect, of short-ened life in the habitual drunkard, is due to this degeneration of tissue, which is gradually, but infallibly brought about by alcoholic excess. Even the very blood, the beginning of all tis-sues, is affected in a similar way, as we might expect.

There is no doubt that in excessive doses, alcohol, if it be a food at all, is a very bad one, and we must remember that the drunkard does in fact test its capacity to act as food ; for by his habits he so impairs his appetite that he can take very little, if any ordinary food.''

And Dr. James Edmunds, a distinguished English phy-sician, said, in a course of lectures given in New York, in 1874, on the medical use of alcohol and stimulants for women :

"Now recollect that food is that which puts strength into a man, and stimulant that which gets strength out of a man ; so that when you want to use stimulant, recollect that you are using that which will exhaust the last particles of strength with a facility with which your body would not otherwise part with them. If a man takes a pint of brandy, what do we see ? It intoxicates, it poisons him. Of course you know *intoxicant* is a modification of the Greek word *toxicon*. The man who is intox-icated is poisoned ; we simply use a Greek instead of a Saxon word for it. We see a man intoxicated. What are the pheno-mena we see then ? A man lies on his back snoring, helpless, senseless. If you set him up, he falls down again like a sack of potatoes. If you try to rouse him you get nothing out of him but a grunt. Is that the effect of a stimulant, do you think ? I should think it is the effect of a paralyzer that you have—mind, and body, and nerve and muscle, all equally and uni-formly paralyzed right through . . . Alcohol in a large dose is a narcotic poison, which paralyzes the body and stupefies the mind. If a man takes a somewhat larger doze, what do you see then ? You see that snoring and breathing come to an end—you see that the soft, flabby pulsations of the heart cease ; that the spark of life goes out, and the man can not be resuscitated. In fact, there are more men killed, so far as I know English statistics, more men poisoned in that way by alcohol than are poisoned by all other poisons put together. We have a great horror of arsenic and fifty other things ; the fact is, that all these other things are a mere bagatelle in relation to the most

direct, absolute, immediate, and certain poisonings which are
caused by alcohol."

Dr. Stephen Smith, of New York, says, in a Paper on
" Alcohol: Its Nature and Effects :"

"An agent which has no properties necessary to the normal
condition of the body, and which is capable in many ways
of perverting its functions, cannot be continuously employed
for any considerable period without seriously impairing organs
and tissues. Alcohol, it is seen, at first, causes a relaxation of
the arteries, and consequent excited action of the heart, followed by depression when the effects entirely pass off. The
frequent repetition of this act creates an unnatural condition
of these organs, and a tendency to require its continuance by
the constant resort to the original means. Relief to depression
is found only in a renewal of the potion, and every unusual
mental or physical strain is sought to be sustained by the same
means. Alcohol thus in time becomes the habitual resort of the
individual whose physiological condition has been originally
perverted by its subtile influence. Meantime, the *arteries* and
minute vessels become permanently dilated ; and the *heart*, subjected to excessive work, is enlarged, its orifices are increased
in size, and its valves overstretched; fatty degeneration follows,
and sudden death. Thus, the entire circulatory system is
thoroughly perverted from its normal condition and function.
The integrity of the *blood* is impaired, and nutrition is perverted. These results can but be followed by corresponding
changes of a degenerative character in other tissues and organs
throughout the entire system. The *liver* at first enlarges and
then undergoes slow contraction, producing the well-known
'hobnailed' or drunkard's liver; or it may change to a condition of fat; or, finally, grape-sugar may be developed within
the body, causing fatal diabetes. The *kidneys* may change to fat,
or undergo contraction, giving rise to Bright's Disease. The
lungs have their vessels enlarged and weakened, and are thus
rendered very susceptible to fatal pneumonia during the cold
season. The *nervous system* undergoes various degenerations,
the vessels are changed, the membrane thickened, the tissue of
nerve centre and cord is gradually changed, and thus the functions are changed, causing epilepsy, paralysis, insanity. The
digestive organs gradually lose their tone, less and less substantial food is taken, constipation is obstinate, assimilation becomes more and more impaired. Then follows muscular debility, with its attendant feebleness and want of endurance. Fi-

nally, all the tissues lose their native integrity, waste exceeds
supply, and the victim of alcoholic poisoning falls an easy prey
to some intercurrent disease, or sinks into hopeless senility." *

Dr. Willard Parker, of New York, perhaps the highest
Medical authority in America, said, at a meeting of the
" American Association for the Cure of Inebriates : "

"What is alcohol? The answer is—a poison. It is so regarded by the best writers and teachers on toxicology. I refer
to Orfila, Christison, and the like, who class it with arsenic,
corrosive sublimate, and prussic acid. Like these poisons, when
introduced into the system, it is capable of destroying life without acting mechanically. The character of alcohol being established, we investigate its physiological and pathological
action upon the living system. It has been established that,
like opium, arsenic, prussic acid, &c., in small doses, it acts as
a mild stimulant and tonic. In larger doses, it becomes a powerful irritant, producing madness, or a narcotic, producing
coma and death." †

In a recent letter to the writer, Dr. Parker says:

" The character of alcohol is settled : it is a *foreign substance*
to the body when in a physiological state; it is like a mote in
the eye.

" It produces *disease* of the system, like other poisons, and
when not too long used, the system will regain health or rid itself of the poison, as in other cases.

" Its fearful heredity is now admitted, *vide* Elam's Problems.
Arrest the use of intoxicants and you lay the axe to the root of
Insanity, Idiocy, Pauperism, etc."

The foregoing are not mere opinions, but are necessary
deductions from facts which have come under the observation of men who have had to do with every known form of
physical disease and weakness. What a sad record those
facts furnish! We present here merely a page from the
immense volume in which they are written.

Dr. Hardwicke, coroner for Central Middlesex, commenting upon a paper before the late Social Science Congress

---

* Centennial Temperance Volume, pp 256, 257.
† Proceedings of the First Meeting, 1870, p. 2.

by Dr. Norman Kerr, who stated the annual alcoholic death-rate of Great Britain at 120,000, spoke of the astonishment which he felt at one time when he traced the history of those dying in the district over which he was officer of health. He said:

"He found hardly any deaths attributed to alcohol, and he knew this must be quite inaccurate. He found the causes of death returned as disease of brain, heart, or liver, sun-stroke, etc. When he ascertained the truth, he found that alcohol was the cause of more deaths than all other causes put together— that is, at certain ages. Between twenty-five and fifty he found something like thirty to fifty per cent. had been really killed by alcohol. Dr. Kerr, though he was the first to place this matter on a scientific basis, had been wonderfully cautious and exact, and he was convinced that that gentleman's estimate of 12),000 dying annually from their own intemperance or the intemperance of others was under rather than over the truth. Dr. Kerr's conclusions were staggering, but from his own experience, both as the medical officer of health of a large borough and as coroner, holding 1,500 inquests a year, he was convinced that the estimate would ultimately be found to be an under-estimate. If such fatality, which really was after all preventible, were to occur among sheep and pigs, the country and the farmers would be all up in arms; but no one seemed to care for the slaughter of human beings by alcohol. Health officers, too, ought to call the attention of their vestries to the great death-rate through alcohol."

Dr. Hooper, in his "Physician's Vade Mecum," says:

"It has been ascertained that in men peculiarly exposed to the temptation of drinking, the mortality before thirty-five years of age is twice as great as in men following similar occupations, but less liable to fall into this fatal habit. It has also been shown that the rate of mortality among persons addicted to intemperance is more than three times as great as among the population at large. At the earlier periods of life the disproportion is still greater, being five times as great between twenty and thirty years of age, and four times as great between thirty and fifty. The annual destruction of life among persons of decidedly intemperate habits has been estimated at upwards of 3000 males, and nearly 700 females, in a population of nearly 54,000 males, and upwards of 11,000 females addicted to intemperance. (That is, of males the death-rate is 55 per 1000 per annum, and of females 63 per 1000 per annum, while the general death-rate of the whole country and at all ages, is only 23 per 1000). The greater number of these deaths are due to delirium tremens and diseases of the brain, and to dropsical affections supervening on diseases of the liver and kidneys."[*]

Mr. Wakely, former Coroner for Middlesex, says:

"I have seen so much of the evil effects of gin, that I am inclined to become a Teetotaller. Gin is the best friend I have; it causes me to hold more inquests than I otherwise should hold; and I have reason to believe that from 10,000 to 15,000 die in this metropolis annually from the effects of gin, upon whom no inquests are held."[†]

"In Scotland, in 1823, the whole consumption of intoxicating liquors amounted to 2,300,000 gallons; in 1837 to 6,776,735 gallons. In the mean time crime increased 400 per cent., fever 1,600 per cent., death 300 per cent., and the chances of human life diminished 44 per cent."[‡]

In the Twenty-third Registration Report of Massachusetts (pages 61, *et seq.*) will be found instructive tables, selected and digested by Dr. Edward Jarvis, from the result of the investigations of Mr. Neison, Actuary of the Medical, Invalid, and General Life Insurance Company of London. It is necessary to premise, in order to appreciate the full force of the tables, that under the designation " General Population" are of course included both the temperate and intemperate ; and that the latter designation includes " only such as were decidedly addicted to drinking habits, and not merely occasional drinkers or free livers." The same general result is displayed in several ways, thus :

" Rate per cent. of Annual Mortality :

Among Beer Drinkers....................................4,597
    Spirit Drinkers.....................................5,995
    Mixed Drinkers. 6,149
General Population, Males...............................2,316
     Females..2,143

" If the death rate of the general population be constantly

represented by 10, for purposes of comparison, then the death rate among the intemperate between the ages of 15 and 20 would be represented by 18; between 20 and 30, by 51; between 30 and 40, by 42; between 50 and 60, by 29, and so on.

"If we take 100,000 intemperate persons and 100,000 of the general population, starting at the age of twenty years, we shall find there will be living at successive periods as follows:

Age.	Intemperate.	General Population.
25	81,975	95,712
30	64,114	91,577
35	50,746	86,830
40	39,671	82,082
50	21,938	70,066
60	11,568	56,355
70	5,076	35,220
80	807	13,169

These tables preach their own sermon." *

Said Dr. Willard Parker, in remarks made in New York not long since:

"Another point settled is that a drinking family dies out in three or four generations. Take one of your best families, and let them commence when twenty, and go on with this drinking; in the third or fourth generation the family becomes extinct. * * * Now, as our statistics show, and from following up these matters, we find that out of the children that are born in these New York slums, over ninety per cent. die during the first year—ninety per cent.; that leaves ten per cent. Now take the ninety per cent., and place them against those who attain the good, substantial middle age or old age, and when you strike the balance it makes a very bad balance for us. Drunkards beget drunkards, and they beget a race that is soon to be destroyed." †

DRINKING AND MENTAL DISEASE, AND HEREDITARY RESULTS.—Few things are more sad to contemplate than are mental aberration and decay. To detect and remove the causes of such ruin would be to confer inestimable good

* Alcohol and the State, pp. 35, 36.
† See preceding remarks on Moderate Drinking; also for many facts and references, Dr. Hargreaves' "Alcohol, what it Is and what it Does."

upon the race. Says a widely acknowledged authority on this subject:

" While we must admit hereditary influence to be the most powerful factor in the causation of insanity, there can be no doubt that intemperance stands next to it in the list of efficient causes: it acts not only as a frequent exciting cause where there is hereditary predisposition, but as an originating cause` of cerebral and mental degeneracy, as a producer of the cause *de novo.* If all hereditary causes of insanity were cut off, and if the disease were thus stamped out for a time, it would assuredly soon be created anew by intemperance and other excesses. A striking example of the effects of intemperance in producing insanity has recently been furnished by the experience of the Glamorgan County Asylum. During the second half of the year 1871, the admissions of male patients were only 24, whereas they were 47 and 73 in the preceding and succeeding half years. During the first quarter of the year 1873, they were 10, whereas they were 21 and 18 in the preceding and succeeding quarters. There was no corresponding difference as regards female admissions. There was, however, a similar experience at the county prison, the production of crime, as well as of insanity, having diminished in a striking manner. Now the interest and instruction of these facts lie in this—that the exceptional periods corresponded exactly with the last two 'strikes' in the coal and iron industries, in which Glamorganshire is extensively engaged. The decrease was undoubtedly due mainly to the fact that the laborers had no money to spend in drinking and debauchery, that they were sober and temperate by compulsion, the direct result of which was that there was a marked decrease in the production of insanity and of crime.

" If men took careful thought of the best use which they could make of their bodies, they would probably never take alcohol except as they would take a dose of medicine, in order to serve some special purpose. It is idle to say that there is any real necessity for persons who are in good health to indulge in any kind of alcoholic liquors. At the best it is an indulgence which is unnecessary ; at the worst, it is a vice which occasions infinite misery, sin, crime, madness, and disease. Short of the patent and undeniable ills which it is admitted on all hands to produce, it is at the bottom of manifold mischiefs which are never brought home to it. How much ill work would not be done, how much good work would be better done, but for its baneful inspiration. Each act of crime, each suicide, each outbreak of madness, each disease, occasioned by it, means an infinite amount of suffering endured and inflicted before matters have reached that climax." *

Dr. Austin Flint, in his "Principles and Practice of Medicine," says:

" The deleterious influence of alcohol on the mental is not less marked than on the physical powers. The inebriate exemplifies a variety of the forms of mental derangement, called dipsomania, from which recovery is extremely rare. The perceptions are blunted, the intellectual and moral faculties progressively deteriorate, until at length the confirmed inebriate, miserably cachetic in body and imbruted in mind, has but one object in life, namely, to gratify the morbid cravings of alcohol."

Dr. M. H. Romberg, of Germany, says, in writing of the "Diseases of the Nerves and Brain :"

" The diseased condition of the blood and its vessels exerts an undoubted influence on the mind. The affections of the mind, such as vertigo, dizziness, fear, terror, etc., are caused in a great measure by the continued use of spirituous liquors and other narcotics, taken into the blood, that inflamed the blood-vessels of the nerves and brain." " The state of the blood is depraved by these poisons, and it (thus depraved) reacts upon the brain. It affects the nerves of the eye so as to make it see sights that do not exist ; and upon the nerves of sound, so as to make them hear sounds that do not exist, such as boiling, screeching, hammering, cutting, etc. After a time the mind becomes clouded, and sopor, and paralysis, and death intervene."

Says Dr. John Higginbottom : " Alcohol is particularly destructive to the brain and nervous system, and consequently, to the mental and physical powers of the whole body. Drunkenness and insanity appear so near akin, that drunkenness has been called voluntary insanity, and we often find that such voluntary insanity terminates in involuntary and incurable insanity."

Dr. Morel, after seventeen years' experience in Insane Asylums in France, says:

" There is always a hopeless number of paralytic and other

---

* Responsibility in Mental Disease. By Henry Maudsley, M. D., pp. 283–285.

insane persons in our (French) hospitals, whose disease is due to no other cause than the abuse of alcoholic liquors. In one thousand, upon whom I have made especial observation, not less than two hundred owed their mental disorder to no other cause."

Behics, in a "Report on the Physical Causes of Insanity in France," says that of eight thousand and eight hundred male lunatics, and seven thousand one hundred female lunatics, thirty-four per cent. of the men, and six per cent. of the women were made insane by intemperance.

Motet, another French writer, speaking of cases of insanity examined by him, says: "Among eight thousand seven hundred and ninety-seven cases of the insane from physical causes, three thousand and forty-four were drunkards."

Dr. Hiram Cox, while Physician to the Probate Court of Cincinnati, Ohio, examined upwards of four hundred cases of insanity, previous to sending them to the State Asylum, and found that "two-thirds of their number became insane from drinking the poisonous liquors sold at the doggeries and taverns of our city and county." *

Dr. James Edmunds, in his Lecture on the "Medical Use of Alcohol and Stimulants for Women," before cited, says:

"It is admitted by every one that alcohol is the cause of more than half the insanity we have. I am not so familiar with the facts on this subject here as I should naturally be at the other side of the Atlantic. . . . I know this: that Lord Shaftesbury, the chairman of our commission on lunacy in England, has said, in a parliamentary report on the subject, that six out of ten lunatics in our asylums are made lunatic by the use of alcohol. It is a fact which can not be disputed that diseases of the liver, diseases of the lungs, diseases of the tissues of the body, are induced directly by the use of alcohol, and that as a general rule you may say that where you have alcohol used most largely and most frequently, there these diseases and degenerations in the tissues of the body become most marked. I could give you very authoritative facts bearing upon this mat-

---

* Cited by Story, in "Alcohol, its Nature and Effects," pp. 227–238.

ter from sources which are not open to the imputation of any
kind of moral bias, as the utterances of some of our temperance
friends may be open to."

And Dr. B. W. Richardson, in an Address at the Royal
Albert Hall, in London, in 1877, said:

"We know, now, scientifically, that alcohol excites the mental power unduly, then depresses it into melancholy, and so
often brings it to complete aberration, that in some of our institutions for the insane as many as 40 per cent. of those who enter per year are made to enter from this one simple cause alone."

Dr. F. R. Lees, has collected statistics showing that:

"The number of deranged people in a country corresponds
very closely with the amount of strong drink they consume.
Till the introduction of fire-water among the American Indians,
insanity was unknown. In Cairo, comparatively teetotal, there
is one insane person to every 30,714 of the inhabitants. In
Spain, comparatively sober, the consumption of alcohol being
only one gallon per head per annum, there is one insane person
in every 7,181. In Normandy, consuming two gallons, one in
every 551. In England, consuming two and a half gallons, the
proportion is one in every 430 of the inhabitants." *

According to Royce: "When the duty on spirits was
removed in Norway, in 1825, between that time and 1835
insanity increased 50 per cent., and idiocy 150 per cent."

Sweden consumes 25,000,000 gallons of spirits, though it
has but 3,000,000 population—of whom but half are of an
age to drink—and the consequence is that insanity, suicide
and crime are fearfully common among them." †

"In Prussia," says Dr. Finkelburg, member of the Prus-
sian Commission of Public Health, "one person in 450 is
insane. The cause is chiefly the abuse of alcoholic
liquors."

"Of 490 maniacs," writes the Bishop of London, "in one
hospital, 257 were deprived of reason by drinking. . . .

* Prize Essay on the Liquor Traffic, p. 199.
† Smith's Temperance Reformation, p. 60.

Of 780 maniacs in different hospitals, 392 were deprived of reason in the same way." *

Says Dr. Townson, of Liverpool: "It is part of my duty to examine pauper lunatics in considerable numbers, and into the history of each I have to inquire, and my conviction is this, that five out of every six of the lunatics of the workhouse have been reduced to that condition by intemperance." †

The chief horror of this great evil considered in a mental or moral point of view, is, that these deplorable consequences of drinking are not confined to those who use the intoxicating cup, but are entailed upon their offspring through several generations. Morell, who has made the study of the causes of human deterioration a specialty, cites many cases of children of inebriates cursed in later years with the hereditary bent of excessive alcoholism, leaving one insane asylum for another, and ending in marasmus, general paralysis, a perfectly brutal condition, and the utter extinction of reason and conscience. "I constantly find," he says, " the children of drunkards in the asylums for the insane, in prisons and houses of correction. The deviation from the normal type of humanity shows itself in these victims by the arrest of the development of their constitutional system as well as by a vicious intellectual disposition and cruel instincts." ‡

Maudsley also says:

"A host of facts might be brought forward to prove that drunkenness in parents, especially that form of drunkenness known as dipsomania, which breaks out from time to time in uncontrollable paroxysms, is a cause of idiocy, suicide or insanity in their offspring." §

Dr. Brown says, in his "Hereditary Tendency to Insanity: "

* Deterioration and Race Education, pp. 428, 429.
† Teetotaller's Companion, p. 287.
‡ Cited by Royce, p. 426.
§ Responsibility in Mental Disease, p. 43.

" The drunkard injures and enfeebles his own nervous system, and entails mental disease upon his family. His daughters are nervous and hysterical, his sons are weak, wayward, eccentric, and sink insane, under the pressure of excitement, from some unforeseen exigency, or of the ordinary calls of duty. This heritage may be the result of a ruined and diseased constitution ; but it is much more likely to proceed from that long continued nervous excitement, in which pleasure was sought in the alternate exaltation of sentiment and oblivion which exhausted and wore out the mental powers, and ultimately produced imbecility and paralysis, both attributable to disease of the substance of the brain." *

Dr. Ray, one of the highest authorities in America on the subject of insanity, says:

" Another potent agency in vitiating the quality of the brain is habitual intemperance, and the effect is far oftener witnessed in the offspring than in the drunkard himself. His habits may induce an attack of insanity where the predisposition exists; but he generally escapes with nothing worse than the loss of some of his natural vigor and hardihood of mind. In the offspring, however, on whom the consequences of the parental vice may be visited, to the third if not the fourth generation, the cerebral disorder may take the form of intemperance, or idiocy, or insanity, or vicious habits, or impulses to crime, or some minor mental obliquities." †

Dr. Storer, of Boston, alluding to certain statements made by Dr. Day, Superintendent of the Washingtonian Home, in that city, says:

" Reference has been made by the doctor to the dire effects so often seen by medical men in the persons of the children of those addicted to habits of intoxication—epilepsy, idiocy, and insanity, congenital or subsequently developing themselves, with or without any apparent exciting cause. He has not, however, I think, sufficiently held up to the victims of this baleful thirst the terrible curse they thus deliberately entail upon their descendants."

And the Massachusetts Board of State Charities, in their

---

* Cited in Teetotaller's Companion, pp. 288-9.
† Mental Hygiene, p. 44.

Report for 1866, in speaking of "one prolific cause of the vitiation of the human stock," say :

That prolific cause is the common habit of taking alcohol into the system, usually as the basis of spirits, wine or beer. .'. . The basis being the same in all, the *constitutional effects* are about the same. The use of alcohol materially modifies a man's bodily condition; and, so far as it affects him individually, it is his own affair; but if it affects also the number and condition of his offspring, that affects society. If its general use does materially influence the number and condition of the dependent and criminal classes, it is the duty of all who have thought and care about social improvement to consider the matter carefully, and it is the special duty of those having official relations with those classes to furnish facts and materials for public considera- tion. It is well known that alcohol acts unequally upon man's nature; that it stimulates the lower propensities and weakens the higher faculties. . . . and represses the functions which manifest themselves in the higher or human sentiments which result in *will*. If the blood, highly alcoholized, goes to the brain, its functions become subverted ; the man does not know and does not care what he says or does. If this process is often repeated . . . the man is no longer under control of his volun- tary power, but has come under the dominion of automatic functions, which are almost as much beyond his control as the beating of his heart. Any morbid condition of body frequently repeated becomes established by habit . . . and makes him more liable to certain diseases, as gout, scrofula, insanity and the like. This liability or tendency he transmits to his children just as surely as he transmits likeness in form or feature . . . Now the use of alcohol certainly does induce a morbid condition of body. It is morally certain that the frequent or the habitual overthrow of the conscience and will, or the *habitual weakening* of them, soon establishes a morbid condition, with morbid appetites and tendencies, and that those appetites and tenden- cies are surely transmitted to the offspring."

"Again, it is admitted that an intemperate mother nurses her babe with alcoholized milk ; but it is not enough considered that a *father* gives to his offspring certain tendencies which lead surely to craving for stimulants. These cravings once indulged, grow to a passion, the vehemence of which passes the compre- hension of common men."

Dr. A. Mitchell, one of the Scotch Lunacy Commis- sioners, stated in evidence before the Select Committee on

Habitual Drunkards (1872) that "it is quite certain that the children of habitual drunkards are in a larger proportion idiotic than other children, and in a larger proportion themselves habitual drunkards; they are also in a larger proportion liable to the ordinary forms of acquired insanity."

Professor Laycock, of Edinburgh, states that 80 per cent. of the children brought up in workhouses (whom he reckons 300,000) prove failures when sent out into the world; most of them either swell the criminal class, or return to the workhouse, or become inmates of asylums: and he adds:

"I might, if time allowed, point out how drunken, vicious imbeciles, tainting their offspring to the third and fourth generations, serve to fill our asylums to overflowing. Dr. Carpenter quotes the testimony of Sir W. A. F. Browne, the first medical Lunacy Commissioner for Scotland, to the following effect: 'The children of drunkards are deficient in bodily and vital energy, and are predisposed by their very organizations to have cravings for alcoholic stimulants. If they pursue the course of their fathers, while they have more temptations to follow and less power to avoid, they add to their hereditary weakness, and increase the tendency to idiocy or insanity in themselves and their children." *

And Dr. Willard Parker, than whom there is no higher authority in America concerning all matters pertaining to Medical Science, says, in a tract entitled " Remarks on the Hereditary Influence of Alcohol : "

"Of all agents, alcohol is the most potent in establishing a heredity that exhibits itself in the destruction of mind and body. Its malign influence was observed by the ancients long before the production of whiskey or brandy, or other distilled liquors, and when fermented liquors or wines only were known. Aristotle says, 'Drunken women bring forth children like unto themselves,' and Plutarch remarks, 'One drunkard begets another.' Lycurgus made drunkenness in women infamous by exhibitions, and Romulus made it punishable with death, because the habit was regarded as leading to immorality, which would compromise the family integrity. But although the

---

* Smith on the Temperance Reformation, pp. 62-3.

broad features of alcoholism were appreciated by the ancients, later and more exact investigations have thrown more light upon the subject.

"The hereditary influence of alcohol manifests itself in various ways. It transmits an appetite for strong drink to the children, and these are likely to have that form of drunkenness which may be termed paroxysmal; that is, they will go for a considerable period without indulging, placing restraint upon themselves, but at last all the barriers of self-control give way; they yield to the irresistible appetite, and then their indulgence is extreme. The drunkard by inheritance is a more helpless slave than his progenitor, and the children that he begets are more helpless still, unless on the mother's side there is engrafted upon them untainted stock.

" But its hereditary influence is not confined to the propagation of drunkards. It produces insanity, idiocy, epilepsy, and other affections of the brain and nervous system, not only in the transgressor himself, but in his children, and these will transmit predisposition to any of these diseases. Pritchard and Esquirol, two great authorities upon the subject, attribute half of the cases of insanity in England to the use of alcohol. Dr. Benjamin Rush believed that one-third of the cases of insanity in this country were caused by intemperance, and this was long before its hereditary potency was adequately appreciated. Dr. S. G. Howe attributed one-half of the cases of idiocy, in the State of Massachusetts, to intemperance, and he is sustained in his opinion by the most reliable authorities. Dr. Howe states that there were seven idiots in one family where both parents were drunkards. One-half of the idiots in England are of drunken parentage, and the same is true of Sweden, and probably of most European countries. It is said, that in St. Petersburg most of the idiots come from drunken parents. When alcoholism does not produce insanity, idiocy, or epilepsy, it weakens the conscience, impairs the will, and makes the individual the creature of impulse and not of reason. Dr. Carpenter regards it as more potent in weakening the will and arousing the more violent passions than any other agent, and thinks it not improbable that the habitual use of alcoholic beverages, which are produced in such great quantities in civilized countries, has been one great cause of the hereditary tendency to insanity. In a work on the 'Diseases of Modern Life,' Dr. Richardson remarks: ' The solemnest fact of all bearing upon the physical deteriorations and upon the mental aberrations produced by alcohol, is, that the mischief inflicted by it on man, through his own act, can not fail to be transmitted to those who descend from him, while the propensity to its use descends also, making the evil interest compound in its totality.' But this is not stating the case as strongly as the truth demands. There is not only a propensity transmitted, but an actual disease of the nervous system, which not merely manifests itself in a propensity, but in an *uncontrollable impulse.* I have been acquainted with several men, having brilliant and cultured minds, who inherited the vice, and they have stated to me that there were times when the impulse to drink strong liquor was perfectly irresistible, and that no offer could be made them which would dissuade them from yielding to it.

"The researches of Morel on the causes of the formation of degenerate varieties of the human race, indicate the influence of the continuance of morbid action through succeeding generations, and its power to finally cause extinction of the family; and it will be noticed how large a share drunkenness holds in the chain of causation. He gives the history of one family as follows: First generation—the father was an habitual drunkard, and was killed in a public-house brawl. Second generation —Hereditary drunkenness, maniacal attacks, general paralysis. Third generation—The grandson was strictly sober, but was full of hypochondriacal and imaginary fears of persecutions, etc., and had homicidal tendencies. Fourth generation- Feeble intelligence, stupidity, first attack of mania at sixteen, transition to complete idiocy and extinction of family. After all, may we not fairly entertain the question whether drunkenness was not the whole cause, or almost the whole, of all the other abnormal characteristics of the case? The thirteenth annual report of the New York Prison Association contains the genealogy of a family called 'Juke,' who have become historical, as affording one of the most illustrative cases on record of the hereditary influence of alcohol and vice. My own experience has furnished me with many similar examples. To instance one: A merchant came to me for medical advice. He was a man in good circumstances, but was in the habit of getting intoxicated every night before retiring. His mother also drank habitually, and died of paralysis. He had two brothers and three sisters; he was the second brother and child. The oldest brother died a paroxysmal drunkard; that is, he had, all his life, periodical fits of drunkenness. My patient was sober, and a successful merchant, but was always in a state of mental discomfort, and was suspicious and jealous to the most unreasonable degree. The third brother and child died a drunkard. The fourth child, a sister, was an inmate of a lunatic asylum. The fifth child was intolerable on account of her eccentricity. The

17

sixth child, also a female, died of consumption. The second son, my patient, married a woman of fine physical and mental organization. They had two sons; the elder was associated with his father in business, and was an energetic man, but exceedingly excitable, and although not an habitual drinker, was a slave to his other animal appetites. The other child was, when only five years old, very unmanageable and exceedingly prone to vicious habits. He was constantly running away, and would take every opportunity to commit theft. He was, in reality, a moral idiot. Here, in spite of the restoring influence of the fine mental and physical organization of the mother, we see the effects of alcohol cropping out in the third generation, exhibiting unmistakable characteristics of the paternal side of the family; traceable, in fact, to the habitual alcoholic indulgence of the grandparents. Thus, we do not always see the *worst* effects of the hereditary influence of alcohol, because of the frequent mingling of good blood with that which is tainted; but in the most squalid portions of our large cities we often see the hereditary tendency of alcoholism exhibited in aggravated forms. There many of the children are born of parents tainted on both sides, and these are brought into the world with constitutions so enfeebled that a large percentage of them die the first year, and those that live are unsound in mind and body. Indeed, from my own observations and the testimony of others, I am led to the conclusion, that by far the larger share of mental disease, poverty, and crime is the direct heritage of alcohol; that it also is the cause of a great share of our bodily diseases, and is a powerful element in shortening the average duration of life in certain localities or among certain classes."

CHAPTER IV.

History of the Means Employed in various Ages and Nations to remove Intemperance; viz.: Antidotes to Intoxication, Infliction of Personal Penalties on Drunkards, Moderation Societies, Total Abstinence Societies, Coffee Houses, Inebriate Asylums, Education, License of the Sale of Intoxicants, Prohibition, Local Option.

THE manifest and acknowledged evils of intemperance have led, in all ages of the history of the vice, to experiments and efforts for avoiding those evils, and to many attempts to restrain or destroy their cause. The former have been characterized by no dislike of the intoxicants themselves, no purpose of diminishing their use, but rather have been incited by strenuous desire and effort to increase individual ability to consume the contents of the bowl with impunity. The latter have aimed either at the restricted use of intoxicants, or at a total avoidance of them; based in the one case on an acknowledgment of the necessity of refraining from over-indulgence, and in the other, on the folly and danger of any indulgence whatever. The purpose of this chapter is to simply narrate facts, not to deduce theories from them; to classify the efforts and outline their history, not to argue as to their worth or their folly.

For convenience they will be considered in the following orders: Antidotes to Intoxication; Infliction of Personal Penalties on Drunkards; Societies to Diminish the Use of Ardent Spirits; Total Abstinence Societies; Coffee Houses; Inebriate Asylums; Education; License of the sale of Intoxicants; Prohibition; Local Option.

I. ANTIDOTES.—Attempts to neutralize the intoxicating power of alcoholic beverages, in order that thereby drinkers might sit longer at their cups, were not uncommon in ancient times. "Athenæus, arguing on the fondness of the Egyptians for wine, says: "This is a proof, that they are the only people amongst whom it is a custom at their feasts to eat boiled cabbages before all the rest of their food, and even to this very time they do so. And many people add cabbage seed to potions which they prepare as preventives against drunkenness. And wherever a vineyard has cabbages growing in it, there the wine is weaker." *

And Eubulus says, somewhere or other:

> "Wife, quick! some cabbage boil, of virtuous healing,
> That I may rid me of this seedy feeling!"

And so Alexis says:

> "Last evening you were drinking deep,
> So now your head aches. Go to sleep:
> Take some boiled cabbage when you wake,
> And there's an end of your headache!"

A like power is attributed by the same author, to bitter almonds: "Bitter almonds were considered a preservative against intoxication, it being said of the physician of Tiberius, that he could carry away the contents of three bottles, if thus fortified, but speedily became drunk if deprived of his almonds." †

Redding ‡ attributes to Pliny the saying that "the drunkards of his day took pumicestone before they set to at a drinking bout in honor of Bacchus," in order that the intoxicating power of large quantities might be neutralized. The Greeks held that the amethyst was an antidote against intoxication; and the adorning of "Queen Elizabeth's Cup," profusely with amethysts, is supposed to have been suggested by this notion." §

---

* Athenæus, Vol. I. p. 56.
† Ibid, Book ii. 39.
‡ History of Wines, p. 10.
§ Biblical Temperance, by John Mair, M. D., p. 185.

Morewood gives the following examples :

"The inhabitants of Jesso, an island of Japan, although great drinkers of a fiery compound, seldom become intoxicated on account of their free use of the oil of the *todo-noevo*, a species of seal, said by the Jesuit priests to be an infallible preventive of inebriety." "In the isle of Skie, the root Carmel or Knaphard—*Argatilis Sylvaticus*,—was used to prevent drunkenness." *

It is not known, to the writer, save as the above citations claim it, that any of these supposed antidotes in any way neutralized the power of the beverages that had been imbibed. But as the drinking of large quantities was a feat in which many of the ancients were ambitious to excel,—as will be seen by referring back to the accounts of drinking customs, especially to those of the Greeks and Romans,— it is not unreasonable to believe that these and other supposed preventives were sought for and used quite as extensively as the above quotations claim, and more generally in ancient times than we have now any means of knowing.

II. PERSONAL PENALTIES FOR DRUNKENNESS.— Among the measures adopted for the suppression of intemperance, may be mentioned the infliction of personal penalties on the drinker. In some instances these penalties have aimed at the reformation of the drunkard; in others, where resulting in death, they have been intended to serve as warnings and restraint to those who might witness or otherwise know of their infliction.

The earliest examples of measures of this kind, of which we have any record, are found in the laws and history of the Israelites, as preserved in the Old Testament Scriptures. In the law, it is written :

"If a man have a stubborn and rebellious son, which will not obey the voice of his father, or the voice of his mother, and that, when they have chastened him, will not hearken unto them : . . . . they shall say unto the elders of his city, this our son is stubborn and rebellious, he will not obey our voice ; he is

* History of Inebriating Liquors, pp. 248, 582.

a glutton and a drunkard. And all the men of his city shall stone him with stones, that he die: so shalt thou put away evil from among you; and all Israel shall fear."—*Deuteronomy* xxi. 18–21.

So also in the regulations made for the guidance of the priesthood, in order that they might not repeat the folly of Nadab and Abihu (whose offence we have described in the previous chapter):

"The Lord spake unto Aaron, saying: Do not drink wine nor strong drink, thou, nor thy sons with thee, when ye go into the tabernacle of the congregation, *lest ye die.* That ye may put difference between holy and unholy, and between unclean and clean. And that ye may teach the children of Israel all the statutes which the Lord hath spoken unto them by the hand of Moses."—*Leviticus* x. 8–11.

CHINA.—In treating of intemperance in China in ancient times, allusion was made to the demoralization and ruin that had come to the land of Yin on account of the intemperance of rulers and people. This led to the promulgation of an imperial edict about the year 1120 B. C., the full text of which is preserved in the " Shoo-King," or history. It is there called, " The Announcement about Drunkenness." In many parts it is vague and somewhat contradictory in its wording, but there is no ambiguity in its description of the evils of drunkenness, nor in its threatening the penalty of death on those who persist in the use of intoxicants. Some of the native commentators on the decree seek to soften down and explain away its more harsh features, but its positive declarations cannot be changed by criticisms which were written many centuries after the proclamation of the decree. Especially is this true of the mandate with which the " Announcement " closes:

" If you are told that there are companies who drink together, do not fail to apprehend them all and send them to Chow, where I will put them to death. As to the ministers and officers of Yin, who have been led to it and been addicted to drink, it is not necessary to put them to death; let them be taught for a time. If they keep these lessons, I will give them bright distinction. If you disregard my lessons, then I, the one man, will show you no pity. As you cannot cleanse your way, you shall be classed with those who are to be put to death. O Fung, give constant heed to my admonitions. If you do not manage right, your officers and the people will continue lost in drink." *

At the present time intemperance among the Chinese, although quite extensively prevailing, is a solitary rather than a convivial or social vice, as to be seen in public in an intoxicated condition is sure to be followed by the infliction of severe penalties. According to the statement of an observant traveller, ordinary offences are dealt with promptly but mildly, while it is one of the laws of the Empire that "A man, who, intoxicated with liquor, commits outrages against the laws, shall be exiled to a distant country, there to remain in a state of servitude." †

INDIA.—The debauchery in ancient India, sanctioned by the religious sacrifices which necessitated inordinate drunkenness on the part of those who participated in them, and especially on the part of the priests, became so alarming as to threaten the speedy destruction of the nation. Providentially there was raised up at this time a new leader, who, being accepted as the religious and moral lawgiver of the people, placed before them the "Institutes or Hindoo Law," in which the severest penalties were pronounced on those who should tamper with intoxicants. The period in which Manu, or as he is sometimes called, Menu, flourished, was probably near the ninth century B. C., although his translators differ quite widely on this point. From Sir William Jones's translation of the "Ordinances of Manu," edition of 1825, the following extracts are made, to show how thoroughly and severely the great lawgiver dealt with the vendors and partakers of intoxicants: "Never let a priest eat part of a sacrifice performed by those who sell fermented

---

* Legge's Chinese Classics, Vol. III. pt. 1.
† Dobell, Vol. II. p. 239.

liquor," chap. iv., v. 216. In chap. vii. vs. 46, 47: "Intoxication" is named as one of the ten vices for which a King
"may lose even his life." "Money due for spirituous
liquors, the son of the debtor shall not be obliged to pay."
"A contract made by a person intoxicated, is utterly null."
viii. 159, 163.

Drinking spirituous liquor is one of the six faults which
bring infamy on a married woman. A wife who drinks any
may at all times be superseded by another wife. A superseded wife, who, having been forbidden, addicts herself to
the use of intoxicating liquors even at jubilees, must be fined
six racticàs of gold. ix. 13, 80, 84. Sellers of spirituous
liquors are classed with gamesters, revilers of Scripture, etc.
They shall be instantly banished from the town. "Those
wretches," it is said, "lurking like unseen thieves in the
dominion of a prince, continually harass his good subjects
with their vicious conduct."

"A soldier or merchant drinking arak, or a priest drinking
arak, mead, or rum, are all to be considered respectively as
offenders in the highest degree, except those whose crimes are
not fit to be named. For drinking spirits, let the mark of a vinter's flag be impressed on the forehead with a hot iron. With
none to eat with them, with none to sacrifice with them, with
none to read with them, with none to be allied by marriage
with them, abject and excluded from all social duties, let them
wander over this earth; branded with indelible marks, they
shall be deserted by their paternal and maternal relations,
treated by none with affection, received by none with respect:
Such is the ordinance of Manu."—ix. 235–239.

Drinking forbidden liquor, is mentioned among the offences which wise legislators must declare to be crimes in
the highest degree. Smelling of any spirituous liquor is
considered as causing a loss of class. Eating what has
been brought in the same basket with spirituous liquors is
an offence which causes defilement. xi. 55, 68, 71.

"Any twice-born man, who has intentionally drunk spirit of
rice through perverse delusion of mind, may drink more spirit
in flame, and atone for his offence by severely burning his body;

or he may drink boiling hot, until he die, the urine of a cow,* or pure water, or milk, or clarified butter, or juice expressed from cow-dung: or, *if he tasted it unknowingly*, he may expiate the sin of drinking spirituous liquor, by eating only some broken rice or grains of *tila*, from which oil has been extracted, once every night for a whole year, wrapped in coarse vesture of hairs from a cow's tail, or sitting unclothed in his house, wearing his locks and beard uncut, and putting out the flag of a tavern-keeper. Since the spirit of rice is distilled from the Mala, or filthy refuse of the grain, and since *Mala* is also a name for sin, let no Brahmin, Cshatriya, or Vaisya drink that spirit." xi. 91–94.

"When the divine spirit, or the light of holy knowledge, which has been infused into the body of a Brahmin, has once been sprinkled with any intoxicating liquor, even his priestly character leaves him, and he sinks to the low degree of Sudra." xi. 98.

The souls of men who have given way to passions, pass at death into the bodies of those "addicted to gaming or drinking."

"A priest who has drunk spirituous liquor, shall migrate into the form of a smaller or larger worm or insect, of a moth, of a fly feeding on ordure, or of some ravenous animal." xii. 45, 56.

In modern India, the East India Company, though largely encouraging distillation as a means of revenue, were forced by the prevalence of intemperance, to issue laws against drunkenness, by ordaining in 1754, for the first offence, admonition ; for the second a fine of five shillings ; and persons of rank were to pay in proportion to their station, as it was expected they should be examples to others.

GREECE.—Athenæus, Book XIII. section 566, gives several instances of the determined manner in which the ancient Greeks dealt with drunkenness, especially when it occurred in the higher classes in society. At Athens the

* "Cow's urine was probably a metaphorical name for 'rain-water' originally—the clouds being cows metaphorically." *Haug's Essays*, p. 242.

Archons of the Court of Areopagus, were made inspectors
of the public morals, and authorized to rigorously punish
intemperance. To have even dined at a public house dis-
qualified one for a seat in that renowned senate and court.
For an Archon to be intoxicated was a capital offence, punished with death. Isocrates is quoted as saying of this
period in the history of Athens, that not even a servant in
the city would be seen eating or drinking in a public house.
The Spartans, according to Plutarch, made their servants
drunk once a year, in order that their children might see
how foolish and contemptible men looked in that state.
Plato says that the vice of intemperance was effectually
rooted out of the republic of Sparta; and that if any man
found another in a state of intoxication, he was, under the
stern laws of Lycurgus, brought to punishment, and even
though he might plead that the feast of Bacchus ought to
excuse him, his defence availed him nothing. Zeleucus,
the Locrian, enacted a law punishing with death any man
who should drink wine, unless by a physician's prescription.
The Massilians had a law that no woman should drink anything stronger than water.* Pittacus of Mitylene, made a
law that he who, when drunk, committed any offence
against the laws, should suffer a double punishment, one
for the crime itself, and the other for the intoxication which
prompted him to commit it; † and the law was applauded
by Plato, Aristotle and Plutarch, as the height of wisdom.

ROME.—Pliny is authority for the statement that women
in Rome were forbidden to drink wine in the year 650,
B. C., ‡ and that the penalty for disobedience was death,
the same as the penalty for adultery;—and the reason
given was, that wine was a certain incitement to lewdness.
An instance is cited by him of a wife killed by her husband
for drinking wine, and that the husband was absolved from

---

* Athenæus II. Bk. X. c. 33.
† Macnish, Anatomy of Drunkenness, Chapter XIII.
‡ Natural History, Bk. XIV. chap. 13.

tho murder by Romulus. Another case is cited by him, as having occurred four hundred years later, in which a certain Roman lady for having in her possession the keys of the wine cellar, was starved to death by her family. It was the usage, he says, for women to be compelled to salute all their male relatives with a kiss in order that it might be ascertained if they smelt of *temetum*, wine being called by that name.

"Whence," Pliny says, "our word *temulentia*, meaning drunkenness. Another case is cited by him in which the judge Domitius, rendered the decision that a certain woman who claimed to have been allowed to take wine as a tonic or medicine, took more than was requisite for health, and must therefore be sentenced to lose her dowry. The Roman Censor, whose office corresponded in many respects to that of the Athenian Areopagite, had, like the latter, " a general supervision of the morals of the people—was empowered to punish, and did punish drunkenness with excessive severity—was required to be himself, a man of rigidly abstemious habits, and was liable to expulsion from the order for a single violation of the laws relative to sobriety. These censors turned drunken members out of the senate without the least mercy, and branded them with perpetual infamy. They would allow them no place of honor or profit in the government." *

MAHOMETANS.—There is no doubt that the early followers of Mahomet interpreted the Koran as absolutely prohibiting the use of intoxicants. A few instances are here adduced to show how those in authority among them regarded the matter. The Caliph Omir, on learning from his general " that the Mussulmans had learned to drink wine during their invasion of Syria, ordered that whoever was guilty of this practice should have fourscore stripes upon the soles of his feet. The punishment was accordingly inflicted, and many were so infatuated, although they had no accusers but their own conscience, as voluntarily to confess their crime, and undergo the same punishment." †

* The War of Four Thousand Years, p. 123.
† Ockely's History, quoted by Morewood, pp. 41, 42.

The Sultan, Soliman the First, caused melted lead to be poured down the throats of those who disobeyed the precepts of the Koran against wine. In 1795, the Sultan Abdelbrahman published an ordinance against the use of Merissah, an intoxicating beverage, under penalty of death.

GERMANY.—In the middle of the eighth century, the emperor Charlemagne, in giving a constitution to his nobles, which conferred on them many privileges of great value, charged them not to sully by intemperance, that which had been conceded to their valor, and the many services which they had rendered. To himself and to his heirs he reserved the right of punishing disobedience to this injunction, in the person of the grantee or his successor. His precautions in regard to drunkenness extended to all classes in society, and the sentence of excommunication was made the penalty of disobedience. The clergy were especially aimed at in many of the regulations, degradation from office and corporal punishment, according to the rank of the offender, being prescribed as the penalty for the offence, even of going inside a tavern. Tippling in any manner was prohibited by penal laws of great severity. The soldier found drunk in camp was restricted to water as a beverage until he admitted the enormity of his offence and publicly sued for pardon. Judges were not allowed to hold courts unless perfectly sober; witnesses and suitors could not appear in the halls of justice if intoxicated, and priests were not suffered to offer drink to penitents.

About seven hundred years later, Frederick III. ordered " all electors, princes, prelates, counts, knights, and gentlemen to discountenance and severely punish drunkenness." Officials of lower rank also established societies and orders, through the regulations of which they sought the moral elevation of the people, and by means of which they imposed fines and imprisonment on the intemperate.* The church authorities also imposed heavy penalties on the in-

* Authorities quoted by Samuelson, pp. 105, 106.

temperate among the different orders of the clergy;—but of this further on.

ENGLAND.—Macnish states,* that by two acts passed in the reign of James I., drunkenness was punished with a fine, and, failing payment, with sitting publicly for six hours in the stocks, and that this law remained in operation till its repeal in 1828, previous to which time the ecclesiastical courts could take cognizance of the offence, and punish it accordingly.*

Rev. William Reid says that, " In the time of Oliver Cromwell, the magistrates in the north of England punished drunkards by making them carry what was called 'The Drunkard's Cloak.' This was a large barrel, with one head out and a hole through the other, through which the offender was made to put his head, while his hands were drawn through two small holes, one on each side. With this he was compelled to march along the public streets." †

In the present century, in the Naval Discipline, the following rule was in force:

"Separate for one month every man who was found drunk, from the rest of the crew: mark his clothes 'drunkard;' give him six-water grog, or, if beer, mixed one half water; let them dine when the crew had finished; employ them in every dirty and disgraceful work, etc. This," says Macnish, " had such a salutary effect, that in less than six months not a drunken man was to be found in the ship. The same system was introduced by the writer into every ship on board which he subsequently served. When first lieutenant of the Victory and Diomede, the beneficial consequences were acknowledged—the culprits were heard to say that they would rather receive six dozen lashes at the gangway, and be done with it, than to be put into the ' drunken mess' (for so it was named) for a month." ‡

SCOTLAND.—By a law of Constantine II., king of Scotland, passed at Scone, A. D. 861, " Young persons, of both sexes, were commanded to abstain from the use of in-

---

* Anatomy of Drunkenness. Chap. xiii.
† Temperance Cyclopædia, p. 282.
‡ Macnish, chap. XIV.

toxicating liquors. Death was the punishment, on conviction of drunkenness." [*] Scottish laws against intemperance in the last century are said to have imposed the following penalties on drunkenness:

"Whosoever shall drink to excess, shall be liable, each nobleman, in £20 Scots; each baron in 20 marks; each gentleman, heritor, or burgess, in 10 marks; each yeoman in 40 shillings Scots, *toties quoties;* each minister in the fifth part of his year's stipend: and that the offender, unable to pay the aforesaid penalties, be exemplarily punished in his body, according to the demerit of his fault." [†]

ANCIENT MEXICANS.—Prescott says of the Aztecs:

"Intemperance, which was the burden, moreover, of their religious homilies, was visited with the severest penalties; as if they had foreseen in it the consuming canker of their own, as well as of other Indian races in later times. It was punished in the young with death, and in older persons with loss of rank and confiscation of property." [‡]

SWEDEN.—Macnish quotes from "Schubert's Travels in Sweden," the following synopsis of the laws against intemperance in force in Sweden during the first half of the present century:

"Whoever is seen drunk, is fined for the first offence, three dollars, for the second, six, for the third and fourth, a still larger sum, and is also deprived of the right of voting at elections, and of being appointed a representative. He is, besides, publicly exposed in the parish church on the following Sunday. If he is found committing the offence a fifth time, he is shut up in a house of correction, and condemned to six months hard labor; and if he is again guilty, to a twelve months punishment of a similar description. If the offence has been committed in public, such as at a fair, an auction, etc., the fine is doubled; and if the offender has made his appearance in a church, the punishment is still more severe. Whoever is convicted of having induced another to intoxicate himself, is fined three dollars, which sum is doubled if the person is a minor. An ecclesiastic who falls into this offence loses his benefice; if it is a layman

* War of Four Thousand Years, p. 165.
† Reid's Cyclopædia, p. 282.
‡ Conquest of Mexico. Vol. I. p. 35.

who occupies any considerable post, his functions are suspended, and perhaps he is dismissed. Drunkenness is never admitted as an excuse for any crime; and whoever dies when drunk, is buried ignominiously, and deprived of the prayers of the church. It is forbidden to give, and more explicitly to sell, any spirituous liquors to students, workmen, servants, apprentices and private soldiers. Whoever is observed drunk in the streets, or making a noise in a tavern, is sure to be taken to prison and detained till sober; without, however, being on that account exempted from the fines. If he is without money he is kept in prison till he works out his deliverance. Twice a year these ordinances are read aloud from the pulpit by the clergy; and every tavern keeper is bound under the penalty of a heavy fine, to have a copy of them hung up in the principal rooms of his house." *

SOCIETY ISLANDS.—The law at Huahine is this:

"If a man drink spirits till he becomes intoxicated, (literally poisoned,) and is then troublesome or mischievous, the magistrates shall cause him to be bound or confined; and when the effects of the drink have subsided, shall admonish him not to offend again. But if he be obstinate in drinking spirits, and when intoxicated becomes mischievous, let him be brought before the magistrate and sentenced to labor, such as road-making, five fathoms in length and two in breadth. If not punished by this, let him make a plantation fence, fifty fathoms long. If it be a woman that is guilty of the crime, she shall plait two large mats, one for the king, and the other for the governor of the district, or make four hibiscus mats, two for the king and two for the governor, or forty fathoms of native cloth, twenty for the king and twenty for the governor." †

BIRMAN EMPIRE.—Alomphra, the founder of the Birman Empire, made intoxication punishable with death.‡

NORTH AMERICA.—Both in the British Possessions, and in the several portions of the United States, intemperance which is manifest in rioting, in trespassing on the rights of others to the extent of interfering with their business or destroying property, is,—as is probably true in all civilized

<hr>

* Macnish, note to chap. XIII.
† Ellis' Polynesian Researches, Vol. II. p. 433.
‡ Two Years in Ava, p. 307.

countries,—treated as a misdemeanor. But some of the early laws of the new world treated drinking as an offence, and vigorously punished it, long before it had resulted in acts of violence to others. We give a few of the more curious, and now nearly forgotten examples.

"Robert Coles, of Rockesbury" (Roxbury) "Massachusetts Bay Colony, seems to have severely taxed the ingenuity of the Court of Assistants, (answering in composition and functions to the present Governor and Council of Massachusetts,) to find a variety of punishment for his determined drunkenness. For his first offence he was brought into court, in 1631, and fined '5 marks.' The next year his second offence was dealt with when he was fined 'xx shillings.' In 1633 he was again before the court, when he was 'fined x£, and enjoyned to stand with a white sheet of paper on his back, whereon *a drunkard* shall be written in great letters, and to stand therewith so long as the court thinks meete for abusing himself shamefully with drink.'

"Again, March 4, 1633–4, he is brought before the court, where his case is thus disposed of: 'It is ordered that Robte Coles, for drunkenness by him committed at Rockesbury, shall be disfranchised, weare aboute his necke, and soe to hang upon his outward garm't, a D, made of redd cloath, and sett upon white; to contynue this for a yeare, and not to leave it off att any tyme when hee comes amougst company, under the penalty of xls. for the first offence, and v£ the second, and after to be punished by the Court as they thinke meete; also he is to wear the D outwards, and is enjoyned to appear att the nexte General Court, and to contynue there till the Court be ended." *

Sept. 6, 1636, at a Quarter Court, at Boston : "Peter Bussaker was censured for drunkenness to be whipped and to have twenty stripes sharply inflicted." † In Plymouth Colony, drunkards were sentenced to pay a fine, to sit in the stocks, to be whipped, and as the extreme penalty, to be disfranchised.‡

One of the earliest laws of Pennsylvania, that adopted at Upland, Dec. 1682, provided that, "Drunkenness, encour-

---

* Records of the Governor and Company of Massachusetts Bay, Vol. I. pp. 90, 93, 107, 112.

† Ibid, p. 177.

‡ Plymouth Colony Records, Vol. I. pp. 12, 36, 100, 132.

agement of drunkenness, drinking or pledging of healths, should be punished by fine and imprisonment."*

Recently, Dr. H. A. Hartt, in a paper read before the Episcopal Church Congress in Boston, and by subsequent agitation and discussion in the columns of the secular press, has sought the cooperation of various classes in community,—the drinkers and the abstinent, the so-called moderates and even the liquor sellers, in a revival of the old methods,—in spirit, if not in detail,—to make all drunkenness an offence against the law, to be punished by imprisonment and other penalties. So far it has no substantial encouragement.

ECCLESIASTICAL PENALTIES.—The Epistles of the New Testament verify the evidence collected from other sources, that intemperance prevailed to an alarming extent in the Gentile world, especially in the cities of Greece and Rome, where the first efforts were made to establish the Christian Religion. They show that the drinking habits of the people greatly hindered the progress of the Gospel, and that exhortation and rebuke were frequently given to the early converts on account of the tendency of their former habits, and the influence of the constant examples before them of indulgence in the use of wine. Hence they were warned to avoid drunkenness, not to keep company with drunkards, and to remember that no drunkards can inherit the Kingdom of God. Rom. xiii. 12. 1 Cor. v. 11; vi. 10. Hence the injunction that a bishop " must not be given to wine," literally " must not sit down at wine," " must not be in company with wine," " must wholly avoid it." 1 Tim. iii. 2 ; Titus i. 7. The Fathers both of the Western and Eastern Churches tell sad tales of its influence in tempting and overcoming both clergy and laity. Taverns,—the keepers of which were, on account of their vile traffic, esteemed more lightly than men who followed any other occupation†—

---

* Proud's History of Pennsylvania, p. 71.
† Neander, Memorials, p. 100.

became the resorts of the clergy to such an extent as to call for the passage of the following law, in the fourth century :

"If any one of the clergy be taken eating in a tavern, let him be suspended, except when he is forced to bait at an inn upon the road." *

In the seventh century intemperance was so prevalent, and the disposition to indulge in it on all occasions was so general, that at the Synod of Trullus, "the clergy and laity were commanded not to partake of the feasts of the Bacchanalia; on pain, the former, of deposition, the latter of excommunication." At about the same period, the emperor Justinian found it necessary to "forbid monks to enter places where liquor was sold, under pain of chastisement upon conviction before a magistrate, and of expulsion from their monasteries." †

Bridgett quotes from a letter written in the eighth century by St. Boniface to the Abp. of Canterbury : (as see the section on "Intemperance in England.") "It is reported that in your dioceses the vice of drunkenness is too frequent," etc., closing with reference to the ancient decrees that "a bishop or a priest given to drink should either resign or be deposed." One of these decrees, that of A. D. 569, referring to the priests, reveals the fact that malice as well as hospitality, sometimes led the unfaithful clergy to tempt and even to force others to become intoxicated : "He that forces another to get drunk out of hospitality must do penance as if he had got drunk himself. But he who out of hatred or wickedness, in order to disgrace or mock at others, forces them to get drunk, if he has not already sufficiently done penance, must do penance as a murderer of souls." Other penalties are thus set forth:

"If a bishop or any one ordained has a habit of drunkenness,

*Law Book of the Ante-Nicene Church, quoted by Ritchie in his *Scripture Testimony against Wine.* p. 150.

† War of Four Thousand Years, pp. 154, 155. See also various chapters in Mosheim's Ecclesiastical History, and in Neander's History of the Christian Church.

he must either resign or be deposed. If a monk drinks till he vomits, he must do thirty days' penance; if a priest or deacon, forty days. If a priest gets drunk through inadvertence, he must do penance seven days; if through carelessness, fifteen days; if through contempt, forty days; a deacon or monk, four weeks; a sub-deacon, three; a layman, one week."

In 1255, Walter, Bishop of Durham, forbids "those in holy orders that they be not drunkards, nor keep taverns, lest they die an eternal death."

In 1536, Henry VIII. as "Defender of the Faith," issued this injunction:

"The said dean, parsons, vicars, curates, and other priests shall in no wise, at any unlawful time, nor for any other cause than for their honest necessity, haunt or resort to any taverns or ale-houses; and after their dinner and supper they shall not give themselves to drinking and riot." *

Abp. Plunkett, speaking of the Irish clergy in the latter part of the 17th century, says:

"While visiting six dioceses of this province, I applied myself especially to root out the cursed vice of drunkenness, which is the parent and nurse of all scandals and contentions. I commanded also, under penalty of privation of benefite, that no priest should frequent public houses or drink whiskey." †

A Scotch Presbytery, in 1637, sentenced a drunkard "to stand in sackcloth two Sabbaths; and to paye four markes penaltye." ‡

As we have seen in a former chapter (the citation from Barbour's Statistics,) that full fifty per cent. of all the cases of Church discipline, are occasioned by intemperance, so we may refer to the history of all Christian sects for legislation on this subject, for various devices of penalties, even where, as in the case of Protestant sects, generally, the withdrawal of fellowship and watch care is the extreme punishment that can be inflicted. As no evil has been a greater

---

* Jeaffreson's Book about the Clergy, I. p. 90.
† Discipline of Drink, pp. 77, 114, 135, 140, 166.
‡ Reid's Cyclopædia, p. 281.

foe to the existence and prosperity of churches, so none has suggested the necessity for a greater number of expedients for making church members feel the shame and disgrace of drunkenness.

III. MODERATION SOCIETIES.—The organized efforts to diminish drunkenness were at first characterized by allowing the use of all known intoxicants, in moderation ; and afterwards, by the exclusion of the use of distilled liquors, and the permission of moderate indulgence in fermented drinks. So far as we have been able to ascertain,— although personal efforts for, and commendations of moderate drinking, are noticeable in the history of most remote times,—the organization of societies for this purpose, is comparatively modern.

In 1517 the first society of this kind, so far as the writer knows, was established in Germany. It was called the " Order of Temperance," and was well supported by the nobility, clergy and gentry. Its founder, Sigismond de Dietrichstein, had especially in view, the cultivation of temperate habits among the highest classes, who were fast becoming dissolute by the exactions of the habit of social drinking ; and it was the chief aim of this society to put an end to the custom of pledging of healths, a practice then carried to such an extreme that intoxication was sure to be an accompaniment of the most casual meeting of friends.

In 1600, Maurice, the Landgrave of Hesse, established another society, the fundamental rule of which was, " That every member of the society pledges himself *never to become intoxicated.*" To guard against the violation of this pledge, it was ordered by the society that no member " should be allowed more than seven goblets of wine at a meal, and that not more than twice a day ! " A third society, of a similar character, was called the " Ring of Gold," and was established by the Count Palatine, Frederick V.

The members of these societies pledged themselves to ob-serve the rules for two years.*

We find no further trace of Temperance Societies of any kind, for nearly two hundred years; for the next mention is of an organization of farmers, in the New World. The Lansingburg, N. Y., *Federal Herald*, of July 13, 1789, contains this item :

"Upwards of two hundred of the most respectable farmers of the County of Litchfield, Connecticut, have formed an *association* to discourage the use of spirituous liquors, and have determined not to use any kind of distilled liquors during their farming work the ensuing season." †

Drinking habits, as we have already seen, were at that time prevailing to an alarming extent among the people of the United States. The French War had done much to demoralize the people, and the War for Independence had increased the tendency to intemperance. Patriots and philanthropists were sounding the alarm and putting in motion the arguments and moral influences which were to make America the birth-place of all subsequent organized work against this great evil.

As early as 1737, Benjamin Lay, an illiterate sailor, but a true lover of his kind, put forth a little work in which he sought to awaken the disgust of his countrymen against the use of rum, by detailing the filthy manner of its manufacture, as he had observed it during his voyages to Barbadoes. He made bitter complaint that "we send away our excellent provisions and other good things, to purchase such filthy stuff, which tends to the corruption of mankind; and they send among us some of the worst slaves when they cannot rule them themselves, *along with their rum,* to

---

* The Teetotaller's Companion, by Peter Burne, p. 314. History of the Temperance Movement, by Samuel Couling, p. 24. War of Four Thousand Years, p. 168.

† Centennial Temperance Volume, p. 423.

complete the tragedy, *i. e.*, to slay, to destroy the people of Pennsylvania, and to ruin the country."*

On the eve of the Revolution, viz., in 1774, a pamphlet with the following title was published in Philadelphia: "The Mighty Destroyer Displayed, in some Accounts of the Dreadful Havock made by the mistaken Use as well as Abuse of Distilled Spirituous Liquors. By a Lover of Mankind."† It was no doubt from the pen of the eminent American philanthropist, Anthony Benezet; and is a wonderful production for the thoroughness of its presentation of the subject. It is largely made up of extracts from the writings of Drs. Hales, Hoffman, Cheyne, Short, Lind, and Buchan, on the physiological evils of the use of distilled spirits; combats the notion that they are needed in extreme hot or cold countries; enforces the moral argument as set forth in 1 Cor. viii. 13; and pleads for Total Abstinence, against the mistaken notion that some use of intoxicants is necessary. And further, it attacks the use of *fermented drinks*, quoting from Dr. Buchan: "There are few great ale drinkers who are not phthisical, nor is that to be wondered at, considering the glutinous and almost indigestible nature of strong ale."

In 1777, Dr. Benjamin Rush, of Philadelphia, at that time "Surgeon-General of the Army for the Middle Department," wrote and published, by request of the "Board of War of the American Army," a pamphlet entitled, "Directions for Preserving the Health of Soldiers," in which he took strong ground against the use of Alcoholic Liquors.

In 1778, Benezet issued another pamphlet, to which he appended his name, entitled: "Remarks on the Nature and Bad Effects of Spirituous Liquors, collected by Anthony

* Samson Shorn, and his Locks Renewed: or the History of Spirituous Liquors in Pennsylvania. By Rev. George Duffield, Jr. Pp. 17, 18.

† A copy may be found in the Library of the Historical Society of Pennsylvania, bearing the autograph of its supposed author.

Benezct." * In this pamphlet, he adopts the opinion of Dr. Cheyne : " Water alone is sufficient and effectual for all the purposes of human want in drink ; strong liquors were never designed for common use."

In 1785 Dr. Rush published an " Enquiry into the Effects of Ardent Spirits upon the Human Body and Mind." " By spirits," he said, " I mean all those liquors which are obtained by distillation from the fermented juices of substances of any kind." To this little book, which was republished in large editions, as early as 1794, 1804, and 1811,—and which is still a standard authority on the subject,—and to the personal efforts of its author, we are largely indebted for general enlightenment on the evils of intemperance, and for the early organizations for its suppression.

In 1788 he appeared before the " Philadelphia Annual Conference of the Methodist Episcopal Church," then in session in the city of Philadelphia, and " made an earnest and animated address on the use of ardent spirits, taking the broad ground then so strongly occupied by the Conference, and since so signally taken and maintained by the Temperance Reformation : ' *that total abstinence is no less the demand of our nature than it is the rule of our safety,*' and he besought the Conference to stop the use as well as the abuse of spirit drinking." †

Two years later (1790), a volume of Sermons on Intemperance was anonymously published in Philadelphia. They were evidently the production of a physician, and have generally been attributed to Dr. Rush.

The session of the General Assembly of the Presbyterian Church, at Philadelphia, in 1811, was visited by Dr. Rush, its ministers and elders were presented by him with 1000 copies of his treatise, and personally urged to take some steps that would show their desire to put a stop to drunkenness. The necessary limits in which the history of the

---

* There is a copy in the Friends' Library, Philadelphia.
† Report of the Pennsylvania State Temperance Union, 1871, p. 393.

movement must be kept in these pages, forbids extended extracts from this " Enquiry," but justice seems to demand that at least this much should be quoted, to set forth its scope, and the forcible manner in which the subject was treated :

" The effects of ardent spirits on the *body* are—1. A decay of appetite. 2. A consuming of the liver of the drunkard, like the vulture preying on that of Prometheus. 3. Jaundice and dropsy. 4. Hoarseness and consumption. 5. Diabetes. 6. ' Rum-buds' in the face, descending to the limbs in the form of leprosy. 7. A fœtid breath. 8. Spontaneous combustion. 9. Epilepsy. 10. Gout in all its various forms of swelled limbs, colic, palsy, apoplexy. 11. Madness. Its effects on the *mind* are—1. To impair the memory. 2. To debilitate the understanding. 3. To pervert the moral faculties. 4. To produce falsehood, fraud, uncleanness and murder. In folly, it causes a man to resemble a calf in stupidity ; an ass in roaring ; a mad bull in quarrelling ; a dog in fighting ; a tiger in cruelty : a skunk in fetor ; a hog in filthiness, and a he-goat in obscenity."

In 1805, the Paper Makers of Philadelphia, " associated themselves together for the purpose of improving their art, and ameliorating the condition of *worthy* unfortunate journeymen and their families. "They soon found that the excessive use of strong drink was almost the only cause of the misery and poverty which they had occasion to relieve, and they at once sought to restrict this evil. The conclusion at which they arrived was, "to use every possible endeavor to *restrain* and *prohibit* the use of ardent spirits in their respective mills." *

In 1806, Dr. Rush found an ardent and efficient co-worker, in Rev. Ebenezer Porter, of Washington, Connecticut, who preached and published a sermon on " The Fatal Effects of Ardent Spirits." In this Sermon we have probably the first attempt to set forth the statistics of the consumption of ardent spirits in the United States.

In March, 1808, Dr. B. J. Clark, a physician in Moreau,

---

* Duffield, p. 25.

Saratoga county, N. Y., alarmed at the increase of intemperance in the place of his residence, sought the advice of his pastor, Rev. Lebbeus Armstrong, of the Congregational Church, to whom he communicated his conviction : " We shall all become a community of drunkards in this town, unless something is done to arrest the progress of intemperance." Developing his plan of a Temperance Organization, his efforts was seconded by his pastor and others, and resulted in the organization on the 30th day of April, in the same year, of " The Temperate Society of Moreau and Northumberland." A Constitution was adopted and received the signatures of forty-three members. It provided for four meetings during the year, and defined its purpose and the means of effecting it, in the following provisions :

"Article IV. No member shall drink rum, gin, whiskey, wine, or any distilled spirits, or composition of the same, or any of them, except by advice of a physician, or in case of actual disease ; also, excepting wine at public dinners, under penalty of twenty-five cents ; provided that this article shall not infringe on any religious ordinance.

" Sec. 2. No member shall be intoxicated, under penalty of fifty cents.

" Sec. 3. No member shall offer any of said liquors to any other member, or urge any other person to drink thereof, under penalty of twenty-five cents for each offence."

" Art. XI. It shall be the duty of each member to accuse any other member of a breach of any regulation contained in Article IV., and the mode of accusative process and trial shall be regulated by a by-law."

" This little feeble band of temperance brethren, held their quarterly and annual meetings in a country district schoolhouse from April, 1808, onward for several years, without the presence of a single female at their temperance meetings." *

Another society was organized in April, 1809, at Greenfield, in the same county, on a similar basis.†

---

* History of the Temperance Reformation, by Rev. Lebbeus Armstrong, pp. 18–28.

† Centennial Temperance Volume, p. 27.

In the "National Temperance Advocate," for January, 1881, it is stated that at a recent Temperance meeting in the city of New York, Rev. Dr. S. Irenæus Prime, of the "New York Observer," said:

"I hold in my hand a printed sermon on intemperance, preached by my father, Nathaniel S. Prime, Nov. 5, 1811, from the words, 'Who hath woe?' etc.; 'they that tarry long at the wine.' That was before I was born. It was preached before the Presbytery of Long Island, and it appeals to ministers and all others to discourage the use of intoxicating drinks. All the arguments now in use are employed in this sermon, the same objections answered, and the same appeals are made. This was fourteen years before Dr. Beecher's famous sermons were preached. The Presbytery that listened to Mr. Prime's sermon requested its publication; it adopted a resolution recommending their people to refrain from offering ardent spirits or wine as an act of hospitality. The session of the church, and then the church of which Mr. Prime was pastor, Freshponds, adopted a similar resolution, and a total revolution in the habits of the community was the result."

The *Advocate* adds: "In 1812 Mr. Prime removed from Long Island, and, after preaching in Saratoga County, was settled, in 1813, in Cambridge, Washington Co., New York, where he at once organized the farmers of his congregation into a temperance society."

Perhaps the most important of the early organizations, certainly the most influential as an example for other localities, was the one organized in Boston, Mass., in 1813. The causes which led to this new society may be traced to Dr. Rush's visit to the session of the Presbyterian General Assembly, in 1811. At that session committees were appointed to consider the evil and its remedy. The same year the Associated Churches of Connecticut and Massachusetts took up the subject, and appointed committees to "cooperate with those of the General Assembly of the Presbyterian Church." The immediate result was that it was resolved to discountenance the use of ardent spirits at the public meetings of these bodies; and a more extended result was the effect produced on subordinate bodies, and the creation of a strong public opinion on the necessity of organized work by means of secular societies, for the arrest of the great evil.

"The Consociation of the Western District of Fairfield County," Connecticut, at a meeting in Oct., 1812—

" *Voted*—That we cordially approve of the doings of the General Association of Connecticut, on this subject, at their session in June last, and will, as far as practicable, comply with their recommendations; Particularly, "1. That the customary use of ardent spirits shall be wholly discontinued, at all future meetings of this body." "4. That we will endeavor to influence the members of our respective churches, and other well-disposed persons in our congregations, to contribute for the purchase and gratuitous distribution of well-written tracts on the subject; particularly one by Dr. Rush of Philadelphia, and report our progress in this undertaking, to the next annual meeting of this body." It was also *voted*—"That Mr. Swan, Mr. Humphrey and Mr. Bonney, be a committee, to draft and print an Address, respecting the intemperate use of Ardent Spirits."

This address, which bears the imprint, "New Haven, 1813," makes mention of the fact that, "The late formation of a general society in this State, for the promotion of good morals, promises to be a powerful engine to put down dram-shops and arrest the progress of intemperance;" and recommends the citizens to become "members of this society, and to establish branch societies in their respective parishes." * What special service was rendered to the Temperance cause of this Connecticut State Society, and to what extent, if at all, branches of it were planted in various parts of the State, the writer is not informed.

Through the mutual influence of a Congregational and Presbyterian Alliance, and under the counsels of the Hon. Samuel Dexter, a distinguished lawyer, who said that he would pay all the taxes of Boston and of the State of Massachusetts, if he might have the profit on the traffic in spirituous liquors, † "The Massachusetts Society for the

---

* "An Address to the Churches and Congregations of the Western District of Fairfield County," pp. 3, 27.

† Autobiography of John Marsh, D.D., p. 12.

Suppression of Intemperance," was formed, Feb. 5, 1813. Its object was, "to discountenance and suppress the too free use of ardent spirits, and its kindred vices, profaneness and gaming; and to encourage and promote temperance and general morality." This society held special meetings as occasion might require, and an annual meeting at which a sermon or address was delivered before the Society, by some person elected for the purpose. According to Dr. Marsh: "The Society did little but observe an anniversary and have a sermon preached, after which preacher and hearers would repair to tables richly laden with wine; and was therefore without efficacy in rooting out the evil." * And Dr. A. P. Peabody, in a recent article in the "Cambridge *Tribune*," thus speaks of it:

"It had among its members the foremost men in Church and State, including the chief-justice of the Commonwealth, the president of Harvard College, Hon. Nathan Dane, and other persons of like standing and character. The members of this society were probably, without exception, perfectly temperate men, most of them opposed to the use of distilled spirits, but, perhaps, none of them scrupulous as to the moderate use of wine They soon experienced the truth of the adage, 'Do that you may know.' One of the original members told the writer of this article that at the earlier meetings, held at private houses, the then usual display of decanters appeared on the sideboard, and was not suffered to remain a mere show: that, when a meeting was to take place at his house, he took care to have his sideboard generously replenished; that the incongruity of such indulgence with the work in hand struck him at the last moment, and induced him to lock up the decanters; and that the members, taking kind and grateful notice of his procedure, resolved informally, but unanimously, to drink no more at their meetings. This society, while it held the foreground, directed its efforts mainly against the use of distilled spirits, and, it must be admitted, in favor of light wines and home-made fermented liquors. We well remember a receipt for making currant wine printed on the last leaf of one of their widely-circulated annual addresses."

The Annual Reports of the Society show, however, that

---

* Ibid, p. 12.

it was not without influence in encouraging the organization of other State, County and Town Societies, and that these in time led to a successful effort for a National organization. In 1833 it changed its form of organization, and became "The Massachusetts Temperance Society," the members "pledging themselves that they will not use distilled spirit as drink, nor provide it as an article of refreshment for their friends, nor for persons in their employment." The organization is still in existence, but does little more than to hold an annual meeting for the purpose of legally holding and managing its funds.

In 1818 a Society "to check and discourage the use of ardent spirits," was organized in Darby, Delaware County, N. Y.

In 1826, a national organization, called the "American Temperance Society,"—ten years later merged into the "American Temperance Union,"—organized in Boston, under a pledge of "Total Abstinence from Ardent Spirits." Rev. J. L. Hanaford states that a by-law of this society at the time of its original organization, read thus:

"Any member of this Association who shall be convicted of intoxication, shall be fined two shillings, unless such act of intoxication shall take place on the 4th of July, or on any regularly appointed military muster." *

And Dr. Marsh says: "In the early stage of the temperance reform, some friends of the cause in Boston thought best to establish a brewery to furnish men who would abstain from ardent spirits with beer. They did so, and soon failed, sinking some $20,000 in the business. How happened it? Why, they were honest men, and made honest ale ; whereas other brewers, using drugs and poisons, were able to undersell them, and compelled them to sell at such a price that they could not sustain the business." †

Rev. Edwin Thompson, for many years, and at present one

---

* Address delivered at Lynn, Mass., April 24, 1864, p. 6.
† Frauds in Intoxicating Liquors. The Sin of Drunkard Making. New York 1856, n. p. See also " Letter to the Friends of Temperance in Massachusetts, by Justin Edwards." Boston, 1836. pp. 11, 12.

of the most active and efficient advocates of the Temperance cause in Massachusetts, informs the writer that he was personally acquainted with one of the parties in this enterprise, and that the brewery was erected at Roxbury, in 1828 or 1829. In the broad light now pouring upon us, these rules and measures seem impolitic and absurd, but they were the honest attempts of earnest men, struggling in efforts to suppress intemperance. And not wholly in vain, for in the first annual report of the National Society, it was shown that State organizations had been established in New Hampshire, Vermont, Illinois and Indiana ; and of local auxiliary societies, " thirteen in Maine, twenty-three in New Hampshire, seven in Vermont, thirty-nine in Massachusetts, two in Rhode Island, thirty-three in Connecticut, seventy-eight in New York, six in New Jersey, seven in Pennsylvania, one in Delaware, one in Maryland, five in Virginia, two in North Carolina, one in South Carolina, one in Kentucky, one in Ohio, and two in Indiana; making a total of two hundred and twenty-two in the Union."*

And in 1833, there were not less than 5,000 societies in the United States warring against the use of ardent spirits, having a membership of 1,500,000, of whom 10,000 had been drunkards ; 4,000 distilleries had been stopped ; 6,000 merchants had given up the sale of distilled liquors, while on over 1,000 vessels their use had been abandoned. †

In 1833 a large number of Senators and Representatives in the United States' Congress, organized a Temperance Society, on the basis of Total Abstinence from the use of Ardent Spirits.

In 1828, Moderation Societies begun to exist in Canada, the first being established in Montreal. Until 1835 kindred societies multiplied rapidly throughout the Dominion.

Turning again to the old world, we find that in 1760, the inhabitants of the town of Leadhills, Scotland, alarmed at the prospect of the re-opening of the malt distilleries,

---

* War of Four Thousand Years, p 199.

† Centennial Temperance Volume, p. 446.

resolved, in view of the evil effects of the use of distilled liquors on health and morals, and regarding the use of grains in the distilleries as a " principal cause " of a recent " famine," " to discourage to the utmost of our power, by all public methods, that pernicious practice, being determined to drink no spirits so distilled ; neither frequent, nor drink any liquor in any tavern or ale-house that we know sells or retails the same." *

How long this resolution was in force, and what influence it exerted, is unknown. Sixty-eight years later, Mr. John Dunlop, who had been engaged for some years in the religious education of the young, and in Bible and missionary societies, in the west of Scotland, paid a visit to France, where he was sadly affected by the fact that the latter country was far in advance of Scotland in general morality. Knowing that his country made great boast of its soundness in religious faith, and of its zeal for the exclusion of heresies, he was led to inquire into the cause of such a strange anomaly, and was forced to the conclusion that the love of the Scotch for whiskey and other distilled drinks, accounted for it. He at once made himself familiar with the laws and methods of the American Temperance Societies, and at a religious meeting in Glasgow, urged the necessity of steps being taken in Great Britain for the suppression of national intemperance. He had little difficulty in demonstrating that intemperance was a growing evil in Scotland, but the differences as to the respective modes of life of the British and Americans seemed so great to his audience that they were quite unanimous in their opinion that what had wrought a good work here, could not possibly accomplish anything there. A year later, he again visited Glasgow, and after spending two days in personal interviews with many clergymen and others, held a conference with about twenty influential gentlemen, before whom he laid such facts as he had been able to collect with regard to the

* Edinburgh Magazine, April, 1760. Quoted on p. 347 of Reid's Temperance Cyclopædia.

extent of intemperance in their country, a detailed account
of the American Temperance Societies, and a proposal for
a system of similar associations and pledges for Scotland.

A long discussion followed, and much interest was manifest, but no one committed himself to a desire or willingness
to try the experiment of organized effort, until a clergyman,
—the only one present,—who had taken the precaution,
before leaving his home, to prejudge the case by preparing
a resolution, rose, and in a solemn manner offered the following :

" *Resolved,* That this meeting tenders its best thanks to Mr.
Dunlop for his address with reference to the sin of drunkenness,
but it is the opinion of the meeting that *no Temperance Association will ever work in Scotland.*"

No one seconded it ; on the contrary, several strongly
condemned it, and Mr. Dunlop was requested to continue his
investigations, and report at a subsequent meeting. A
month later, having carefully fortified himself with facts
and arguments, he again visited Glasgow, to give a public
lecture on " The Extent and Remedy of National Intemperance." But no religious society was willing to allow
him the use of either church or chapel for such a purpose.
At last, securing a suitable place, he sent courteous notices
to the clergy, requesting them to announce his meeting and
its purpose. Nearly all of the ministers threw the notices
aside, thinking that the project was vain and foolish. And
of the few who complied with his request, one afterwards
acknowledged that during the reading of the announcement,
he kept his eyes doggedly fixed on the paper, not daring
to look either to the right or to the left, lest he might be
drawn to laugh outright, in case any of his audience
should show symptoms of risibility !

The lecture, however, was a success, both in the numbers who heard it, and in the effect produced. Quite a
number of Divinity students who were in attendance, were
disposed, as were many others in the audience, to receive the
discourse with levity ; but soon, as facts which they could

not gainsay, began to be marshalled before them, they were awed into quiet, and then to earnest attention. But even then, the dislike of innovations, possibly the special aversion to being taught by America, and the fear of failure, caused even the most interested to refuse to start an organization in Glasgow. "Let the experiment be first tried elsewhere," they said, " and if it shall succeed there, we will venture on a trial here." The Divinity students before referred to, came forward and offered to become members whenever it should be deemed best to organize in that city. Mr. Dunlop therefore retired to his home in Greenock, and started a society there. Meanwhile Miss Graham and Miss Allan, his personal friends, succeeded in organizing a Female Temperance Society at Maryhill, a small village, near Glasgow. Their organization was created October 1st, 1829, four days earlier than the one at Greenock. " And the two societies," says Mr. Dunlop, " beginning immediately to flourish and do extraordinary good, proved the means, under Providence, of showing to the gentlemen now interested, in Glasgow and elsewhere, that there was a possibility of the plan succeeding if persevered in."*

On the 12th of November, 1829, the " Glasgow and West of Scotland Temperance Society," was formed. Its platform was thus expressed in Article II. of its Constitution.

" That the society shall consist of all who, under the conviction that intemperance and its attendant evils are promoted by existing habits and opinions in regard to the use of intoxicating liquors, and that decisive measures for effecting a reformation are indispensable—do voluntarily agree to relinquish entirely the use of ardent spirits, except for medicinal purposes; and although the moderate use of other liquors is not excluded, yet as the promotion of temperance in every form is the specific design of the society, it is understood that excess in these necessarily excludes from membership."†

* The Early Heroes of the Temperance Reformation. By William Logan. pp. 26–40.
† Couling, pp. 36, 37.

At the first annual meeting, the society—

"Reported that during the year they had circulated 425,300 tracts, in addition to 20,200 temperance pamphlets, which had been printed at the Glasgow press. The society in Edinburgh had also circulated 40,000 tracts, the society in Greenock, 9,000, Dundee, 4,000, Perth, 4,000; and it was estimated that the total number of temperance tracts, and larger publications, issued in Scotland, during the year, was considerably more than half a million. The number of members in Glasgow, was reported to be 5,072, while in all Scotland it was stated that there were not fewer than 130 societies, and 25,478 members. The balance sheet showed an income of £347. 11s. 2½d., and an expenditure of £575. 18s. 7½d., leaving a balance due to the treasurer of £168 7s. 5d." *

In 1829, a combination of circumstances, notable among which was the recent reduction of the duty on spirits, so increased the consumption of intoxicants, and the consequent evils, in Ireland, that three ministers of Belfast convened a public meeting to consider what could be done to check the evil, and especially to diminish the traffic in ardent spirits on the Sabbath. It was decided at that meeting that Rev. Dr. Edgar should prepare and present through the public press, an appeal to the public. While engaged in preparing his address, he learned through an American clergyman, the nature, progress and beneficent results of the American movement, and embodying these facts in his address, accompanied with strong recommendations for the trial of the experiment, gave it to the public through the columns of the secular papers, on the 14th of August.

On that same day, Rev. G. W. Carr, a minister in New Ross, in the south of Ireland, who had also been studying and advocating the American system, organized a Temperance Society among his own people; the following being their pledge :

"We, the undersigned members of the New Ross Temperance Society, being persuaded that the use of intoxicating liquor is, for persons in health, not only unnecessary but hurtful, and that the practice forms intemperate appetites and habits; and that while it is continued, the evils of intemperance can never be prevented—do agree to abstain from the use of distilled spirits, except as a medicine in case of bodily ailment; that we will not allow the use of them in our families, nor provide them for the entertainment of our friends; and that we will, in all suitable ways, discountenance the use of them in the community at large."*

" In less than twelve months it was reported that there were sixty societies in existence with about 3,500 registered members. Of their further progress we have no information, except that the Solicitor-General stated in 1831, that there were then ' upwards of 15,000 members of Temperance Societies in Ireland.' "†

In the Fall of 1829, Mr. Henry Forbes, a merchant of Bradford, Yorkshire, England, while on a business visit to Glasgow, attended one of the public meetings of the Temperance Society, and at once became interested in its work. Procuring a number of the tracts then being circulated by that society, he put them in circulation in Bradford, and on the 2nd of February, 1830, organized there the first Temperance Society in England. Advocates of the cause from Scotland and Ireland, assisting in the work, other societies were formed,—at Warrington in April, Manchester in May, Liverpool in July, and Leeds in September,— so that at the close of the year there were about 30 societies, with an aggregate of 10,000 members. An organization was effected in London, in June, 1831. The pledge was:

" We agree to abstain from distilled spirits, except for medicinal purposes, and to discountenance the causes and practice of intemperance."

The London Society, in their first annual report, stated that " 55 auxiliary societies had been formed, and that nearly 100,000 of the publications of the Society had been printed in London alone." Societies had also been formed in the army and navy.

---

* Teetotaller's Companion, p. 320.   † Couling, pp. 29, 30.

A pledge adopted by the Society in Blackburn, in 1831, will give an idea of the general reservation which the temperance men of that day made of their right to moderate indulgence.

" We, the undersigned, believing that the prevailing opinions and practices in regard to the use of intoxicating liquors are most injurious, both to the temporal and spiritual interests of the people of this place, and that decided means of reformation, including example as well as precept, are loudly and imperatively called for, do voluntarily agree, that we will totally abstain from the use of ardent spirits, except for medicinal purposes; that if we use other liquors it shall be in great moderation; and that we will never use them in any inn or house in which they are sold, *except when necessary for refreshment in travelling or transacting business from home*, in order that, by all proper means, we may, to the utmost of our power, discountenance the causes and the practices of intemperance." *

The pledge adopted by the Preston Society in 1832 was similar to this.

In Sweden, the first Society on the basis of total abstinence from distilled liquors was organized at Stockholm, in 1831. The societies continued to increase, though encountering great opposition, till in ten years they numbered about 500.

Recently several attempts have been made in the United States to revive Moderation Societies, and in a few localities such organizations have been created. The plan of Dr. Hartt, before referred to, looks no farther than to the prevention of "excessive drinking." Dr. Crosby, of New York, organized the " Business Men's Society for the encouragement of Moderation." This society seeks to accomplish its purpose by means of pledges, four in all, which are offered to the choice of those who would be enrolled in membership. The first is a pledge of total abstinence from all intoxicants for such length of time as the signer may designate; the second is a pledge to abstain from all intoxicants except wine and beer, and these to be

---

drunk only at meals; the third is a pledge, not to drink any intoxicating liquors till after 5 o'clock in the afternoon of any day; and the fourth, pledges the signer not to drink as a beverage, any intoxicating liquors at the expense of any other person, nor to invite another to drink. It is reported that during the first year of the existence of the society it gave 22,616 pledges, being 5,661 of the first, 4,100 of the third; and 12,855 of the fourth. Many, it is said, who began with the third or fourth, changed them for the first.* It is believed that many who have taken the first, were already enrolled in other total abstinence organizations, but renewed their pledges here for the purpose of influencing others.

In May, 1878, the Pontifical sanction was obtained for "the erection of the Sodality of the *League of the Cross for the Suppression of Drunkenness* in St. Lawrence's Church, 64 E. 84th St., New York," and in Feb. 1879, "power to affiliate similar societies," was also obtained from the Pope. This Society offers its members the option of two pledges: one of Total and the other of Partial Abstinence. Those who take the partial pledge, are forbidden by Rule seventh of their Laws to drink what is allowed them, "in bar-rooms, saloons, beer-gardens, or such like places, or at picnics, or excursions."

"Those who take the Partial pledge may at any time change it for the Total; but those who take the Total, must keep it for at least a year before they may change it for the Partial." "The pledge of Total Abstinence binds those who take it to abstain from all intoxicating drinks, beer of every kind, no matter by what name it may be known, cider, cordials, bitters, etc. etc." "Men who take the pledge of Partial Abstinence, bind themselves not to exceed in the day of 24 hours, three glasses of porter, ale, beer of any kind, or cider, or three wine-glasses of wine, or two wine glasses of brandy, whiskey, gin, or rum, and not to drink this in places forbidden by Rule 7. Women who take the pledge of Partial Abstinence are limited to two glasses of beer, ale, etc., or one glass of brandy, whiskey,

---

* "The Signal," Chicago, April 29, 1880.

etc., or two glasses of wine, and under restrictions similar to those of the men.

" If any exceed the quantity the rule allows, or violate Rule 7 in their use of it, they must renew their pledge before the Director, give up their old card, receive a new one, dated on the day on which they renewed their pledge, and undergo another probation." *

IV. Total Abstinence Societies.—Dr. Lees places at the head of his chapter on "Ancient Teetotalism," † the following quotation from the Medico-Chirurgical Review : " Without contradiction, in Every Age of the World there has been a Total Abstinence Movement." And then, in confirmation of this statement, gives historic proof by citations from the records of antiquity. China, India, Persia, Egypt, Greece, all contribute their testimonies. The Bible also speaks for the work among the Israelites, and of the Apostolic demands in Christian antiquity. Having already alluded to the latter in speaking of Ecclesiastical Penalties, no attempt will be made here to go over the ground so well covered by Dr. Lees ; but a brief mention will be made of some of the early traces of Total Abstinence not mentioned by him, and of their causes or chief incitements, before attempting to trace the origin and extent of this effort in modern times.

The author just named has quoted from the "Modern Universal History," to the effect that eleven hundred years before Christ, a Chinese emperor issued a decree forbidding the use of wine in his dominion. Beyond question the " Announcement about Drunkenness, " (already quoted from, a few pages back,) is the decree alluded to. This prohibition did not aim at Total Abstinence, but only at checking such excessive use of what the document calls " spirits," as rendered the drinker wholly unable to attend to his business. The subsequent mention of Buddah, of

---

* The Crusade or League of the Cross for the Suppression of Drunkenness, 1879, pp. 15, 18, 19–21.

† Works, Vol. II. pp. 1–15.

his five special laws, and particularly of the fifth : " Walk steadily in the path of purity, and drink not liquors that intoxicate or disturb the reason," and that his opinions rapidly spread and prevailed " over China " as well as over other countries, is the more reliable intimation of the fact of Total Abstinence in China, and of its causes. Buddah, —which means, " The Wise,"—was born in India nearly six hundred years before Christ. He founded a religion in which Total-Abstinence was imperatively demanded of the priesthood, and intoxication was forbidden to the laity. The vow taken by the young priest included this : " I will observe the precept or ordinance that FORBIDS THE USE OF INTOXICATING DRINKS, which lead to indifference to religion." He was instructed that it was his duty, and must be made his rule, " to proclaim first the reward to be received for the giving of alms, and then to enforce the precepts." " *But there is no reward to him who gives intoxicating liquors.*" * This religion, high in its aim, and rigid in its requirements, made early and rapid spread in China, and for nearly two thousand years, having most hearty reception, taught and secured Total Abstinence.

If the reader will turn back to the account of Intemperance in Persia, in the previous chapter, it will be noticed that,— strange as the inconsistency may seem,—while the gods were to be plentifully supplied with inebriating drinks, drunkenness among the people was attributed wholly to the hostile evil powers, and that even to simulate intoxication was regarded as sinful. These teachings, to a people so devout as the Zoroastrians were noted for being,—teachings that were repeated on every religious occasion,—must have made the ancient people strong in their efforts to wholly abstain.

Mahomet's conquest of Arabia, and the spread of his religion over other countries, was also the triumph and establishment of Total Abstinence as the rule for millions

---

* Hardy's Eastern Monachism, pp. 24, 80–82.

of human beings, as is too well known to be disputed, however true it may be that in later years, many of his followers have fallen away from his precepts.

The Romans enforced total abstinence on the women of their nation, guarding against its violation in a great variety of ways. Athenæus quotes Polybius as saying in his sixth book:

> "It was forbidden to women to drink wine at all. However, they drank what is called Passiun; and that is made of raisins, and when drank is very like the sweet Ægosthenite and Cretan wine, on which account men use it when oppressed by excessive thirst. And it is impossible for a woman to drink wine without being detected: for, first of all, she has not the key of the cellar; and in the next place, she is bound to kiss her relations, and those of her husband, down to cousins, and to do this every day when she first sees them; and besides this, she is forced to be on her best behavior, as it is quite uncertain whom she may chance to meet: for if she has newly tasted wine, it needs no informer, but is sure to betray itself." *

Turning to modern times, where the principle of Total Abstinence has been maintained through the agency of pledges and organizations, the first instance of which we have reliable information, is the case of Micajah Pendleton, of Virginia, who, witnessing the lamentable effects of drinking on his neighbors, and desiring to fortify himself in all possible ways against becoming a victim to the evil, drew up, and signed a Total Abstinence pledge in the early part of the year 1800. Desiring to associate others with himself in the good work,—though not, so far as is now known, attempting any formal organization,—he induced many of his neighbors to sign with him; and it is said that others, in different parts of the State, followed his example. Had he been a man possessing the gifts of popular address, he would no doubt have become the leader in organized work; but lacking these, and depending exclusively on his personal influence, manifest wholly in a

* Book X. c. 56.

private way with his neighbors, his work was limited.*

Following soon after this, as we have seen, the Pledges
and Organizations to secure community against the evils
of the use of ardent spirits, were introduced, and held
exclusive possession of the temperance field for more than
a quarter of a century. In 1833 these began to give way
to Total Abstinence organizations, which in a few years
obtained entire control of the temperance work. A
variety of causes led to this radical change.

(I.) The first, and most significant, was the fact that little
or no reform was produced in the inebriates who had been
induced to sign the half-way pledge, and that the allowed
moderate use of fermented drinks to the young was rapidly
recruiting the army of drunkards. This inevitable tendency of an allowed moderate indulgence by those seeking
to reform, had been predicted in the Address issued by the
Fairfield Association in 1813.

"The first remedy which we would suggest," it said, "particularly to those whose appetite for drink is strong and increasing, is a total abstinence from the use of all intoxicating
liquors. This may be deemed a harsh remedy, but the nature
of the disease absolutely requires it. People often form resolutions of breaking off from the use of spirits by degrees. . . .
For the drunkard, or the almost drunkard to think of reforming by degrees, is perfectly idle. If he should attempt and ever
begin to reform by taking a little less and a little less, daily, he
would most certainly relapse, in a very short time. To parley
with the enemy in this way, is just about the same thing as
surrendering at discretion. . . . Could we make our voice
heard by all these persons throughout the United States, we
would entreat them to avoid the gulf into which they are plunging. We would say, friends and fellow-citizens, move not a
step further in your downward course. Conquer by total abstinence from strong drink, that perilous habit, which, if persisted in, will prove your undoing. Escape from the enemy of
your souls and bodies, while you may. Make no half-way efforts. Be resolute, be persevering, and you will obtain a glori-

ous victory. Say not, you cannot break off all at once. This very plea proves the greatness of your danger. Have you come to love liquor so well, that you cannot do without it?" *

Experience demonstrated the truth of this prediction the more rapidly the moderation movement spread, and the more numerous its converts became. As early as 1825, Rev. Joshua Leavitt, of Connecticut, was exposing the evils of moderate drinking, and advocating Total Abstinence, in the columns of the "Christian Spectator." He was seconded by Rev. Calvin Chapin, who in January, 1826, commenced a series of able papers on "Total Abstinence the only Infallible Antidote," in the "Connecticut Observer." In April, 1826, the "National Philanthropist," a weekly paper in aid of the temperance cause, was commenced in Boston, and had for its motto : "Temperate Drinking is the Down-hill road to Intemperance." The next year, both the Massachusetts and the New Hampshire Medical Societies passed resolutions declaring it to be their profound conviction that water was the only proper beverage for man." †

In 1832, Dr. Edwards, in his report as corresponding Secretary of the American Temperance Union, said that of the many reformed drunkards who had joined the various Temperance Societies, some had gone back to drunkenness, and that in most cases "they had done so without breaking their pledge, having become intoxicated on other than distilled liquors." This statement being challenged, much investigation followed. Gerrit Smith, of Peterboro', N. Y., instanced "numerous reformed drunkards who had gone back on cider. Others reported lapses on wine. Others on beer. The basis of all these drinks was found to be alcohol, generated in fermentation and not in distillation; and hence the conclusion was, that if men would have the reform progress, and our children saved, the pledge must embrace all intoxicating drinks." It was also declared to be a growing

---

* An Address, etc., pp. 25, 26.
† Centennial Temperance Volume, pp. 435–40.

conviction that "sound and stable-minded temperance men were becoming satisfied that they were far better off without fermented drinks than they were with them."

A circular of inquiry on this subject, addressed to a large number of intelligent gentlemen, brought replies of a most decided character, placing wine, cider and malt liquors under the ban as deleterious articles to the human constitution. No testimony was stronger or more influential than that of Professor Hitchcock, of Amherst College. Said he :

"I have watched the reformation of some dozens of inebriates, and have been compelled to witness the relapse of many who had run well for a time. And I say, without any fear of contradiction, that the greatest obstacle to the reformation of drunkards is the habitual use of wine, beer, cider and cordials by the respectable members of community; as in very many, I believe in most cases, intemperate habits are formed, and the love of alcoholic drinks induced, by the habitual use of these lighter beverages. I rejoice to say that a very great majority of the several hundreds of clergymen of my acquaintance, are decided friends of the temperance cause, and both by preaching and practice inculcate total abstinence from all that can intoxicate as a beverage." *

This testimony had great weight, especially as in 1830, Prof. Hitchcock had said in one of his Lectures to the students of Amherst : "I should consider it *extremely injudicious* and even *Quixotic*, for any temperance society to require total abstinence from the milder stimulants." †

Dr. Marsh says of this period :

"Alcohol was diffusing itself through all the veins of society, in fermented drinks. Breweries sprang up as by enchantment. Distillers turned their whiskey over to the wine factor. The happy family saw their hopes blasted in the return of the reformed father, through hard cider and drugged beer, to drunkenness. Even temperance men were seen intoxicated on sherry and porter, and the youth of the land were lawfully plunging, amid the exhilarations of champagne, into the vortex of ruin." ‡

* Marsh, pp. 42, 43.

† Centennial Temperance Volume, p. 442.

‡ Introduction to the American edition of Anti-Bacchus, 1840, p. 15.

Mr. Pleasants says : " The most deplorable apostacy was common in all quarters of the Union, and the enemies of sobriety, when a man was seen more than ordinarily drunk, were wont to say, ' there goes a member of the Temperance Society.' " *

What was true in America was also alarmingly true in England, and elsewhere abroad. A member of the Society in Bradford, England, said·:

"Here the first moderation society was formed ;. and here there was no want of zeal, talent, or piety in the working of that system; and yet, in five years, we did not succeed in reforming one solitary drunkard." ·

Of the Preston Society it was reported that : " By visiting the members at home it was discovered that numbers of them got drunk, not with ardent spirits, but malt liquor." Of the Society of Halifax it was said : " No drunkards were reclaimed, and not many of the members reduced their daily consumption of wine or porter."

Rev. Dr. Edgar, one of the pioneers of the movement in Ireland, testified :

" We have seen as plainly as light can show it, that all plans which we have hitherto adopted for putting an end to intemperance, have been to a melancholy extent unavailing. They have employed only a portion of the means which the gospel prescribes; and hence not sufficiently strengthened precept by example. They have said to the drunkard : ' We will wean you off by degrees from your intemperate habits ;' and thus, with the best intentions, they have contributed to the drunkard's doom. They have said to the temperate : ' We will allow you to drink moderately '—without inquiring into the nature of the drink employed; and thus they have contributed to support and patronize *the school in which drunkards are trained.* They have unconsciously conducted the temperate man forward through all the stages of free drinking, till he is temperate no more." †

(II.) Not only was the pledge against the use of ardent spirits defective in principle, in ignoring the fact, that drunkenness is produced by all beverages which contain

* War of Four Thousand Years, p. 194.
† Teetotaller's Companion, pp. 324, 325.

alcohol, whether distilled or fermented, but it was practically a discrimination against the poor and in favor of the rich, in that it forbade the former and allowed the latter.

Dr. Macnish intimated this in his " Anatomy of Drunkenness," where he accused the members of temperance societies of practising on themselves a delusion, and showed that while they follow the rules of the society they are certainly habituating themselves to intemperance, by the "inconsistency of allowing their members to drink wine and malt liquors, while they debar them from ardent spirits. They do this," he says, " on the ground that on the first two a man is much less likely to become a drunkard than upon spirits—a fact which may be fairly admitted, but which, I believe, arises in some measure from its requiring more money to get drunk upon malt liquors and wine than upon spirits."*

Dr. Marsh relates that when, in 1831, he was endeavoring to hold a Congressional Temperance meeting in Washington, he called on Senator Grundy, to secure his aid, who replied to him that "he would speak; but, if he did, he should be an ultra, for he should go against wine; he had no idea of calling upon the laboring population to give up their ardent spirit and leave the more refined and wealthy to drink their wine, when he knew it was equally a source of drunkenness."† An anecdote, showing that even the poorest of the poor were aware of this discrimination, is told of an ignorant negro, who was employed by a clerical member of the old society, to cut wood for him :

"On visiting him to see what progress he was making, the clergyman saw a jug among the chips, and said : 'What is here? Rum ?' ' Oh! yes, massa,' said the negro; 'but if I could buy wine as you do, I would not have this vile stuff.'"

The following incident will also show how it was regarded by the intemperate generally. It is said that somewhere about the year 1830, a man in Philadelphia,

---

* Chapter XV. † Marsh, Autobiography, p. 31.

who had taken the old pledge, called a meeting of the friends of Temperance, for the purpose of awakening public interest in the cause.

"Singularly enough, many noted topers were among the audience; and as soon as the hour had come, one of these moved that a certain person, at that time a very intemperate man, be called to the chair. As the nominee, though a hard drinker, was a very popular man, the motion was quickly seconded and easily carried. As soon as the chairman had taken his seat, some one in the crowd offered the following resolution:

"' *Whereas,* The object of all drinking is to produce intoxication in the cheapest and most expeditious manner possible; and whereas, the substitution of the more costly drinks, such as wine and beer, has a tendency to increase the expense of the operation without lessening the disposition to drink, therefore,

"' *Resolved,* That we recommend to all true friends of temperance to quit the use of every other intoxicating beverage except whiskey, rum, gin, or brandy.'

"These were carried by a large majority, and the gentleman who called the meeting together left it amid peals of laughter." *

Such an exhibition of the manifest partiality and injustice of the old plans, had no little influence in leading to a radical change in both theory and practice.

(III.) Another agency in showing the need of total abstinence, was the startling facts brought prominently to light at about the time of this transition period, in regard to the effects of all kinds of intoxicants on the human body in predisposing it to disease, and in preventing its recovery from sudden and violent sickness. The first fact was apparent during the general prevalence of the cholera, in 1832. Physicians were so well aware of the inability of the drinker of intoxicants to resist the attacks of this scourge, that early that year, and before it had reached their locality, they sounded the alarm and called on the people to abstain. In London, England, placards were daily carried through the streets some weeks before a case of cholera had been reported, bearing in large letters the words: " ALL SPIRIT

---

* Centennial Temperance Volume, pp. 446, 447.

DRINKERS WILL BE THE FIRST VICTIMS OF THE CHOL-
ERA."

In New York and Albany, U. S., thousands of posters
with this advice printed on them, were put up with good
effect: "QUIT DRAM DRINKING IF YOU WOULD NOT
HAVE THE CHOLERA." Those who were so unwise as to
disregard the warning, paid the penalty of their infatuated
self-assurance with their lives. Nine-tenths of the victims
of the disease were from the ranks of the intemperate.

M. Huber said of 2,160 persons whom he saw die in twenty-one days in one town in Russia: "It is a most remarkable circumstance that persons given to drinking have been
swept away like flies. In Tifflis, containing 20,000 inhabitants, every drunkard has fallen—all are dead—not
one remains."

In the city of Washington, D. C., the Board of Health,
alarmed at the progress of the disease, issued an order declaring the grog-shops a public nuisance, and forbidding
their continuing their traffic for ninety days. The Boston
Board of Health also published their opinion: "That all
kinds of ardent spirits and other strong stimulants are not
useful in preventing cholera, but that they dispose to its
attack." *

The statistics which were gathered after this visit of the
scourge had ceased, show how wise these warnings and precautions were. A physician writing from Montreal, said:

"Cholera has stood up here, as it has done everywhere,—the
advocate of Temperance. It has pleaded most eloquently, and
with most tremendous effect. The disease has searched out the
haunt of the drunkard, and has seldom left it, without bearing
away its victim. Even *moderate* drinkers have been but little
better off."

Dr. Bronson, of Albany, said:

"Drunkards and tipplers have been searched out with such

---

* Tract No. 3, published by the National Temperance Society
and Publication House, New York: "Cholera Conductors." By
Rev. James B. Dunn.

unerring certainty, as to show that the arrows of death have not been dealt out with indiscrimination."

With regard to those who died in that city, over 16 years of age, the physicians reported, and the Board of Health, vouching for the accuracy of the report, recommended that it be published for general circulation :

```
" Whole number of deaths.............................336
Intemperate......................................140
Free drinkers.....  ...................... ....... 55
Moderate drinkers............................ 131
Strictly temperate............................. 5
Members of Temperance Societies................ 2
Unknown.......  ............................... 3
                                              ___
                                              336    336." *
```

A more fearful illustration of the dangerous evil of intemperance as a physical curse, it would be difficult to find. Other testimonies of like character might be adduced, but these are sufficient to establish the fact that any use of intoxicants disables the human system and makes it an easy prey to this dreadful disease.† The apprehension of this fact had great influence in bringing in the Total Abstinence era in the Temperance cause.

A second illustration of the physiological result of the use of all intoxicants, was that afforded by a series of remarkable experiments by Dr. Beaumont, Surgeon in the U. S. Army, on the stomach of a man in full health and strength. Alexis St. Martin, while in the employ of the American Fur Company, in June, 1822, was accidentally wounded by the discharge of a musket, which produced an opening in his stomach, about two and a half inches in circumference. In 1825, some time after he had fully recovered his health and strength, this aperture still remaining, and the surrounding wound firmly cicatrized to its edges, Dr. Beaumont commenced a series of gastric experi-

---

* Mirror of Intemperance, pp. 125, 126.

† See more full citations in Dr. Hargreaves' " Alcohol, What it Is and what it Does," pp. 313–323.

ments with him, lasting four months. Similar experiments were made four years later, and a more extended series for a year, commencing in November, 1832. Through this aperture in his patient's stomach, the experimenter saw the full effect on the stomach of everything that entered it.* Alcoholic drinks of all kinds were swallowed by St. Martin, and invariably they produced inflammation and disease, acting in every case as poison would act. Many of these experiments were made in Washington, D.C., and many in Plattsburg, N. Y.; and as facts in regard to them were from time to time made known to the public, they could not fail to arrest attention, and while they excited wonder, they also produced a deep impression, especially in those whose attention had been given to the subject of intemperance. Finally, Dr. Beaumont gave a history of his experiments, in which he declared that "The *whole class* of alcoholic liquors, whether simply fermented or distilled, produced very little difference in their ultimate effects on the system." † This declaration, based not on speculation, but on what Dr. Beaumont had seen with his own eyes, as actually going on in St. Martin's stomach, exploded many theories in regard to the harmlessness of the lighter forms of intoxicants, and gave a wonderful impetus to the cause of total abstinence.

(IV.) The attitude of the Christian Church, in its various branches, had no little to do in changing the character of temperance theories and acts. I do not overlook the fact that many ministers stoutly arrayed themselves against the total abstinence movement, but I feel confident that in nearly every ecclesiastical organization where there was any difference of opinion in regard to temperance methods, the friends of total abstinence steadily gained ground, and at last were largely in the majority. And even before this idea gained the ascendant, the immorality of the traffic

---

* See also Dr. Hargreaves' Essay, p. 332.
† Experiments, etc., 1833, p. 50.

was clearly seen, and this clearing of the vision at once helped to still farther sight and to a constantly nearer approach to solid vantage ground. Drunkenness was acknowledged, and had long been, as the chief cause of church discipline, and ministers and churches could not be blind to the fact that drunkenness was not checked by moderation.

The action of several of the leading religious bodies was similar to that of the General Assembly of the Presbyterian Church, in 1831, when it declared it to be "a well-established fact that the common use of strong drink, *however moderate*, has been a fatal, soul-destroying barrier against the influence of the Gospel." This was not long after followed by a thorough discussion of the "Wine question," especially of the relations of that question to the Sacrament of the Lord's Supper. To obtain all possible light on the subject, a premium of $150 was offered for the best essay on the use of alcoholic wine at the Communion. There were numerous competitors, but the award was made to Rev. Calvin Chapin.

Rev. Geo. B. Cheever, D.D., then of Salem, Mass., contributed largely to a wholesome agitation of the public mind by the publication of his little work: "Deacon Giles' Distillery," "one of the most masterly and effective blows ever inflicted on the liquor system up to the date of its publication."

Bishop Hopkins, of the Episcopal Church in Vermont, afterwards an unfortunate defender of American slavery, came out in defence of the work of the rumsellers and the arguments of the drunkards, by giving to the public a book, entitled, "The Triumph of Temperance the Triumph of Infidelity." Arguing in this that the wines mentioned in the Bible are all intoxicating wines, he charged the advocates of Temperance with attempting to do what Christianity itself could not do, and with setting revelation aside as useless ; and so endeavored to prove that the triumph of the modern Temperance cause would be the triumph of Infidelity. Gerrit Smith, Dr. Edwards, and many others took up the challenge, and the churches and the secular world alike became educated aright by the controversy.

It was under these circumstances, and also largely by the circulation of millions of pages of temperance literature, that the public mind became prepared for the advance step, the Total Abstinence Pledge, and organizations for the spread of the Total Abstinence cause.

The first Total Abstinence Society is said by Messrs. White and Pleasants, to have been established " at Hector," (Schuyler County,) "in the State of New York, in 1818." *

For the first eight years, however, it pledged its members to abstinence from the use of distilled spirits only. In 1826 it adopted two pledges, leaving it optional with its members to sign either, as they might prefer. The first was the old pledge, the second required total abstinence from fermented as well as distilled intoxicants. In 1827 the secretary of the society made up a roll of the membership, and to distinguish one class from another, marked the letters " O. P." against the names of those who took the old pledge, and the letter " T." against the names of those who took the pledge against the use of all intoxicants. Before long the latter were called "T-totallers,"—hence the origin of the word " Teetotaller." †

The Fifth Annual Report of the American Temperance Society,—1832, p. 46—said :

"Before the formation of the Hector Temperance Society, more than 8,500 gallons of ardent spirit were annually consumed in the town. Eleven distilleries were in operation. Since that time the consumption of ardent spirit has diminished nine-tenths. Nine of the distilleries have been stopped, and two are now struggling for a doubtful existence. . . . There was scarcely grain enough raised in the town for the supply of its inhabitants ; but the last year it is supposed that 60,000 bushels were sold for foreign consumption."

---

* War of Four Thousand Years, p. 193.
† One Hundred Years of Temperance, p. 129.

Messrs. White and Pleasants also announce that, in 1826, Rev. Thomas P. Hunt,—identified through a long life with the Temperance movement in America, in its earliest, and also in its most advanced stages of effort,—prepared a Pledge of Total Abstinence for the use of children, which, being put in rhyme, made a deep impression on the memory. It was to this effect :

> " We will drink no whiskey, brandy or rum,
> Or anything else that will make drunk come." *

Consecutive efforts in this direction, in America, may be said to date from the City of New York, early in 1833. In February of that year, meetings for the purpose of creating an interest in a National Convention to be convened in May, were held in all parts of the Union. After these meetings, and prior to the Convention in May, Luther Jackson, a missionary in the city of New York, and Secretary of the Eighth Ward Temperance Society, drew up, circulated and published the following pledge :

" We, whose names are hereunto annexed, believing that the use of intoxicating liquors, as a drink, is not only needless, but hurtful to the social, civil and religious interests of men ; that they tend to form intemperate appetites and habits ; and that, while they are continued, the evils of intemperance can never be done away: do therefore agree that we will not use them, or traffic in them ; that we will not provide them as articles of entertainment, or for persons in our employment; and that in all suitable ways, we will discountenance the use of them in the community."

This pledge being industriously circulated, received in a short time over one thousand signatures. It is a fair inference to say that the society of which Mr. Jackson was secretary, adopted this pledge, as at his suggestion they held a grand festival on the fourth of July of that year, conducted strictly on total abstinence principles.†

The National Convention held in Philadelphia, in May,

* Ibid, p. 256.   † Ibid, pp. 211, 212.

1833, composed of 440 delegates, representing nineteen states and one territory, took advanced ground in reference to the great reform. The two leading conclusions which it reached, and which by the generosity of one of its members it was able to publish extensively by the gratuitous distribution of 100,000 copies of its proceedings, were:

"*First*, that the traffic in ardent spirits, to be used as a beverage, was morally wrong, and ought to be universally abandoned: *Second*, that an advance in the cause was demanded, and that it was expedient to adopt the total abstinence pledge as soon as possible." *

For the purpose of diffusing information on these subjects, and exerting a moral influence which would tend to the spread of the principles and blessings of temperance throughout the land, the convention put in operation the " United States Temperance Union," an organization consisting of the officers of the American Temperance Society at Boston and the officers of each of the State Temperance Societies.

The records of "The Massachusetts Society for the Suppression of Intemperance," show that:

"A special meeting of the Society was held June 14, 1833. Mr. Sullivan, Chairman of the Committee, appointed at the last meeting, made a report, which concluded with several propositions for the action of the Society, in the form of resolutions; of which the second affirmed, that *total abstinence* should be a fundamental principle of the proceedings of the Society in this cause. A very interesting debate arose with regard to it, which was continued, at the same place, during six successive appointed meetings of the Society, and was finally terminated by the Society's adopting an article distinctly comprehending the pledge of total abstinence from intoxicating drinks. The President ruled that this pledge did not apply to the old members of the Society, unless they severally subscribed to it." †

In June, 1834, the "Juvenile Branch of the Eighth

---

* Centennial Temperance Volume, p. 449.

† "Extracts from the Records," published in a volume issued by the Society, entitled: "When Will the Day Come?" p. 100.

Ward Temperance Society, on the principle of total abstinence, as a drink, from all intoxicating liquors," was organized in New York; and shortly after, another Total Abstinence Society was started in the Fourteenth Ward of that city. But as early as the month of May, over eight thousand persons, members of the New York Temperance Society, had signed the new pledge.*

During 1835 many local Societies had substituted the Total Abstinence Pledge for the Partial Pledge; and their example being followed by the New York State Society, in February, 1837, a second National Convention was held at Saratoga, in the summer of that year, which declared that " the pledge of temperance henceforth, should be that of total abstinence from *all* intoxicating liquors." †
The United States Temperance Union, organized in 1833, having done little or nothing, was at this Convention superseded by the " American Temperance Union," which at once became a mighty power in the land, and for many years led the great work. By means of lectures, circulation of temperance literature, publication of Journals and papers, it exerted a wonderful and far-reaching influence. The story of its great work is faithfully told by the late Dr. Marsh, for many years its secretary and editor of its Journals, in his Autobiography.

In some cases these changes were not effected without encountering great opposition. The pulpit often helped on the good work, but in many instances it threw obstacles in its way. The old societies, in some localities, denounced the new movement as fanatical and extreme, and in not a few places both the intemperates and the so-called moderates made common cause against the innovation. As a consequence, the church was violently assailed by some of the reformers, prejudice led to the free use of invective. and bad passions were in various ways manifest. But surely,

* War of Four Thousand Years, p. 213.
† Centennial Temperance Volume, p. 453.

slowly and permanently, the Total Abstinence cause advanced, and in a short time, comparatively, occupied the entire field.

"Two thousand societies, formed in New York State in 1837, on the moderation principle, had, in 1839, disbanded, and some 1,200 societies adopting the total abstinence principle, had been organized, with a membership of 130,000; and in many of the towns and villages of the New England States more than half the entire population of the towns were members of societies pledged to entire abstinence; while in the United States and Canada there were fifteen temperance papers, ably conducted, advocating total abstinence."[*]

About this time, some new and wholly unexpected influences for the encouragement of workers in the field of Total Abstinence, were put in operation, and as one of them came from abroad, it may be well to drop the story of the cause in this country, and look for awhile at the progress that was being made in the Old World.

The making of the initial movement in England has been claimed for several localities. Peter Burne, in his "Teetotaller's Companion," p. 328, says:

"Paisley has the honor of being the first to declare for unqualified and uncompromising temperance. Mr. D. Richmond, surgeon, on the 14th of January, 1832, founded and became president of 'The Paisley Youth's Society for promoting Temperance on the principle of Abstinence from all Intoxicating Liquors.' The pledge adopted was as follows : 'We, the undersigned, believing that the widely-extended and hitherto rapidly increasing vice of intemperance, with its many ruinous consequences, is greatly promoted by existing habits and opinions in regard to the use of intoxicating liquors in every form, and believing that it will be calculated to promote the furtherance of true and consistent temperance principles, and of the cause in general, do voluntarily agree to *abstain* from *all* liquors containing *any* quantity of alcohol, except when absolutely necessary [*i. e.* as medicines.]

"A similar pledge was adopted some time in the month of May of the same year, by the temperance society of St. John's, New Brunswick. The palm has been claimed and erroneously

---

[*] Ibid, p. 457.

awarded to the Preston Society, as being the first to declare for entire abstinence; but the first pledge drawn up on that principle in Preston, was not signed till the 23d of August following the establishment of the preceding societies; and even then it was but a *private* pledge, and only signed by two individuals— John King and Joseph Livesey."

It is allowed, however, by all who have written on the history of the cause in England, that to the Preston Society, more than to any or all others, the present position of the Temperance cause abroad is due. Their persistence, zeal and energy in the extension of the principle, entitle them to the credit of being the founders of the cause on a permanent basis. The Total Abstinence Pledge, in Preston, was largely the outgrowth of opposition on the part of the officers of the old Society, to the advocacy of Total Abstinence Principles. When shortly after the organization of the general temperance society on moderation principles, in March, 1832, it was found that members were often drunk on beer, the leaders in the cause were perplexed and discouraged, and some of them determined on denouncing the use of fermented drinks and advocating the principle of total abstinence. This was prominently the attitude of Mr. James Teare, who gave an address of this character at a public meeting held in June of that year. The committee of the society were incensed, and cited Mr. Teare to appear before them to answer to the charge of breach of the rules— ardent spirits only being the article prohibited by the constitution. A few members of the society agreed with the committee that Mr. Teare's proceeding was unwarranted by the laws; while the accused acknowledged the truth of what had been charged against him, and promised that he should repeat the offence at every possible opportunity, and would never advocate any less thorough-going doctrine, in his attempts to further the temperance cause.

Curiosity and reflection being awakened by this discussion, the Total Abstinence cause rapidly grew into favor, especially as, within a few weeks, other speakers adopted and advocated it, and so provoked new consideration and further hearings before the committee. In the midst of the agitation thus caused, Mr. King challenged Mr. Livesey to sign a pledge of Total Abstinence with him, which he accepted, and the " private pledge " before alluded to thus had its origin : " We agree," it reads, " to abstain from *all* liquors of an intoxicating quality, whether ale, porter, wine or ardent spirits, except as medicine."

The following week five others in the old society, also appended their names. Six of the seven who had now signed, were members of the committee, which then consisted of thirty-seven members of the general society. The opinions of their associates were rapidly changing as they noted the effects produced in those who had adopted the new measures, and soon it was proposed in committee to insert the new pledge in the constitution. The discussion brought a diversity of views to light. Some members were for placing the new pledge beside the old one, and working with both ; others for rejecting the new altogether, while others were in favor of the new alone. After numerous meetings, a motion prevailed for the adoption of the new pledge in connection with the old one, and a committee was appointed to revise the private pledge and adapt it to use in connection with the original pledge in the constitution of the society. At the annual meeting in March, 1833, at which about 2000 persons were present, the private pledge, in this modified form, was adopted as a second pledge of the society, for those who wished to subscribe to it :

" We do further voluntarily agree to abstain, *for one year,* from ale, porter, wine, ardent spirits, and all intoxicating liquors, except as medicines or in a religious ordinance."

The advanced teetotallers objected to the limitation of time in the pledge, but their objections were overruled, and they heartily renewed their efforts for success. The new pledge was at once signed by 34 persons, and during the year by 598. From this time the public advocacy of the temperance cause was on the Total Abstinence basis. An effort to diffuse a knowledge of the principles of the new movement, led to a voluntary mission to the adjoining towns, by six of the leading and zealous members, and great good was thus accomplished. Mr. Burne, from whose historical sketch these facts are obtained, relates the following incident as accounting for the introduction of a now universally recognized word, as denoting the thorough-going principle which characterized the new movement :

"In the month of September of the present year, (1833), a new name was found for it by (the late) Richard Turner, a simple, eccentric, but honest and consistent reclaimed drunkard, who at this time had risen to the position of plasterer's laborer. Being in the habit of speaking at the meetings, he is said to have made use of the following provincialisms in a phillipic against the old system : ' I'll hev nowt to do wi' this moderation—*botheration*—pledge; I'll be rect down *tee-tee-total* for ever and ever.'

" ' Well done,' exclaimed the audience. 'Well done, Dicky,' said Mr. Livesey, ' that shall be the name of our new pledge.' The prefix *tee* had before been occasionally employed in Lancashire to express a *final* resolve or event ; thus a thing irrecoverable was sometimes said to be ' *tee*-totally lost,' a perfectly complete piece of work was said to be ' *tee*-totally finished,' and a determination of relinquishment was expressed to ' give up *tee*-totally.' Conveniently embodying the sense of the new principle, it was eagerly adopted to express it ; and being a few times employed in Livesey's Moral Reformer, soon became popularly established." P. 333.

Meanwhile collisions between the moderationists and the teetotallers in the Preston Society, were frequent, and as a consequence spirited debates and sharp action followed. A member of the committee having been reported as being in the habit of giving liquors to some of his customers, it was resolved, after warm discussion, at a meeting in March, 1834, to add the words " Neither give nor offer," to the teetotal pledge. In April following, the young people of the society, impatient of the dissensions caused by the dual character of the old organization, formed an exclusively Total Abstinence Society, under the following pledge :

"I do voluntarily promise that I will abstain for one year from ale, porter, wine, ardent spirits, and all intoxicating liquors, and will not give nor offer them to others, except as medicines, or in a religious ordinance; and I will endeavor to discountenance all the causes and practices of intemperance."

The society started with 101 members, between the ages of sixteen and twenty-five. During the year their numbers increased to 998. In less than a year, viz., at the annual meeting in March, 1835, the old society repudiated their " great moderation" pledge, by voting that those members who had subscribed to it only, should have three months' notice to take either another step in advance and adopt teetotalism, or step out of the society altogether.*

These several stages in the history of the Preston Society, have thus been somewhat minutely traced, because they bring into light the persistence and zeal of those to whose missionary spirit and labors we trace the permanent establishment of the Total Abstinence cause throughout the world, America as well as Europe being indebted to them for general success in this great Temperance work, on the only safe basis of individual security against intemperance. They were instrumental in establishing societies in Manchester, Birmingham, Lancaster, London and Garstang, in 1834, and hence indirectly influential in securing the many organizations which came into being on account of the good work done in these various localities.

At Birmingham a physician confronted them and appointed a meeting in which he proposed to " explode the folly of total abstinence." He was given a hearing, and answered. The formation of the teetotal society followed, and the opponent of the movement expressed his willingness to sign the pledge.

In London, Mr. Livesey presented himself at the office of the " Moderation Society," and offered his services to aid them in their general work; but as his labors were directed

---

* Teetotallers' Companion, p. 334.

against all intoxicants, he met with no encouragement, and was compelled, in order to get a hearing, to hire a place for his lecture on his own responsibility, and to go through the streets bell in hand, announcing the meeting in the fashion of a town crier. This meeting was followed up by others, and a society was formed in the summer of 1835, which was called " The British Teetotal Temperance Society." It met with great success, enrolling 3,000 members in ten months. At Garstang the difficulty of obtaining a place in which to hold their meetings, was overcome by the erection of a wooden building for that express purpose, probably the first building in the history of the world that was put up for a Total Abstinence hall. Mr. Teare named it " The Temperance Lighthouse."

At the close of the year 1835, it was estimated that 48,000 persons had signed the teetotal pledge in England, and that 2000 drunkards had been reclaimed. In July, 1836, the first Conference of the British Association was held at Preston, delegates from twenty-seven societies being present, when it was

"*Resolved*, That no society which, after three months from that date, retained the old pledge, should be considered a branch of the association ; and that the only pledge of the association should be the following : 'I do voluntarily declare that I will abstain from wine, porter, ale, cider, ardent spirits, or any other intoxicating liquor ; and that I will not give or offer them to others, except as medicines, or in a religious ordinance ; and that I will discountenance all the causes and practices of intemperance."

For a long time the supporters of total abstinence, in England, had far less opposition from the public at large, than they received from the old societies. For about fourteen years these Moderates obstructed the work in England ; noticeably so in London.

There was, however, outside opposition.

" Sometimes drunken men were employed by the publicans to disturb the meetings, and to annoy the advocates. At other times, the opposition took another form, and men were made to suffer pecuniarily for their teetotalism. The following handbill, for example, was issued at Oswestry :—

'SWEENY NEW COLLIERY.

" ' The proprietors of the above colliery have come to the conclusion not to employ any teetotaller; therefore none need apply.

" ' The proprietors conceive that this resolution is a duty which they owe to the agricultural interests of the country, as well as to the welfare of the public in general. February 19, 1883.'

" 'In consequence of this resolution, 80 teetotallers who were already employed in the colliery, received their discharge, after one week's notice."

But the cause went forward in spite of opposition, so that in 1838, the "New British and Foreign Temperance Society," (a new name for the British Teetotal Temperance Society, mentioned above,) reported:

"In five years only, we have some hundred thousand members. In North Wales alone, about one hundred thousand; amongst whom there are thousands of reclaimed drunkards. Amongst the advocates we can now enumerate at least 400 ministers of religion, of all denominations, who have espoused our cause."

In all England the number of teetotallers "amounted to 400,000."

Until 1839, the pledges in the New British and Foreign Society were of two kinds; the one containing the " neither give nor offer" clause, was called the long pledge; the other, without this clause, was called the short pledge. In March of this year a majority of the committee of the Society had resolved to substitute the American pledge, for both the long and short pledges; but before this could be done it must be submitted to a meeting of the society, composed of delegates from the different auxiliaries. The American pledge was as follows:

" We, the undersigned, do agree that we will not use intoxicating liquors as a beverage, nor traffic in them; that we will not provide them as an article of entertainment, or for persons in our employment; and that, in all suitable ways, we will discountenance their use throughout the community."

The society held its meeting in May. The motion to adopt the American pledge as a substitute for the other pledges, created a long, exciting, and unpleasant debate, and much contradictory and unsatisfactory action. The proposed change was several times defeated, but finally, at the annual meeting, on the 21st of May, the American pledge was adopted by a large majority, but not till many who were opposed to it, including Earl Stanhope, the president of the society, had withdrawn. These formed another society, in June, on the basis of the short pledge, called "The British and Foreign Society for the Suppression of Intemperance."

In 1842, after many unsuccessful attempts in that direction, these two societies were dissolved, and their members immediately formed "The National Temperance Society."

A great impetus was given to the cause in London in the summer of 1843, by the visit of Father Matthew, who during his stay of six weeks, administered the total abstinence pledge to 69,446 persons, extending his visit to other cities and towns,—notably and with great success to Manchester, where 84,000 persons took the pledge in three days. He succeeded during his brief stay in England, in giving the pledge to 180,000 persons, exclusive of those pledged in London. A World's Temperance Convention, held in London, in August, 1846, was also of great service to the cause. A year later it was estimated that in all England there were about 1,200,000 pledged teetotallers. Societies rapidly multiplied; clergymen of all denominations enlisted in the cause; temperance papers were numerous and well patronized, and temperance literature in various forms, was widely distributed and profitably read.

In 1861, Mr. Tweedie, the publisher, reported to a member of Parliament that there were " at least 4,000 temperance societies in the United Kingdom, and not less than 3,000,000 teetotallers." Of the present number and variety of agencies for the advance of the general cause in England, more will be said further on. The facts already narrated have been chiefly drawn from Coulin's History of the Temperance Movement, and Burne's Teetotaller's Companion.

In Scotland, as in England and America, the dangerous indulgence in lighter intoxicants, led to the abandonment of the pledge of abstinence from the use of ardent spirits only and to the substitution of the pledge of abstinence from all intoxicants. The first instance of this in Scotland, is said to have been at Dunfermline, in Sept. 1830. A coffee-house being about to be established in that place, a meeting of the Temperance Society was called to consider what should be done to encourage and assist the enterprise. At that meeting it was made known that the society's committee had agreed to allow the coffee-house keeper to sell *porter and ales.* To this, opposition was offered, though probably without success in preventing the committee's action from being endorsed by the society; as on the following evening, the opponents of the measure met to consider how they could counteract its influence. Their deliberations resulted in the formation of a new society under the following:

" We, the subscribers, influenced by the conviction that temperance is best promoted by total abstinence from all intoxicating liquors, do voluntarily consent to relinquish entirely their use, and neither to give nor receive them upon any, save medical cases—small beer excepted, and wine on sacramental occasions.

" We likewise agree to give no encouragement or support to any coffee-house, established, or receiving countenance from, any temperance society, for the sale of intoxicating liquors.

"Upon these principles we form ourselves into a society, to be called, 'The Dunfermline Association for the promotion of Temperance by the relinquishment of all intoxicating liquors.'"

The reservation in regard to " small beer,"—if that was an intoxicating beverage,—debars us from calling this a strictly Total Abstinence Society, although it would seem, both from the " conviction " on which the signers acted, and the name chosen by them, that it was their intention to make such an organization. Perhaps the most that can justly be said is, that it was in advance of all previous action. Sixty persons united with this society in a few days after its organization.

It was nearly two years before this example of progress was imitated elsewhere. On the 14th of January, 1832, "The Paisley Youths' Society for promoting Temperance and the principle of abstinence from all Intoxicating Liquors," was instituted; and on the following day, an organization was effected at Glasgow, called the "Wadeston Total Abstinence Society." Of the radical character of these societies, there is no doubt. A few days later, the society at Greenlaw adopted the following rule:

"VIII. Finally, that as some wish the 'other liquors,' the moderate use of which is allowed in the second article, placed upon the same footing with ardent spirits, as best suiting their peculiar views and circumstances, the society do not think this prejudicial to the cause. And looking upon the temperance and total abstinence principles as parts of one great whole, provision is here made for acting upon the latter. All, therefore, who do so, shall be considered members of this society; and those wishing to avail themselves of this article shall be required to sign the following declaration:

" We do resolve that so long as we are members of this association we shall abstain from the use of distilled spirits, wines, and all other intoxicating liquors, except for medicine and sacramental purposes. Adherence to this principle will be notified by prefixing a * to the name."

How far this may have been imitated by other societies, we have no information; but it was probably almost an isolated case, as we are assured that many leaders of the Temperance work in Scotland, viewed the Total Abstinence movement "with alarm, and hesitated not to denounce it as a rash and dangerous procedure, most certain to alienate friends, and thereby damage the movement." " In several instances they sought to modify the constitution of the reformed associations, and again permit the family and household use of intoxicants among their members."*

The Total Abstinence cause made slow progress, in spite of earnest and eloquent efforts in its behalf. In September 1836, a teetotal society was formed in Annan, a small town in Dumfries; and in the same month, at the close of a lecture in Glasgow, it was voted: " That the old society pledge be abandoned, and the society meeting there adopt the clean pledge of the Preston friends, namely,—not to take or give any drinks, of whatever kind, that can cause intoxication." This led, among others things, to a debate for two evenings in a week for three weeks, between the champions of moderation and teetotallism, resulting in a majority of the hearers declaring teetotallism the victor. In 1838 the cause received a new impetus through the efforts of several devoted lecturers and organizers, one of whom, the Rev. Robt. G. Mason, said in September of that year :

" The cause is going on in Scotland as well, perhaps, as in any part of Great Britain. We have, at this moment, no fewer than 70,000 pledges to total abstinence, and nearly double that number materially improved by the influence of our principles. No fewer than 50,000 have been added to our ranks during the last year, and the good cause is daily making new accessions. In one small county, commonly called the 'Kingdom of Fife,' we have fifty separate societies, averaging about 300 each, and going on in a most flourishing manner. I have recently made a tour in the north, where during the short space of nine days, I lectured in all the following places—' Inverary, Huntly, Keith, Fochabers, Elgin, Newin, Campbelton, Cromarty, Fostrose, Inverness, Farres, Cullen, Portroy, Bauff, Aberdour Fraserburgh, Old Deer, Peterhead, and Aberdeen. In nearly all these places I succeeded in forming societies, most of which promise to do well. We have now, 15,000 in Edinburgh ; 12,000 in Glasgow ; 5,000 in Paisley ; 3,000 in Dumfries ; 2,000 in Greenock: 1,500 in Dunfermline ; and 1,200 in Kilcardy ; in addition to several societies which average at 700. I have delivered nearly one hun-

---

* The History of the Temperance Movement in Scotland. Edinburgh, n. d. pp. 10, 12.

dred and fifty public lectures during the past two months, in upwards of eighty different places, and have formed more than one hundred and twenty societies in the past year."

In 1842 Father Mathew made a brief visit to Glasgow, and during his stay administered the pledge to at least 40,000 in that city.

In 1838 " The Scottish Temperance Union," was formed, but unfortunately local jealousies caused a division in a short time, and two Associations, one called the Eastern, and the other the Western, occupied the ground which the original organization could have covered as well. Both of these associations were short lived, and were superseded by the "Scottish Temperance League," in 1844. This is still the national association, and it does its work chiefly by publishing information and supporting temperance lecturers. It has seven agents constantly in the field ; circulates an average of 6,000 volumes of carefully prepared temperance literature in a year, distributes 600,000 tracts, and raises and spends about £8,000 annually.*

It is claimed that a Total Abstinence Society was formed in Skibbereen, County Cork, Ireland, as early as the year 1817. This claim was put forth at the " Temperance Congress of 1862," by Mr. Robt. Rae, Secretary to the National Temperance League, who stated that he then had documents in his possession to show this fact, and that the society continued in active operation until it was absorbed in the more comprehensive movement in 1838. Their records and books were destroyed by fire in 1854 ; but several of the original members who were alive in 1862, maintained that total abstinence was their bond of union from the beginning, and that the first rule of the society was expressed in the following words : " No person can take malt or spirituous liquors, or distilled waters, or anything inebriating, except prescribed by a priest or doctor." †

---

* Coulin, pp. 40–142. Logan, pp. 82-85. Centennial Temperance Volume, p. 42.

† Logan, p. 81.

It may be safely said, however, without detracting from the honor which may belong to this pioneer movement, that no general interest was taken in the cause till about 1838, when a few members of the Religious Society of Friends begun to hold weekly Total Abstinence meetings in the city of Cork. Meeting with very little success, they bethought them that much good might be accomplished, if it should be possible to induce Father Mathew, the Roman Catholic priest at Cork, to sign the pledge and give his influence to the movement. William Martin, of the Society of Friends, waited on the priest to inform him of the conviction of Friends in regard to it. He succeeded in interesting him, and at a small meeting, on the 10th of April, 1838, Father Mathew attended and took the pledge. " If only one poor soul," he said, " can be rescued from intemperance and destruction, it will be doing a noble act and adding to the glory of God ; here goes in the name of the Lord." On the same evening a new society is formed, he is elected president, and commences the advocacy of Total Abstinence in an old school-room in Blackamore-lane. He had tried the moderation pledge, it is said, among his people, and had found it inadequate to meet and arrest the great evil of intemperance. Immediately his influence is felt among the members of his spiritual flock. Large numbers crowded his residence, to whom, kneeling, he administers the following pledge; each person repeating after him: " I promise to abstain from all intoxicating drinks, except used medicinally, and by order of a medical man, and to discountenance the cause and practice of intemperance." Then passing among them, he lays his hand in blessing on the head of each, making the sign of the cross, dismisses them to his secretaries, where their names are registered, and each receives a medal and a card, containing the rules of the society; and so makes room for other groups who are in waiting. Soon the crowds are too great to be accommodated in this manner, and so twice in the week he goes to the Horse Bazaar, where thousands kneel and receive the pledge; the numbers so constantly increasing, that before the end of the year—in eight months—156,000 persons have received the pledge from him in Cork.

His fame extends rapidly, and people flock to him from all portions of the land, so that he yields to solicitations and goes abroad to meet them. Visiting Limerick, the people swarm in immense crowds to meet him, extending for two miles along all the roads leading to the place. Private houses are crowded with those who have not been able to get within sound of his voice during his first day there, and over 5,000 persons remain in the streets over night. For four days he is almost incessantly engaged in giving pledges, and his secretaries register no fewer than 150,000 names. Moving on to other places, he is able to show at the close of the year 1840, that 1,800,000 men and women have on bended knees given him their solemn pledge of Total Abstinence from all intoxicants. And so the work goes on until in November, 1844, he reports that he has registered " in Ireland 5,640,000 adherents of Total Abstinence principles."

Father Mathew had a brother who was a distiller, and a near relative, who is also in the same business, and they write to him while he is thus engaged : " If you go on thus, you will certainly ruin our fortunes." His answer goes back : " Change your trade; turn your premises into factories for flour; at all events my course is fixed. Though heaven and earth should come together, we should do what is right." At one of the meetings in London, previously mentioned, he is reported to have said :

"He had no sectarian object in view. Though a Catholic priest, he had been received in the most cordial manner by clergymen and lay members of the Established Church, by Wesleyans, Dissenters, Quakers, aye, and even Jews: and he had administered the pledge to millions of all sects. He wished to elevate mankind, and to promote the interests of religion, and the good of community, by that greatest of all blessings, sobriety. The people of Yorkshire, where he had administered the pledge to upwards of 100,000 persons, wished to pay him for his services, and presents were offered to him from persons of wealth and high standing in society, but he refused to accept of a farthing. He had expended £300 of his own money since he had been in England, but he did not regret it; and if he had been disposed to favor himself and family, he should not have been a temperance advocate, and converted millions of his own countrymen from drunkenness to sobriety. A brother he dearly loved was the proprietor of a large distillery in Ireland, the bare walls of which cost £30,000 ; and he was compelled to close it and was almost ruined by the temperance movement in that country, and the pledge which the people had taken to leave off drinking whiskey, which had caused so much disorder and bloodshed in his native land. The husband of his only sister, whom he also dearly loved, was a distiller, and became a bankrupt from the same cause. He was sorry to speak of those things, but when he was accused of being instigated to do what he had done to enrich himself, he felt compelled to deny the charge. It had also been intimated that he was making a large profit by the sale of medals—he never profited a shilling, and never would. There were 200 of them sold on Monday for a shilling each. The expenses of the day amounted to £15, and the overplus, if any, would be devoted to the furtherance of the cause of total abstinence."

Through his whole career in this work, Father Mathew so carried himself as to convince all that he was seeking the good of others, was devoted to their best interests, and was self-sacrificing in time, means and energy. At the present time the Irish Temperance League, organized in 1859, is the chief instrumentality for the advance of the cause in Ireland. It has something over 100 local Total Abstinence societies connected with it.

In various parts of the world, Total Abstinence societies are now established. In all the British colonies ; in all the countries of Europe ; in India, China, Japan, Africa ; in all the islands of the seas, the cause has gained a footing ; and this principle has been declared to be the only reliable and efficient means for staying the progress of intemperance.

Returning now to note the progress of the Total Abstinence cause in America, we find that by 1840 great gains had been made. Dr. Charles Jewett, then active in the reform in Massachusetts, said that in that State, "nineteen-twentieths of the clergy were total abstainers; and besides occasional sermons, very many of them gave the liquor-system a blow wherever and whenever they had opportunity." The same might as well have been said of New England at large ; and all other parts of the nation were more or less actively engaged in the work, and to so good purpose, that in ten years the number of distilleries had diminished several thousands. And whereas in 1831, 12,000,000 of people were consuming 70,000,000 gallons of ardent spirits,—an average of six gallons for every man, woman, and child,—in 1840, with the population increased to 17,000,000, the whole amount of distilled spirits consumed, was 43,000,000 gallons, less than three gallons for each person. The number of pledged teetotallers was about 2,000,000, at least 15,000 of whom were reformed inebriates. It seemed as though, in some localities, the limit of success had been reached, and some new impulse was needed, to insure further progress. Suddenly, and from an unexpected source, it came.

Six men, far gone in their love of liquor, formed themselves, in the city of Baltimore, into a club for social tippling. They met in the bar-room of Chase's tavern, where they frequently indulged in what they called " a good time." Meeting together on the night of the 2nd of April, 1840, they learned that a noted temperance lecturer was to speak in the city that evening, and, more in sport than from any better motive, they appointed two of their number to go and hear him, and report. The committee brought back a favorable report, and while repeating the arguments to which they had listened, their landlord, overhearing, broke out in a tirade against temperance lecturers, denouncing them all as hypocrites. One of the six tartly replying, " Of course it is for your interest to cry them down, at any rate," provoked further debate, which being renewed each evening, resulted on the evening of April 5th, in the six

signing a pledge of Total Abstinence, and forming themselves into a temperance society, which they named "The
Washington Society." The names of these six mechanics,
with their trades, were Mr. K. Mitchel, tailor ; J. T. Hoss,
carpenter ; D. Anderson, blacksmith ; G. Steers, wheel-
wright ; J. McConley, coachmaker ; and A. Campbell,
silver-plater.

The following was their Pledge :

" We, whose names are annexed, desirous of guarding against
a pernicious practice, which is injurious to our health, standing
and families, do pledge ourselves, as gentlemen, that we will
not drink any spirituous or malt liquors, wine or cider."

It is said that each member was at once put into office,
as it required the full number to fill the positions which
they had provided for in their constitution. Each member
also became an advocate of the cause, the agreement being
that each should relate his history and the results of his
experience. Their first meeting after their organization at
the tavern, was at the carpenter's shop belonging to one of
their number, and here they continued to assemble for a
few weeks, gaining converts and increasing their membership ; but in a short time, living strictly up to their rule,
that each member should "attend all the meetings and
bring a man with him," the shop was too small for their
accommodation, and they moved to a school-house. Soon
outgrowing this they obtained a church, but this soon failing to hold them, they commenced meetings in the open
air. As they were making it their business to seek out
the drunkard in the day time, as far as practicable, their
meetings rapidly increased in attendance and frequency, and
they soon began to send out speakers by twos and threes
to the neighboring towns and cities, where equal success
awaited them as at home. Their success was marvellous.
The speeches of the reformed men, mainly confined to the
relation of personal experiences, touched the hearts and
consciences of drinking men as no other efforts could touch
them ; and at the close of their first year's work, they had

brought more than 100,000 to sign the pledge, some of whom had been low in their drunkenness.

Shortly after their first anniversary they sent out missionaries, Messrs. Pollard and Wright going to the western part of New York; Mr. Hawkins to New England; and Messrs. Vickars and Small to the western States. Each delegation met with a success beyond their most sanguine expectations; so that in a year from their starting, "it was computed that the reformation had included at least 100,000 common drunkards, and three times that number of tipplers who were in a fair way to become sots." For four years their career was one of almost uninterrupted success, for in 1846 there were not less than 5,000,000 teetotallers, connected with over 10,000 societies, in the United States. With great liberality, all classes of society rendered the reformed men assistance in various ways: by providing them with clothing; hiring halls for their meetings; obtaining employment for such as were able to work; and in numerous ways contributing to the comfort of their homes. The women organized Martha Washington Societies, as auxiliaries to the original organizations, and made the well-being of the homes of the reformed men their special care. The Martha Washington Society of Boston, still survives. The movement even penetrated the legislative halls of the nation, leading to the reorganization of the Congressional Society, on a Total Abstinence basis. It awoke many of the churches to renewed activity in the temperance cause, and led in many places to extended revivals of religion.

Much of fact, and much of speculation, has been written to account for the decline of the Washingtonian movement. The Churches have been blamed for their attitude to the Reformers, and the Reformers for their attitude to the Churches. The difficulty in each particular locality, it may not be easy to specify, but these general observations may perhaps cover the case as a whole: The example set by the parent society, of having their meetings characterized by the relation of personal experiences, rather than by arguments based on the varied relations of intemperance to the individual and to society, was too faithfully followed by the organizations which patterned after it. For a time the novelty of having a meeting wholly conducted by those who had been drunkards, was popular and exciting ; and no doubt thousands of men who feared that they were hopelessly enslaved to their cups, were made strong to attempt reform, by the story and example of those who had been as low as themselves, who would not have been reached in any other way. But necessarily there was great sameness both in the tragic and the lighter character of these personal experiences ; intense excitement never can be made lasting ; appeal to mere feeling cannot be long continued ; and therefore, both to the reformed and to the public at large, sameness became insipid, and at last burdensome.

In many instances the reformed men restricted their membership, and especially the management of their affairs, to themselves. It was unwise, as the bar-room is not a suitable school for business of any kind, and is especially defective in fitting men for wise and orderly management of organizations. In very many cases these managers repelled all advice offered by those who were wise in such matters, and they soon became involved in misunderstandings which ripened into disorganizing difficulties. They could not, or would not, see, that there was any danger of surfeiting the public with personal experiences, and invariably the public interest reached its limit, and could not by a repetition of that which had caused its stagnation, be revived. It was also true, that although many of the clergy heartily sympathized with the movement, some were to be found in nearly every community, who, in the spirit of the Bishop before mentioned, opposed it as fanatical, infidel and extreme. An equally direct attack to that made by Bp. Hopkins, was made by Rev. Hiram Mattison, of Watertown, N. Y. His argument was—

"First—No Christian is at liberty to select or adopt any general system, organization, agencies or means, for the moral reformation of mankind, except those prescribed and recognized by Jesus Christ. But,

"Secondly—Christ has designated his Church as his chosen organization ; his Ministers as his chosen ambassadors or public teachers ; and his Gospel as the system of truth and motives by which to reform mankind. Nor has he prescribed any other means. Therefore,

"Thirdly—All voluntary organizations and societies, for the suppression of particular vices, and the promotion of particular virtues, being invented by man without a divine model or command, and proceeding upon principles and employing agencies, means and motives not recognized in the Gospel, are incompatible with the plan ordained of Heaven, and consequently superfluous, inexpedient and dangerous." In seeking to support this style of argument, he declares (p. 12), that the Temperance Reform has not done half the good that has been awarded it, but has done infinitely more hurt than good, comparing what is professed with what is accomplished, to the reputed and the actual effects of quack medicines, a pretended cure-all, but really killing ten where they cure one. Elsewhere, he expresses his opinion that "God has no attributes that can take side with the popular moral societies of the age." And again, he says ; "but suppose a man is reformed in the popular way ; is he any better in the sight of God ? Has that morality which has reference solely to one's present interest, or public sentiment, one single element in common with the morality enjoined in the Bible ?" *

It was also true that not a few who had at first favored Washingtonianism, unwisely,—though no doubt with great honesty,—sought to make it an immediate instrument for sectarian propagandism, and repelled and soured those who in the first flush of their great victory over the appetite for strong drink, were making this the one idea in their efforts for others, and who regarded everything else as direct interference with their work. To add to this tendency, perhaps in some instances to originate it, philanthropists in other departments of benevolent and humanitarian work, some of whom were eminent especially in their labors to

* "A Tract for the Times ; or the Church, the Ministry and the Gospel, the only means for promoting Moral Reformation. By H. Mattison, Minister of the M. E. Church. 1844."

free the American slaves, and who had been made to feel that too many pulpits apologized for this iniquity, entered into hearty sympathy with the Washingtonian movement, and intensified whatever tendency there might have been produced by any other cause, to suspicion, jealousy, distrust of, if not to open war with, the Churches.

It is a mistake, however, to suppose, as many have, that, as a whole, there was nothing but antagonism between the Washingtonians and the Christian Churches. With few, or no exceptions, save, perhaps, on the part of the Episcopal Church, the General Conventions of the Protestant Churches endorsed and encouraged the movement. The action of the General Conference of the M. E. Church, in 1844, is a fair index of the feeling and sentiments of all the Protestant denominations (except as above noted,) at that time :

"*Resolved*, That we recommend to all our preachers, both travelling and local, and to all our members and friends to give to the Temperance Reformation (now in successful operation in this and other countries,) their unreserved approval, and earnest and liberal support."

It can be said with stronger emphasis, that Washingtonianism was not an irreligious movement. In most all localities in the New England states, where the organization was probably the most perfect, there was hearty co-operation with it by churches of nearly every name. Chaplains were appointed who opened each meeting with religious services, and not a few of the reformed men identified themselves with the churches, some becoming even eminent in the ministry. That some coarse, impractical, skeptical men, had part in the enterprise, and that in certain localities such managed it, repelling others from co-operation with them, is no doubt true; and that local failure and abandonment of the plan is thus to be accounted for, cannot be denied; but the general abandonment of this mode of operation cannot be so explained. It is to be traced, rather, to those general peculiarities of the movement itself,

which have been before mentioned. But Washingtonianism was not a failure, for it rescued thousands from ruin,
and opened the way for more advanced thought and effort
in this great cause. It stopped in its career, simply because
it had reached the limit of the exclusive use of the instrumentalities which it employed. It was a great misfortune
that, as Dr. Jewett, who spoke from personal knowledge,
said :

" Some of the most influential of those reformed speakers, including Mitchell, one of the original five, seeing the extensive
and growing influence of the new method of promoting temperance, came, honestly, no doubt, to regard all other efforts as
useless, and did not hesitate so to express themselves. Temperance sermons, prayers, arguments, and exhortations, which
were not *experiences*, were of no account."

Then again, on the same authority :

" Some of the most prominent of the new disciples, although
they advocated total abstinence, held and advocated zealously
doctrines utterly unsound in many important particulars.
Mitchell, the leading spirit of the group, held that, as Washingtonians, they should have nothing to say against the traffic or
the men engaged in it. He would have no pledge even, against
engaging in the manufacture or traffic in liquors ; nor did he
counsel reformed men to avoid liquor-sellers' society or places of
business. He would even admit men to membership in his
societies who were engaged in the traffic, and in my hearing he
admitted that he had paid for liquor, at the bar, for *others* to
drink, after he had signed the pledge. *He* would not drink
liquor, but if others chose to, that was their business." *

While the Washingtonian movement was in progress,
the cause of Total Abstinence received a powerful impetus
from the labors of Dr. Thomas Sewall, of the Columbian
College, District of Columbia. For upwards of thirty
years, Dr. Sewall had been engaged in pathological
researches, during which time he had many opportunities
of inspecting the stomach of the intemperate after death from
the various degrees and stages of the use of intoxicants.
In 1841, he published the " Pathology of Drunkenness,"

---

* A Forty Years' Fight with the Drink Demon, pp. 136, 144.

illustrated by seven drawings of the human stomach as it appeared in various conditions, which he thus described :

" *Plate* 1—Represents the internal or mucous coat of the stomach in a *healthy state.* It was drawn from one who had lived an entirely temperate life, and died under circumstances which could not have changed the appearance of the organ after death : blood-vessels *invisible.*"

" *Plate* 2—Shows the appearance of the stomach of the *Moderate Drinker*—the man who takes his grog daily, but *moderately*, or who sips his wine with his meals:—blood vessels enlarged so as to be *visible*, and distended with blood."

" *Plate* 3—Represents the *first* stage in habitual drunkenness, or the stomach of the *Hard Drinker* :—internal *coat irritated*—blood vessels *more enlarged.*"

" *Plate* 4—Represents the stomach of the drunkard after a *debauch* of several days :—internal coat highly inflamed, red and livid."

" *Plate* 5—Represents the drunkard's ulcerated stomach :—internal coat *corroded.*"

" *Plate* 6—Represents the appearance of the *Cancerous* stomach :—the coats of the organ are *thickened* and *Scirrhus*, with *corroding Cancer* of the size represented."

" *Plate* 7—Represents the internal state of the stomach after death from *Delirium Tremens*. The mucous coat is covered by a dark brown flaky substance, which being removed, shows the organ to have been in a high degree of inflammation before death. In some points it is quite dark, as if in an incipient state of mortification." *

These drawings were first exhibited to the public, in connection with a lecture on the subject, at Washington, in 1842, to an audience of about 3000. They made a great impression, especially on the minds of the more cultivated people. Their correctness was attested by such eminent physicians as Drs. Mott, Warren, Homer, and Green ; and many editions of them in enlarged form were published and used with great effect by several eminent lecturers on temperance. The most intelligent approved of them, and desired their more extended use.

---

* " Alcohol, What it Is, and What It Does," p. 339, contains these plates.

"General Scott desired that they might be furnished to every military post. The Hon. Samuel Young desired that they might be hung in every common school in the state. The presidents of the Marine Insurance Companies expressed a wish that they might be put on board of every vessel on the ocean, on our rivers, and on our lakes, counteracting the peculiar temptations to which mariners and emigrants were exposed. Testimonials from lecturers were often of a most affecting character. 'It is very frequently the case,' said one, 'that after all the facts I could present, or the appeals I could make, seem to fall powerless on the ear of the drunkard, his head up and apparently unmoved, when these *pictures* are shown, his cheeks turn pale and his head droops.' 'I have heard,' said another, 'the unfortunate drunkard exclaim, when looking at them,—and particularly at the one representing the stomach after a debauch —'*they look as I have often felt!*'—Missionaries in foreign lands, at Constantinople and other places, were found to be exhibiting the plates with great effect." *

Shortly after the Washingtonian movement had expended its power, viz., in the summer of 1849, Father Mathew made a visit to the United States. Administering the pledge to about 100,000 in the principal cities of New England, he contemplated making an extended tour through the United States, but his health failed, and he was forced to be quiet. A few societies were founded by him while here, at least two of which, one in Philadelphia, and one at East Cambridge, Mass., are in existence to-day. Quite a number were active in 1860, when a reorganization of Total Abstinence work among the Catholics took place, of which it may be well to speak here. This new movement among the Catholics puts the whole superintendence of Temperance work in the hands of their clergy. It commenced in Jersey City, New Jersey: and was rapidly forwarded in 1867, 1868 and 1869 by the Missionary orders, —the Passionists, Jesuits, and Paulists,—who in prosecuting their mission work, not unfrequently had whole congregations respond to their invitations to rise and take the pledge. By 1870 these local societies began to form dio-

* Marsh, Autobiography, p. 115.

cese or State unions, and in 1872 these Unions became a
national body, bearing the name of the Catholic Total
Abstinence Union of America. In 1876 there were in this
national Union 600 societies of over 150,000 members.
" These comprise only those aggregated, to the Union ;
of other Catholic total abstinence societies there are probably 300 working as local societies. The Catholic women,
under the lead of Father Bessoines, of Indiana, organized in
a few localities about 1878. They failed to receive recognition from the National Union, at first, but in 1880 they
were heartily recognized and endorsed. We may safely
estimate the entire number of Catholic total abstinence
societies in this country at not less than 1,000, having an
active membership of 200,000 persons." These societies
erected a Fountain at Fairmount Park, Philadelphia, on
the 4th of July, 1876, at a cost of $60,000.*

Early in the history of the Total Abstinence Pledge, it
was judged by many that neither the mere taking of the
pledge, nor the affiliation of those who signed it with the
simple and imperfect organizations with which they experimented in their efforts for concerted action, were sufficient
for the peculiar needs and exposures of the reformed men.
It was believed that such could be more perfectly banded
together and greatly helped in ways for which their first
organization made no provision, by societies having a wider
aim, and a more imposing and attractive form of admission,
and method of operation. Hence the origin of the so-called
Secret Temperance Societies.

The first in order of time, were the Rechabites. They
organized in the town of Salford, county of Lancaster,
England, in August, 1835, under the name of " The Independent Order of Rechabites," taking their title from the
ancient people mentioned in Jeremiah xxxv. who put themselves under obligation " to drink no wine forever." Membership is strictly confined to total abstainers, and is limited

---

to male persons of healthy constitution and good moral
character, from fifteen to fifty years of age. To the active
members certain pecuniary benefits are secured, while those
who may not desire such consideration, but are willing to
give countenance and moral support to the order, become
honorary members by paying a small annual fee. Originally the entrance fee of active members was the same for
all, irrespective of their age, within the limits before mentioned. Two funds were established and divided into
shares, each member being at liberty to take from one to
six shares in the sick fund, and from one to four shares in
the funeral fund. For every penny per share paid weekly
into the sick fund, two shillings and sixpence per week are
received in time of sickness; and for every five pence per
share paid quarterly into the funeral fund, five pounds are
paid at death. Recently a graduated scale of contribu-
tions, according to age, has been made for new members,
which in time will become the more just rule throughout
the order.

At an early day, 1842, this order was brought to the
United States, and for a time extensively flourished,
numbering at one period not far from 100,000 members.
At present the membership in this country is from 3,500 to
4,400. It is now chiefly confined to Great Britain and its
Possessions and Dependencies. The local organizations are
called Tents; several of these united, form a District; and
the supreme power of the Order is created by a biennial
conference of representatives from the various districts.
The present aggregate of active membership is a little
more than 30,000, and funds are accumulated to the amount
of £140,000.*

The next organization was the result of a consultation
on the part of a few active Washingtonians in the city of
New York. They had noticed that although the Washingtonian movement was making rapid advance in new

---

* Centennial Temperance Volume, pp. 865–867.

fields, there were already many falling away from the pledge, and they desired, if possible, to hit upon some new plan of operations, some more perfect organization, one that should shield the members from temptation, and more effectually elevate and guide them. In a short time the number of those thus solicitous for a new experiment was increased to ten, when it was agreed that a plan of an improved organization should be drafted and copies of a call for a meeting to consider it, should be distributed among forty prominent Washingtonians. The call was headed " Sons of Temperance," and invited those to whom it was sent, "to attend a select meeting," on Thursday evening, Sept. 29, 1842. It further stated:

"The object of the meeting is to organize a beneficial society based on total abstinence, bearing the above title. It is proposed to make the initiation fee, at first, $1, and dues 6¼ cents a week; in case of sickness a member to be entitled to $4 a week, and in case of death $30 to be appropriated for funeral expenses."

In response sixteen person assembled, and after adopting and signing the following resolution, approved of the Constitution which had been submitted.

"*Resolved,* That we now form a society to be called New York Division No. 1, Sons of Temperance."

A second meeting was held on the 7th of October, when it was decided to adopt a form of initiation. Officers were then elected, and sixteen persons were proposed for membership. The pledge then adopted, and unaltered to the present, reads:

"I will neither make, buy, sell, nor use as a beverage, any spirituous or malt liquors, wine or cider."

In November, a circular was prepared for the Temperance press throughout the country, calling attention to the existence of the order, and giving an outline of its purposes and plans.

"The order of the Sons of Temperance," it said, "has three distinct objects in view, which are, as declared in the preamble

22

of our constitution : ' To shield us from the evils of intemperance; afford mutual assistance in case of sickness ; and elevate
our characters as men.' The design contemplates permanent
systematic organization throughout the United States, divided
into three classes—viz., Subordinate Divisions, State Divisions,
and a National Division."

The New York Grand Division *pro tem.*, was appointed
from the members of the Division in New York, with
power to grant charters. On the following January a
sufficient number of past officers having been obtained,
the Grand Division duly organized, taking on itself the
name of " The Fountain-head of the Sons of Temperance
of the State of New York." Propagand work soon after
commenced, and in a few months Divisions had been established in New Jersey, Pennsylvania, Maryland, North
Carolina, Virginia, Connecticut and Massachusetts. In
July a Quarterly Session of the Grand Division of New
York was held, and the membership was announced as
being 1500. Grand Divisions now begun to be chartered,
and as soon as the number necessary for the purpose had
been obtained, viz., in June, 1844, the National Division
was organized. At that time the order was composed of
7 Grand Divisions, 75 Subordinate Divisions, and 6,000
members. The Order increased rapidly in membership
until 1850, when it had 35 Grand Divisions, 5,563 Subor-
dinates, and 232,233 members. From that time there were
many fluctuations in the numerical condition of the Order,
till in 1864 the membership had fallen to 55,736. The
rebellion of the Southern States in 1861, and the political
antagonisms which for several years preceded the conflict
of arms, had much to do with weakening all philanthropic
efforts. Since the close of the war the Order has revived.

In 1876 it had about 2,000 Subordinate Divisions and
90,000 members in North America ; 600 Divisions and
35,000 members in Australia ; and several thousand members in Great Britain and Ireland. Originally membership was restricted to " male persons eighteen years of age
or over." In 1854, " mothers, wives, sisters, or daughters "

of members were admitted as visitors; a rule which was soon modified, so that any "female, sixteen years of age and upwards," might be so admitted. In 1866, women were made eligible to full membership on the same terms as men; but it was left optional with subordinate Divisions, to accept or reject this privilege. It is now the almost universal policy of the Order to admit women to equal rights and benefits. Controversies have also arisen in the Order in regard to the admission of colored people. The National Division, in 1850, decided on a question of appeal from a decision of the Grand Division of Ohio, to affirm that decision, and declared that "the admission of negroes into Subordinate or Grand Divisions under this jurisdiction is improper and illegal." In 1870 it took the position indicated in the following resolution:

"'That the M. W. P. be, and is hereby, authorized to organize separate Grand Divisions for our colored members, when requested by them and approved by the Grand Division having jurisdiction in the State."

This position was abandoned the next year, but in 1872, all conflicting legislation was repealed, and the action of 1870 is now the law of the order.*

An organization called "The Daughters of Temperance," was founded in the city of New York, in October, 1843. They were incited to and helped in their work, by a number of the Sons, with a view of doing among women what that organization was doing among men. In a short time, by reason of some misunderstanding, a division was made in this fraternity, and a second organization was formed, taking the name of the "Original Daughters." They called their societies "Unions," and having spread quite extensively in the United States, they formed, by means of their representatives from the single Unions, "Grand

---

* The above facts are gleaned from an article by Frederick A. Fickart, M. W. S., in "The Sons of Temperance Offering for 1851;" and the Historical Sketch by Samuel W. Hodges, M. W. S., in the Centennial Temperance Volume.

Unions," in several States. The organization was not long-lived, and on the admission of women to full rights and privileges in the Sons of Temperance, the main reason for its continuance ceased to exist.*

" The Juvenile Sons of Temperance," originated in Lehigh County, Pennsylvania, in May, 1845, an example that was soon after imitated in Bethlehem, in the same State. Later,—in December, 1846,—an effort was made to effect a more general organization for boys, in the city of Philadelphia. They were brought together in what was called a "Section of the Cadets of Temperance," and became partly under the control of the Sons of Temperance, and auxiliary to them. In two years they had spread into twenty-two States, numbering in all, one hundred and thirty sections.†

" The Juvenile Sisters of Temperance," an organization of young girls, was established in 1846. How long it existed, we have no means at hand of knowing; but probably it had a brief life.

The " Cadets of Temperance," an organization for boys, was started in Germantown, Pa., in 1846, by earnest workers in the order of The Sons of Temperance. For some time past the organization has been, as to its oversight and patronage, a mixed body, some being under the " Sons," some, as the " Cadets of Temperance and Honor," under the " Temple of Honor," and some, Independent. Statistics are not easily obtained, but it is estimated that their number in the United States is about 10,000, about a fourth part being under the " Sons," a smaller fraction under the " Temple of Honor," and the balance Independent.

The Temple of Honor was originated by prominent and active members of the Sons of Temperance, with a view to supplying a popular need not provided for by the Sons. These were chiefly, a more elaborate and finished ritual,

---

* "The Beauties of Temperance." By Rev. E. Francis, 1847, p. 64.

† Ibid, p. 58–62.

such as would give the reformed man so much satisfaction as would prevent his seeking fraternal relations with organizations possessing attractive ceremonials, but not in their principles or practices favorable to the maintenance of his integrity as a total abstainer. After many unsuccessful attempts to induce the National Division of the Sons of Temperance to adopt degrees, signs, and other peculiar methods of working, employed by the older so-called Secret Societies, it was determined by a few who believed that such an advance would be of service to the Temperance cause, to test its worth. In June, 1845, they organized a Society called the "Marshall Temperance Fraternity," which in November, they changed to "Marshall Temple, No. 1, Sons of Honor," and again changed in December, to "Marshall Temple of Honor, No. 1, Sons of Temperance." Their membership then numbered forty-five. Before the last of February, 1846, they had instituted eleven Temples in New York City, and one each, in New Jersey, Maryland and Massachusetts. The twelve Temples in New York City, by representatives, met on the 21st of February, and created the "Grand Temple of Honor of the State of New York," which they resolved, "Shall be the supreme power of the Order till the National Division shall take upon themselves that power." After several unsuccessful efforts to induce the National Division to give official recognition to the new order, it was determined in 1849, at a session of the National Temple, which had been organized in 1846, to "make the Temple of Honor an entirely independent organization." At the same session it was voted to prepare a new degree, to be called the "Social Degree," to which the wives, sisters and daughters of Templars, should be eligible. In 1850 the privileges of this degree were extended to "all ladies of good standing;" in 1855 the name of the Social Degree was changed to "Social Temple." In 1852, as the order had spread beyond the limits of the United States, the name of the highest power in the order was changed to "Supreme

Council of Templars of Honor and Temperance." At the completion of ten years in the history of the order, there were 20 Grand Temples, and a membership of 12,980. The twentieth year closed just after the end of the rebellion, which had been disastrous to this as to all other Temperance organizations ; and the membership was 10,530. At the thirtieth annual session of the Supreme Temple, in 1876, there were 21 Grand Temples, 357 Subordinate Temples, and a total membership of 16,229. The entire receipts that year were $74,262.59 ; amount paid for benefits, $7,856.17 ; cash on hand in Subordinate Temples, $101,746.16.*

A "Band of Hope," an organization for children, was first formed in Leeds, England, in 1847. The name being attractive, societies soon rapidly increased. The United Kingdom Band of Hope Union was formed in 1855, for the purpose of promoting total abstinence among the young by means of this organization and such other means as may from time to time be available. It is computed that at the present time there are in the United Kingdom, about 6,000 Bands of Hope, with 810,000 members, and 35,000 adults as officers, members of committees, and workers in the movement ; of whom 7,000 are honorary speakers. Children of both sexes, seven years old and upwards, may become members.† There are Bands of Hope in America, but no statistics concerning them are available.

The "Independent Order of Good Samaritans," and " Daughters of Samaria," was organized in the city of New York, in February, 1847. Originally intended for white men, it took its first advanced step during the first year of its existence by opening its doors to colored people, granting them equal rights and privileges with all others ; and in 1848, admitted women to full membership and privileges. It is a beneficial society ; and is persistent in its efforts to reclaim the inebriate, however often he may fall from his

---

* Centennial Temperance Volume, pp. 625–651.
† Ibid, pp. 858, 863.

promise to abstain. It has seen prosperous days, when its numerical strength was satisfactory to its most ardent members, and it has been again and again brought low. Its organization is into Lodges, Grand Lodges, and a Supreme body called the National Grand Lodge ; but the latter has no definite bounds, but extends its jurisdiction wherever the order exists. Its colored members prefer to keep by themselves in their Subordinate and Grand bodies, but unite with all the others in the Supreme Lodge. It has organizations in many of the States of the American Union, and in Africa. Its present membership is about 14,000. It has also a Juvenile Branch, numbering nearly as many in its membership as the adult.*

" 'The Good Templars," originated in central New York, in 1851. For the first ten years their growth was not rapid, although they extended over quite a large territory. Their membership increased to about 75,000. After the close of the war, in 1865, the order spread rapidly in all the States, and in 1868 numbered about 400,000 members. It was then introduced into various parts of Great Britain, and from thence to Australia, India, China, Japan, Africa, and other foreign countries. Its largest membership was in 1875, when it reached 735,000. In 1876, a portion of the foreign membership seceded ; but a conference held in September, 1886, agreed on a basis of reunion, honorable to all concerned, which will probably be ratified at the coming annual session of the two bodies in May, 1887. In that event the entire membership will be about 620,000. The Good Templars claim the following as the peculiarities of their organization and purpose : The equality of woman in all the work and honors of the order; no discrimination as to race or nationality; the total abstinence pledge binding during life; the prohibition of the traffic in intoxicants ; the reform of inebriates, and the protection of the young from falling into the snares of temptation ; a perfect and equit-

---

* Ibid, pp. 734–737.

able system of finance, by means of which all subordinate lodges may be self-sustaining, and may support the State Grand Lodges, and these in turn support the Right Worthy Grand Lodge, the Supreme Head of the Order. The following is the Platform of the Order:

"1. Total abstinence from all intoxicating liquors as a beverage.

"2. No license, in any form, or under any circumstances, for the sale of such liquors to be used as a beverage.

"3. The absolute prohibition of the manufacture, importation, and sale of all intoxicating liquors for such purposes—prohibition by the will of the people, expressed in due form of law, with the penalties deserved for a crime of such enormity.

"4. The creation of a healthy public opinion upon the subject by the active dissemination of truth in all the modes known to an enlightened philanthropy.

"5. The election of good, honest men to administer the laws.

"6. Persistence in efforts to save individuals and communities from so direful a scourge, against all forms of opposition and difficulty, until our success is complete and universal."

Although all lodges are allowed to admit to membership all persons of twelve years old, and upwards, some have always desired an organization under the auspices of the Good Templars, which would receive younger persons to membership, and train them from their earliest years in the principles of total abstinence. To meet this need, an organization called the "Juvenile Templars," was established some years ago. It is now established in 68 States and countries. Full statistics have not been obtained, but the recent Reports from 35 jurisdictions, show a total of 478 Temples, and 28,694 members. The pledge of the organization contains obligations against the use of all intoxicating liquors, tobacco and profanity. These Temples are managed in very much the same manner as are the British Bands of Hope.

"The British American Order of Good Templars," was an offshoot from the Canadian Grand Lodge of Good Templars, in 1858. In 1866 the name was changed by dropping the word American, in order that the operations of the fraternity might be extended beyond the provinces. It then extended to Newfoundland, Bermuda and New Zealand, and subsequently to Great Britain, Australia, Queensland, Tasmania and Manitoba. In 1876 it consolidated with the Free Templars of S¹. John, in Scotland; the Independent Order of Free Templars in England; and the United Templar Order in Great Britain and Ireland and South Africa, and formed the " United Temperance Association." Its membership is unknown to the writer.*

" The Dashaways," so called from the resolution of their founders to dash away the intoxicating bowl, were organized in San Francisco, Cal., in January, 1859. They spread rapidly through California and Oregon, but we have no information of their present condition and numbers.

The same must also be said of an organization formed in Chicago, Illinois, in 1860, and called " The Temperance Flying Artillery." " Its members were chiefly young men, whose ardor and activity soon organized bands in almost every town and city in Illinois."

" The Friends of Temperance," an organization for white persons only, was organized in Petersburgh, Virginia, in November, 1865. It has since spread into eleven States, and has about 20,000 members.

" The Sons of the Soil," an organization for colored persons only, was organized in Virginia, in 1865. It has a large membership.

" The Vanguard of Freedom," an organization for the children of the Freedmen in the South, was organized in 1868, and has spread into nearly every Southern State.

There is also an organization, chiefly, if not exclusively operating in the Southern States, called the Knights of Jericho, but we fail to get statistical information in regard to it.

The " Sons of Jonadab " was instituted at Washington D. C., in 1867. It is chiefly distinguished from the before

---

* Centennial Temperance Volume, pp. 782-785.

mentioned Orders, in punishing all violations of its total abstinence pledge by expulsion for life. It numbers about 3,000 members.

" The Royal Templars of Temperance " was, for nearly seven years, a local organization in the city of Buffalo, N. Y., originally formed for the purpose of enforcing the law against the sale of intoxicants on the Lord's Day. In 1877, it reorganized, adding a beneficiary fund for the benefit of its members. Up to the present time its field of operations is on the American Continent. It has 533 Select Councils, and a membership of about 20,000. Its aim is to promote " the cause of temperance, morally, socially, religiously and politically.

" The United Friends of Temperance," was organized at Chattanooga, Tennessee, in November, 1871. It is composed exclusively of white persons.

"The United Order of True Reformers," was introduced into the Southern States in 1872, for the benefit of the colored people. It numbered at one time about 45,000 members. Many of its members are now enrolled with the Good Templars, and the original organization has about 10,000 members.

Side by side with the earliest of these secret organizations, the American Temperance Union continued its work for several years, planting and encouraging State societies and other local associations, publishing and distributing valuable temperance literature, and in various ways keeping the temperance sentiment alive and active throughout the country. The corresponding Secretary, Rev. John Marsh, said in the Appendix to his " Half Century Tribute," in 1851, that the Society had at that time distributed 4,964,733 copies of books, and tracts, beside hundreds of thousands of Journals and papers. Several States approved of placing in District School Libraries the volumes of Permanent Temperance Documents, made up of Addresses, Statistics and Reports of the various societies, and thus millions of the people were instructed in right views of the enormity of the evils of Intemperance. Political views of the duties of Temperance men were also advanced through the nation, and made themselves felt in securing more thorough and radical laws on the subject in several States.

But as early as 1854, the efforts of northern men for this cause, became wholly inoperative in the southern sections of the country ; and shortly after this the political excitement arising from the growing prominence of the slavery question, absorbed attention at the north, and pushed all Temperance enterprise to one side. The rebellion following on, very little was done, though the organization continued to hold its annual meetings, till the war ceased.

In August, 1865, a Fifth National Convention was held at Saratoga Springs. Twenty States and the Canadas were represented by 326 delegates. Several able papers on various phases of the Temperance Reform were read, and as a result of the discussions and deliberations, it was determined to make a new departure in the cause, suited to the new condition of the country at large ; one that should secure co-operation on the part of the various open and secret organizations ; enlist the sympathies and efforts of Sunday-schools and churches ; and combine, as far as possible, the many moral instrumentalities throughout the nation and the continent, with the political discussion of the subject, and with wise municipal and legislative action. To this end committees were appointed, one to organize a new National Temperance Society, and the other to provide for and locate a National Publication House ; and each State was requested to organize a State Society, on the broad platform which was to distinguish the National Society, and to become so in harmony with and auxiliary to it, as to have a true union of purpose, and concentration of effort in the all-important cause. At the first meeting of the two committees, assembled at the same time, under power conferred, it was deemed wisest to attempt but one organization, and the two were merged into one committee, and acted together in their deliberations, which, in October, 1865, resulted in the organization of " The National Temperance Society and Publication House," with the following object and pledge :

" The object shall be to promote the cause of total abstinence from the use, manufacture, and sale of all intoxicating drinks as a beverage. This shall be done by the publication and circulation of temperance literature, by the use of the pledge, and by all other methods calculated to remove the evil from the community.—No person shall be a member of this society, who does not subscribe to the following pledge,—namely : We, the undersigned, do agree that we will not use intoxicating liquors as a beverage, nor traffic in them ; that we will not provide them as an article of entertainment or for persons in our employment; and that in all suitable ways we will discountenance their use throughout the country."

Shortly after this organization had been perfected, the Executive Committee of the American Temperance Union, passed the following : " *Resolved*, That the work of the Union be suspended after the 1st of December, 1865, and that its periodicals, documents, tracts, stereotype plates, and good will be transferred to the National Temperance Society and Publication House."

The Twentieth Annual Report of this organization, made in June, 1885, gives a brief statement of various branches of its work since the society was created. $105,719 have been spent in copyrights, stereotyping and engraving its many valuable standard publications, now numbering 1,383 books, tracts and pamphlets; including 138 carefully selected books for Sunday-school libraries. It publishes *The Youth's Temperance Banner*, having a monthly circulation of 116,000 copies, and a total of 27,640,000 copies since the publication first commenced. It also publishes *The National Temperance Advocate*, monthly, containing the latest information of the state of the cause throughout the world, and replete with facts and arguments arranged and presented by the ablest writers. The total number of copies of this journal issued to May, 1885, is 2,085,695.

The assets of the Society are now $34,000, and its sales of publications amounted, in 1885, to over $51,500. It is in hearty accord with, and enjoys the confidence of, the various Temperance Organizations in the land, and does a work of far-reaching and incalculable importance and influence.

Passing by the almost innumerable local Societies which have from time to time sprung up in various parts of the United States, some of them merely experimental and short-lived, and many still in existence and prosperous in States, Counties and Districts, there are a few of more extended purpose and influence which should be mentioned here.

The first is the so-called Reform Club movement. This originated with Mr. J. K. Osgood, a reformed man of Gardiner, Maine, in 1872. The first Club numbered about 100 members, all reformed men. In a few months the movement became popular throughout the State, Clubs were rapidly organized, and in a year they had a membership of from 15,000 to 20,000. From Maine the reform spread to New Hampshire, Vermont and Massachusetts. Women are admitted to membership on equal terms with men, and the platform of principles and methods, embraces these: first, total abstinence; second, reliance on God's help in all things ; and third, missionary work to induce others to sign the pledge.

In 1874, Mr. Francis Murphy entered the field, confining his operations chiefly to the Western States, though doing much in Pennsylvania ; he has been instrumental in inducing tens of thousands to sign the pledge, and to organize themselves into Reform Clubs.

Dr. Henry A. Reynolds also commenced the same work in 1874, at Bangor, Maine, where he organized a Club composed wholly of men who had been intemperate to a greater or less extent. In a year he had organized many such Clubs throughout the State. His labors were then extended to Massachusetts and other New England States, and subsequently to the West. Great success in organizing attended his labors.

THE WOMEN'S WORK.—In December, 1873, the women of a small town in Southern Ohio, organized themselves into a Praying Band, for the purpose of inducing the keepers of saloons, and other drinking places, who seemed to be beyond the reach of the imperfect law of that State, to give up the sale of intoxicants. Their mode of operation was to visit such drinking places as they could obtain access to, and there pray and sing. If the doors were closed against them they knelt on the sidewalks; and were so persistent in their efforts, that the liquor sellers abandoned their business, signed the pledge, and in many cases, became Temperance Missionaries. Their example and its success, was soon imitated in Illinois, Indiana, Wisconsin, Iowa, Nebraska, Kansas, California, Oregon, Maryland, Massachusetts, Pennsylvania and New York. It has crossed the Atlantic, and been a power in England, Scotland, India, Japan and China.

In the spring of 1874, the women who had been " Crusading," as they called it, all winter, called conventions in their respective States, for the purpose of organizing for systematic work. At first they called their new Societies, " State Temperance *Leagues.*" Soon, however, they changed the name to " Unions." A National meeting, attended by delegates from sixteen States, was held in Cleveland, Ohio, in November, 1874, at which time they took to themselves the name of the " Woman's National Christian Union," perfected plans of organization intended to reach every hamlet, town and city in the land, and issued addresses to the women of the country, the girls of America, and to women across the sea. During the first year of this organized life they added six State organizations to their numbers, and established a monthly paper, the " Woman's Temperance Union." Messrs. Osgood, Murphy and Reynolds have been employed by them in several States, and their work is prosecuted with great wisdom and zeal.*

---

* Centennial Temperance Volume, pp. 716–768; 687-704.

Growing out of the Woman's Crusade, was a general awakening of the Churches to a clearer apprehension of their duty, and a deeper sense of responsibility on this great subject. In May, 1874, representatives from several denominations in various parts of the country, assembled at Pittsburgh, Pennsylvania, and organized the "National Christian Temperance Alliance." Its object is, "To bring the influence of the whole Christian Church and all friends of humanity to bear directly and steadily against every part of this 'vile liquor system,' until the principles of total abstinence and prohibition shall universally prevail." In accomplishing this, it aims, "not to effect an organization outside and independent of the Church of Christ, but to organize and unite the Churches themselves in aggressive temperance work."* The latest reports, confessedly incomplete, show a membership, in 1885, of 70,360.

In Great Britain the work is carried on by ten National Total Abstinence Associations, and a great number of District Unions; by ten Religious Temperance Organizations; two Medical Associations; six Associations seeking to Advance the Cause by Legislative Action; three Temperance Insurance and Benefit Societies; and three Societies for Providing Substitutes for Drinking Houses and Indulgences.†

V. COFFEE HOUSES.—One other agency, suggested by this last mentioned effort, is worthy of separate notice, since it has been employed in many countries, and with uniformly good results. Coffee-houses, Friendly Inns, Holly Tree Inns, all uniform in their purpose to provide cheap, attractive and wholesome restaurants, where all classes, and particularly reformed men, can resort without being tempted by intoxicants, have been established as aids to the temperance work.

The first coffee-house of which we have any knowledge, was opened in Paris, in 1643, not in the interest of the

---

* Ibid, p. 750.   † Ibid, pp. 807, 8J8.

temperance cause, but as a novelty and an addition to the attractions furnished by other drinks. The first coffee-house in England, was at Oxford, opened by a Jew, in 1650. Two years later one was opened in London. For a long time there was a great prejudice against the use of coffee, and its odor was said to be a nuisance and unwholesome. A duty of 4d. was in 1660, laid on every gallon made and sold, and in 1663 it was directed by law that all coffee-houses should be licensed. In a broadside against coffee, published in 1652, it is said: " To cure drunkards it has got great fame." * Coffee-houses were opened in Vienna in 1683, in Augsburg in 1712, and in Stuttgart in 1713. Buxton, in his " How to Stop Drunkenness," says that there are at present "1400 coffee-houses in London." Of these, a late number of the " *Leeds Mercury* " says that " twenty-three are operated by The London Coffee Tavern Company. They are frequented by 14,000 to 15,000 customers per day, or upwards of 5,000,-000 per annum. The statistics for one week—the first in December last year—were 78,104 cups of coffee, tea, or cocoa, 3,256 pounds of meat, 5,656 basins of soup, and 10,153 loaves of bread."

In the United States, a nearly simultaneous effort for their establishment was made in 1874, wherever the Woman's Crusade extended its work. In the cities of Boston, Cleveland, Chicago, Baltimore, and Philadelphia, particularly in the last named, they have been successful. In Philadelphia there are two large establishments, at one of which 2,600 persons, and at the other 1,400 lunch daily. They are located on the chief business thoroughfares, are furnished with good and wholesome food, at a low price, and are attractively fitted up. " Quite a variety of nutritious and substantial dishes are provided, and each at the uniform price of *five cents*. The main feature—the coffee—is, however, preserved. A full pint mug of the best Java

* Club Life of London, by John Timbs, Vol. II. p. 1, 4, 15.

(equal to two ordinary cups), with pure rich milk and white sugar, and two ounces of either wheat or brown bread, all for *five cents,* is the everyday lunch of many a man who, but for this provision, would be found at the dram-shop." *

VI. INEBRIATE ASYLUMS.—Impressed with the conviction that drunkenness is often a disease, and is therefore to be treated from a physiological standpoint,—for reasons which will be obvious to those consulting pages 247–258,— "Asylums for the Cure of Inebriates" have been established in several States; but notwithstanding the good which they have accomplished in many cases, the experiments have been sadly interfered with and in many instances have ceased. The several asylums have been crippled, and in many cases have failed, says Dr. Willard Parker, in a letter to the writer, "Not because they had not true merit, nor because the idea upon which they were founded was unsound or untenable, but because the manufacturers, venders, and consumers of liquor were arrayed against them as a part of a Temperance Movement. All these were sustained by venal politicians, in their opposition; hence it was impossible from the want of effective legislation to control the inebriate, and shut him off from spirits *absolutely.* The asylum established by this City's Commissioners was a failure. It was opposed by a large class of voters, and the inmates could procure, and were gladly helped to liquors, to bring reproach upon the humane purpose."

VII. EDUCATION.—As a preventive measure, one of the wisest and most hopeful instrumentalities now being employed, is the education of the young in the Public Schools, Academies and Colleges, in the Nature and Effects of Alcoholic Beverages. Drs. Lees and Richardson, of England, have each prepared books for this purpose. The one, the "Text-Book of Temperance," the other, "The

---

* Centennial Temperance Volume, pp. 303, 304.

Temperance Lesson Book;" and "The National Temperance Society and Publication House," has published
a "Catechism on Alcohol," "Juvenile Temperance
Manual," "The Temperance School," and "Alcohol and
Hygiene," each prepared by Julia Colman. The necessity
for such instruction is obvious, and its results in the hands
of faithful teachers must be far-reaching and salutary.
Already some of these books are used in Public Schools in
various parts of Maine, New Hampshire, Vermont, Massachusetts, New York, Pennsylvania, Ohio, Indiana, Illinois,
and perhaps other States. How far these or kindred text-
books, or special instructions of any kind, on this subject,
are employed in the old world, the writer is not informed.
The following, from an address by Ex-bailie Lewis, of
Edinburgh, given during the year 1879, shows that some
attention is given to the subject in the schools of Sweden :

"In visiting the Swedish public schools, I was particularly
struck with the thorough manner in which physiology. was
taught to the children. I recollect going into one school in
Gothenburg, where there was a large number of scholars, and
the teacher said he would put any question to the scholars I
wished; and I pointed to a large physiological map, and asked
the teacher to put a few questions in regard to that map, and in
reply to the questions a young lad told correctly how butcher
meat and potatoes built up the physical system. I then put
the question, 'In what manner does brandevine or brandy
build up the human system?' and the young boy, with a look
of contempt at my ignorance, answered with a kind of smile,
'Brandy does not build up—it pulls down.' So that you see
we are much behind the educational authorities in Sweden."

VIII. LICENSE LAWS.—Law has in various ways interposed its authority to arrest the evil of intemperance. At
first it attempted to restrain by regulating the places and
the amount of sales. Nearly four hundred years ago,
(1495,) sureties were taken of alehouse-keepers, in England,
against the improper sale of intoxicants. Prior to this the
traffic was in no way interfered with, except by laws which
looked to the avoidance of adulterations and short meas-

ures. It was in this respect treated as ordinary articles of commerce were looked after, everything being in a sense subject to the crown, and yielding a revenue thereto. But at the date above given it was looked upon as a vicious business; and in this light it has been regarded by all subsequent legislation, and dealt with as an exceptional and dangerous trade. It was obvious that it led to immorality and to pauperism, for its second mention in law is in "an Acte against vacabounds and beggars," passed in 1504, wherein Justices of the Peace were authorized to reject and put away common ale selling in towns and places where they shall think convenient, and to take sureties of the keepers of ale houses, of their good behavior." In 1552 we meet with the first attempt to put a price on the license for the privilege of selling, in "An Acte for Kepers of Alehouses to be bounde by Recognizaunces." It is prefaced with a declaration that the reasons which lead to its enactment are: "Forasmuch as intolerable hurts and troubles to the Common Wealth of this Realm doth daily grow and increase through such abuses and disorders as are had and used in common Alehouses and other houses called Tippling-houses." Full power to determine how many such houses should be allowed in the cities, towns and shires, was given to the magistrates. They had absolute control of the trade, except in places where fairs were being held, then all were free to sell who might desire, but at other times only those who were licensed and entered into bonds " to maintain good order in their houses " could sell. For this privilege they "shall pay but twelve pence." The fine for every offence against their bond was put "at twenty shillings."

In 1553, another excise law was passed, the object of which was declared to be:

"For the avoiding of many inconveniences, much evil rule, and common resort of misruled persons used and frequented in many taverns of late, newly set up in very great number in back lanes, corners and suspicious places within the city of

London, and in divers other towns and villages within this Realm."

Under this law none could sell " wine by retail, except by the license of the corporate Magistrates or the Justices of the Peace at the general sessions ; " and except in London these authorities "could only allot two taverns to one town." This limitation was demanded wholly on account of the fact that the taverns had become mere resorts for drinking and the necessarily accompanying licentiousness and riot; a great perversion from their original intent, as is manifest in an Act passed in the reign of James II., which recites that:

"The ancient, true, and principal use of ale-houses was for the lodging of wayfaring people, and for the supply of the wants of such as are not able, by greater quantities to make their provisions of victuals, and not for entertainment and harboring of lewd and idle people, to spend their money and their time in a lewd and drunken manner."

It would seem that various degrees of zeal and wisdom, or the want of it, characterized the Justices to whom the workings of the license law were committed, for it is recorded that:

"Lord-keeper Egerton, in his charge to the Judges when going on circuit in 1602, instructed them to ascertain for the Queen's information 'how many ale-houses the justices of the peace had pulled down, so that the good justices might be rewarded, and the evil removed.'"

In almost every subsequent reign laws further regulating the sale of intoxicants were passed, some for limiting the traffic, and one or two for encouraging it; but all the former, without exception, were based on the ground that for the benefit of the people at large, some restrictions were necessary. Said the Lord-keeper :

"I account ale-houses and tippling-houses the greatest pests in the kingdom. I give it you in charge to take a course that none be permitted unless they be licensed; and for the licensed ale-houses let them be few, and in fit places; if they be in pri-

vate corners and ill-places, they become the den of thieves—
they are the public stages of drunkenness and disorder."

In 1606, a new enactment, with more stringent provisions for regulating the traffic, was created, the grounds for
it being as stated in the law, that:

"The loathsome and odious sin of drunkenness had of late
grown into common use within this realm; being the root and
foundation of many other enormous sins, as bloodshed, stabbing,
murder, swearing, fornication, adultery, and such like; to the
great dishonor of God and of our nation; the overthrow of many
good arts and manual trades; the disabling of divers workmen
and the general impoverishing of many good subjects: abusively wasting the good creatures of God."

A few years later, as little benefit resulted from this
law, owing to the various ways of evading it, it was
amended so as "to put it within the power of a justice of
the peace to convict upon the oath of one witness, or upon
his own personal observation."

But in spite of all regulations and restrictions the
traffic increased; the stronger intoxicants crowded the lighter
drinks aside, until, in 1736, there were over 7,000 houses
in London; an average of one house to every seven in the
place, where gin could be obtained at the lowest prices:
(as has been already related in the previous chapter;)
besides a large number of places where only fermented
liquors could be obtained. To remedy this evil, Parliament
passed the following Act:

"Whereas the excessive drinking of spirituous liquors by the
common people tends not only to the destruction of their health
and the debauching of their morals, but to the public ruin;"

"For remedy thereof—

"Be it enacted, that from December 29th no person shall
presume, by themselves or any others employed by them, to sell
or retail any brandy, rum, arrack, usquebaugh, geneva, aqua
vitæ, or any other distilled spirituous liquors, mixed or unmixed, in any less quantity than two gallons, without first taking out a license for that purpose within ten days at least before
they sell or retail the same; for which they shall pay down £50,
to be renewed ten days before the year expires, paying the like

sum, and in case of neglect to forfeit £100 ; such licenses to be taken out within the limits of the penny post at the chief office of excise, London, and at the next office of excise for the country. And be it enacted that for all such spirituous liquors as any retailers shall be possessed of on or after September 29th, 1736, there shall be paid a duty of 20s. per gallon, and so in proportion for a greater or lesser quantity, above all other duties charged on the same."

This law, known in history as the famous Gin Law, continued on the statute books for eight years, encountering great opposition to its enforcement, and finally becoming a dead letter. Many reasons might no doubt be stated as accounting for this; but it is very certain that, like the old moderation pledge, a large defect was found in its discrimination against intoxicants of a particular grade; while it allowed, if it did not encourage, the use of others which experience, and even previous legislation, had shown to be equally pernicious in their results. Certain it is that the law of 1753, entitled an "Act for regulating the number of public houses, and the more easy conviction of persons selling ale and strong liquors without a license," was an abandonment of such discrimination. This law was in force until 1828, when it gave way to one more elaborate, and containing what were supposed to be more stringent regulations.

In 1830, England entered on a new experiment, by means of which it hoped to diminish intemperance. Licenses, though continued on the sales of distilled spirits, were wholly removed from beer, and free beer shops were permitted without limit.

"The idea entertained at that time," says the London *Times,* "was that free trade in beer would gradually wean men from the temptations of the regular tavern, would promote the consumption of a wholesome national beverage in place of ardent spirits, would break down the monopoly of the old license-houses, and impart, in short, a better character to the whole trade ! . . . . The results of this experiment did not confirm the expectations of its promoters. The sale of beer was increased ; but the sale of spirituous liquors was not diminished." It had been in operation but a few weeks, when Sidney Smith wrote: "The new beer-bill has begun its operations. Everybody is drunk. Those who are not singing are sprawling. The sovereign people are in a beastly state."

In one year the number of beer shops increased 30,000, without any diminution of the spirit stores. In a short time the quantity of distilled liquors consumed was much larger than the gain in the consumption of beer. The official reports to parliament show that:

"During the ten years preceding the passage of the Beer-house Act, the quantity of malt used for brewing was 268,139,-389 bushels; during the ten years immediately succeeding the quantity was 344,143,550 bushels, showing an increase of 28 per cent. During the ten years 1821-1830 the quantity of British spirits consumed was 57,970,963 gallons, and during the next ten years it rose to 76,797,365 gallons—an increase of 32 per cent."

"The licenses for the sale of spirits—of which in 1830, 48,904 were granted—numbered in 1833, 50,828; being an increase of 1924. In Sheffield 300 beer-shops were added to the old complement of public-houses; and it is a striking fact that before the second year had transpired, uot less than 110 of the keepers of these houses had applied for *spirit licenses,* to satisfy the desire for ardent spirits."

There could be no more startling demonstration of the folly on which the Beer Act was based,—the expectation that "free beer," would diminish the demand for ardent spirits. Concerning these beer houses, Lord Brougham said, in 1839, in the House of Lords:

"To what good was it that the legislature should pass laws to punish crime, or that their Lordships should occupy themselves in finding out modes of improving the morals of the people by giving them education? What could be the use of sowing a little seed here and plucking up a weed there, if these beer-shops were to be continued that they might go on to sow the seeds of immorality broadcast over the land, germinating the most frightful produce that ever had been allowed to grow up in a civilized country, and, he was ashamed to add, under the fostering care of Parliament, and throwing its baleful influences over the whole community."

Subsequent reports made to Parliament, and to various Houses of Convocation in the several Ecclesiastical Provinces of England, have all shown that uniform testimony is borne to the fact that the beer-houses are in themselves the sources of poverty, immorality and crime; that they never diminish the demand for ardent spirits, but invariably increase that demand.*

The English Colonists brought the License System with them and incorporated it into their Colonial Laws when they settled in America. Perhaps the earliest attempt at legislation on this subject in this country, was a law against drunkenness, passed by the Plymouth Colony authorities, July 1, 1633: " That the person in whose house any were found or suffered to drink drunk be left to the arbitrary fine and punishment of the Governor and Council, according to the nature and circumstances of the same." †

The first mention of places wherein sales were allowed is in a law of the same colony, passed in 1636 : " That none be suffered to retail wine, strong water, or beer, either within doors or without, except in inns or victualling houses allowed." ‡

Ten years later, the Massachusetts Colony enacted:

"Forasmuch as drunkenness is a vice to be abhorred of all nations, especially of those who hold out and profess the Gospel of Christ Jesus, and seeing any strict law will not prevail unless the cause be taken away," ordered that "no merchant, cooper, or any other person whatever, shall sell any wine under one quarter cask, neither by quart, gallon, or any other measure, but only such taverns as are licensed to sell by the gallon." And it forbade "Any person licensed to sell strong waters, or any private housekeeper, to permit any person to sit drinking or tippling strong waters, wine or strong beer in their houses." §

---

* Teetotaller's Companion, pp. 21, 327. Samuelson, p. 162. Smith's Prize Essay, p. 202. Alcohol and the State. By Robert C. Pitman, LL.D., p. 266.

† Plymouth Colony Records, Vol. I. p. 13.

‡ Ibid, I. p. 31.

' § Massachusetts Colony Records, Vol. II. p. 171.

At Long Island, the people at East Hampton, alarmed at the progress of the evils of intemperance, passed an order at a town meeting held in 1651 :

" That no man shall sell any liquor but such as are deputed to by the town ; and such men shall not let youths, and such as are under other men's management, remain drinking at unreasonable hours ; and such persons shall not have above half-a-pint at a time among four men." *

In 1665, in Massachusetts Colony, and in 1667, in Plymouth Colony, cider is placed among the intoxicants not to be sold without a license. Attempts multiply as the population of the colonies increases, to restrain the evils of drinking by making more stringent regulations, the various laws having such prefaces as these : " Upon complaint of the great abuses that are daily committed by the retailers of strong waters, this Court doth order, etc." 1661. " To prevent the mischiefs and great disorders happening daily by the abuse of such houses, it is further enacted, etc." 1692.

"Whereas, divers persons that obtain license for the retailing of wine and strong liquors out of doors only, and not to be spent or drunk in their houses, do notwithstanding take upon them to give entertainments to persons to sit drinking and tippling there, and others who have no license at all are yet so hardy as to run upon the law, in adventuring to sell without, tending to the great increase of drunkenness and other debaucheries, &c." 1694.

A year later, a law is passed aimed against " divers ill-disposed persons, who the pains and penalties in the laws already made not regarding, are so hardy as to presume to sell and retail strong beer, ale, cider, sherry wine, rum or other strong liquors or mixed drinks ; " and sentences such to be punished at " the whipping post." In 1698, it becomes necessary to pass a law " For the Inspecting and Suppressing of Disorders in Licensed Houses ; " and in 1710, it was ordered that no person shall be licensed to sell liquors

---

* Centennial Temperance Volume, p. 422.

without a " certificate from the Selectmen of the town where they dwell, of their recommendation of them to be persons of sober conversation, suitably qualified and provided for such an employment." The same law also provided that

> " No town, except the maritime towns, shall have more than *one* inn-holder and *one* retailer at one and the same time, unless the Selectmen of the town shall judge there is *need* of more for the better accommodation of *travellers.*"

But this does not seem to have produced the desired result, for in 1811, there was passed " An Act against Intemperance, Immorality, and Profaneness, and for Reformation of Manners," in the preamble to which occurs the sad confession of the failure to regulate and restrain the evils of drinking and the greed of those who are licensed to sell :

> "For reclaiming the over great number of licensed houses, many of which are chiefly used for revelling and tippling, and become nurseries of intemperance and debaucheries, indulged by the masters or keepers of the same for the sake of gain."

And so on through the remainder of the colonial period, there are confessions of failure as more and more stringent regulations are adopted and then abandoned. John Adams writes in his Diary, in 1760 : " Few things, I believe, have deviated so far from the first design of their institution, are so fruitful of destructive evils, or so needful of a speedy regulation, as licensed houses." * What was true in Massachusetts was also true in all the Colonies. The traffic was licensed everywhere and licensed because it was confessedly an evil which it would not do to leave unrestrained ; and all restraints in the way of regulation, were so weak and inoperative, that the license laws were in a continual state of amendment and change. In Pennsylvania, although licenses were granted in 1710, only to those who were " first recommended by the Quarter Sessions to the Governor," it became necessary by additional legislation to protect "minors" against the greed of these honorably recommended men. Shortly after, the Grand Jury of Philadelphia County presented the houses kept by such persons as "a great nuisance," and represented "that there are upwards of a hundred houses licensed, which, with all the retailers, make the houses which sell drink nearly a tenth part of the city."

In 1763, the Governor was petitioned to make such additional regulations as would "prevent youth from committing excesses to their own ruin, the injury of their masters, and the affliction of their parents and friends;" and a little later there are loud and bitter complaints to the authorities,

"That the multiplication of inns, taverns and dram-shops is an obvious national evil, which calls loudly for legislative interference; in no country are they more numerous or more universally baneful."

Since the successful close of the War for Independence the License system has been in some period of their history, the polity of all the State Governments in dealing with the liquor traffic. But without exception, this has been done on the ground that the safety of community requires that the traffic shall be made difficult; and without exception, also, no regulation has made it sufficiently difficult to secure the desired safety. Over one hundred License Laws have been enacted in Massachusetts, and from 1682 to 1879, 342 statutes and changes have been made in Pennsylvania; and still the wisdom of legislators is vainly taxed to revise, and alter and amend. The same is true everywhere; no license law standing long on the statute books without being greatly modified in order to securing its greater efficiency. Why these numerous experiments fail, it will be more pertinent for us to show elsewhere; but that they have not yet done what it was expected they would, is confessed in all communities, and by all legislators.

"The Swedish Licensing Act of 1855," contains some unique features, and ought, therefore, to be briefly described here, before we pass to mention other proposed remedies for the evil of intemperance. Under that law the parochial authorities, or the town councils, fix, annually, the number of places where spirits may be sold at retail, subject to approval by the Governor of the Province. The licenses are of two classes, the one for shops, and the other for public-houses, including restaurants. The former pay for the privilege of selling in quantities of not less than half a kan (three-tenths of a gallon) not to be drunk on the premises, at the rate of eleven cents a gallon; the latter to sell in unlimited quantities, and to be drunk if desired, on the premises, pay seventeen cents per gallon. These licenses are sold by auction, for a term of three years, to those who offer to pay the required tax on the greatest number of kans, estimating beforehand what their sales may be, but not bound to pay for any excess of sales beyond the number actually stipulated to be paid for in their bids. The law also provided that, with certain guarantees, to be approved by the Governor of the Province, the authorities may, without an auction sale, dispose of the whole number of public-house or restaurant licenses, to any company that may organize for the purpose of attending to their distribution. Many of the parishes instruct their authorities to grant no licenses, and in consequence, in a population of three and a half millions of people, there are only 450 licensed places. In Gothenburg, the second city in Sweden, with a population of about 56,000, the authorities fixed on the number of licenses and sold them at auction. After pursuing this method for ten years they were confronted with such a condition of demoralization that the Town Council was impelled to appoint a committee to inquire into the causes of increasing degradation and pauperism. The chief cause they found to be intemperance; and the Council determined on seeking a remedy in this:

" That public-houses should no longer bo conducted by individuals for the sake of profit, but by an association, which should neither bring individual profit to the persons so associated, nor to tho persons who should manage the different establishments."

Such a company was soon organized, and avowed that tho following were its leading objects :

"*First.*—To reduce tho number of public houses.

"*Second.*—To improve their condition as to light, ventilation, cleanliness, etc.

"*Third.*—To make public-houses eating-houses, where warm, cooked food should be procurable at moderate prices.

"*Fourth.*—To refuse sale of spirits on credit or pledge.

"*Fifth.*—To employ as managers respectable persons who should derive no profit from the sale of spirits, but should be entitled to profits from the sale of food and other refreshments, including malt liquors.

"*Sixth.*—To secure strict supervision of all public-houses by inspectors of their own, in addition to the police.

"*Seventh.*—To pay to the town treasury all the net profits of sales of spirits."

This company went into operation in 1865. At once they extinguished one-third of the number of licenses, and at once there were evidences of an improved condition of things. But it was soon apparent that the relief was only temporary ; the " ugly statistics " of pauperism and crime soon showed that these fruits of the traffic were as prolific as before. The fact that the shop licenses were still under the control of the city authorities, was supposed to account for this in part ; and another cause was confessed at last to be found in the unlicensed and free beer shops. The shop licenses were, therefore, transferred to the company, who extinguished some of them, and transferred the remainder to private wine merchants, who, it was claimed, kept their stores " exclusively for the sale of the higher class of spirits and liquors not in ordinary use by the working classes." The 400 free-beer shops were supposed to be wiped out by a change in the law, " placing malt liquors under the same regulations as wine." While the licenses were sold at auction by the authorities, the city received annually about £7,000, or $35,000. In 1875, the Company paid to the city as the net profits on the traffic £35,000, or $175,000. In reply to a statement in the British House of Commons, that the Gothenburg system was not a success, but that drunkenness was on the increase in that city, a Gothenburg paper, *The Hændel's Tidning,* of March 20, 1877, pronounced the statement misleading ; but made the following confession that there was a gain in the amount of liquor consumed :

"The figures for the year October 1st, 1875, to October 1st, 1876, which we lately gave, show a total sale of brän-vin 614,-608 kans, of which on 'selling off shops' 357,445 ; therefore in public-houses 257,163 kans, or 11,000 more than the former year. The sale of spirits of higher class was 52,788 kans, or 1,000 more than last year."

This consumption of intoxicants is nearly six gallons per capita of the population of Gothenburg, and its fruits are manifest in the annual arrests for drunkenness of one in about twenty-six of the population.* In common with all License Laws of modern times, the Swedish law prohibits the sale of intoxicants on Sunday, and also sets a limit to the hours of evening business.

IX. PROHIBITORY LAWS.—As of other efforts to suppress the traffic in and use of intoxicants, so also may it be said of the prohibition of their sale, traces of it are to be found in very ancient times. Du Halde is authority for the following with regard to China : " Under the government of *Yu* or *Ta Yu,* 2207 B. C., an ingenious farmer invented wine from rice. The Emperor, seeing that its use was likely to be attended with evil consequences, expressly forbade the manufacture or drinking of it under the severest penalties ; and even renounced its use himself, and dismissed his cup bearer, lest the princes should be demoralized by it." †

---

* Pitman, p. 216-242.   † Annals of the Monarchs, Vol. I. p. 145.

In Manu's Institutes of Hindoo Law, Book IX. verse 225, is the following: " Sellers of spirituous liquors shall be classed with gamesters, revilers of scripture, etc., and shall be instantly banished from the town." And it is added, v. 226: " Those wretches, lurking like unseen thieves in the dominion of a prince, continually harass his good subjects with their vicious conduct." Picart * assigns as the reason for this prohibition :

" The high sense which the ancient Brahmins entertained for virtue, their strong aversion to anything which might disorder the senses and lead to irregularities. A drink that would extinguish reason must be pernicious, they said, and they felt obliged to inspire their people with similar sentiments."

Al. Henderson, speaking of the houses of entertainment in Rome, " in which all kinds of prepared liquors were sold," says that they became so obnoxious, that "In the reign of Claudius an edict was issued for their suppression." † Morewood, p. 156, says that "Drunkenness in Mysore (South India), from Tari, a liquor extracted from the wild palm tree, increased to such an extent that the Sultan Tippoo issued an order that all the trees be cut down." Partial Prohibition—the prohibition of the sale of ardent spirits—was vigorously maintained in Sweden, in 1753–1756, and again in 1772–1775. ‡

In the early history of America, special emergencies several times occasioned partial, if not absolute prohibition. Mr. W. Fraser Rae, in his recent work, entitled " Newfoundland to Manitoba," gives (p. 19) the following clause from the commission of King Charles I. for the government of the fishermen of Newfoundland, in 1630 :

" That no person do set up any tavern for selling wine, beer, or strong waters, cyder or tobacco to entertain the fishermen : because it is found that by such means they are debauched, neglecting their labour, and poor ill-governed men not only

---

* Religious Ceremonies, Vol. III. p. 274.
† History of Ancient and Modern Wines, p. 104.
‡ Alcohol and the State, p. 308.

spend most part of their *shares* before they come home, upon which the life and maintainance of their wives and children depend, but are likewise hurtful in divers other ways, as by neglecting and making themselves unfit for their labour, by purloining and stealing from their owners, and making unlawful shifts to supply their disorders, which disorders they frequently follow since these occasions have presented themselves."

In 1637, the General Court of Massachusetts made the following order : "In regard to the great abuse in ordinances, it is ordered that no ordinary keeper shall sell either sack or strong water."* In 1676, in a new Constitution of Virginia, " The sale of wines and ardent spirits was absolutely prohibited [if not in Jamestown, yet otherwise] throughout the whole country." †

As early as 1805, the Paper Makers Association of Philadelphia, before referred to, (see chapter II.) declared the principle of prohibition, in these words:

" The quantity of liquor drunk by those who have a propensity for it, will always bear some proportion to the facility of getting it. This fact is sufficiently proved by daily experience, and will refute that silly plea by which retailers attempt to justify themselves, viz.: ' If a man wants liquor he will have it, and if I don't sell it to him another will.' An argument that might as well be used to justify selling opium, or arsenic, to a lunatic." ‡

So in the "Address to the Churches and Congregations," in 1813, the fact is recognized that

" To the great and increasing numbers of taverns and dram-shops, may be traced many of the evils of intemperance. They are at once, causes and effects of these mischiefs. Their very existence proves that the thirst for ardent spirits is already insatiable ; and while they strongly indicate, they greatly increase the disease. . . . It cannot be *safe* to provide so many facilities for hard drinking." p. 22.

As the modern Temperance movement progressed, it was natural that the liquor traffic should appear to those

---

* Records, Vol. I.
† Centennial Temperance Vol. p. 422.
‡ Sampson Shorn, &c., pp. 25, 26.

who were trying to rescue its victims, as an immoral and dangerous business ; and that as this conviction deepened there should be a growing repugnance to its being sanctioned by law. This first made itself manifest by the withholding of licenses, and subsequently by the passage of stringent laws forbidding the sale of intoxicants. As early as 1829, the town of Harwich, Mass., instructed its selectmen not to grant licenses. At once the traders gave up the traffic, but unprincipled men re-opened it, until prosecuted by a committee appointed in town meeting, they abandoned the business. Subsequently other towns in Massachusetts, and several cities and towns in Maine, Vermont, Connecticut, Rhode Island, and other states, forbade the granting of licenses. The legislatures of Connecticut, Michigan and New York submitted the question of license to the popular vote of the people. In Connecticut 200 out of 220 towns elected Temperance Commissioners. In Michigan a majority of the towns voted no License. In New York more than five-sixths of the towns and cities gave overwhelming majorities against License.

In 1832, prohibition was advocated in the columns of "The Genius of Temperance," a weekly paper, and the "Temperance Agent," semi-weekly, both published in New York city. Many of the annual gatherings of the various Protestant churches, "proclaimed the immorality of the liquor traffic and its utter inconsistency with the spirit and requirements of the Christian religion." Near the close of the same year, General Cass, then Secretary of War, issued an order forbidding the introduction of ardent spirits into any fort, camp or garrison of the United States, and prohibiting their sale by any sutler to the troops. Eminent statesmen, jurists and divines gave utterance to their convictions of the immorality of the traffic. Said Rev. Dr. Humphrey, of Amherst College :

"It is plain to me as the sun in a clear summer sky that the license laws of our country constitute one of the main pillars on which the stupendous fabric of intemperance now rests." '

24

Said the Hon. F. Frelinghuysen: "If men will engage in
this destructive traffic, if they will stoop to degrade their reason and reap the wages of iniquity, let them no longer have the
law-book as a pillow, nor quiet conscience by the opiate of a
court-license."

Judge Pratt made the declaration: "The law which licenses
the sale of ardent spirits is an impediment to the temperance
reformation, and the time will come when dram-shops will be
indictable at common law as *public nuisances.*"

Said the Grand Jury of the city of New York, after
recording their deliberate judgment that if drinking were
at an end three-quarters of the crime and pauperism would
be prevented:

"It is our solemn impression that the time has now arrived
when our public authorities should no longer sanction the evil
complained of by granting licenses for the purpose of vending
ardent spirits, thereby legalizing the traffic at the expense of
our moral and physical power." *

The first National Temperance Convention of America
was held this year, and one result of its deliberations was
the avowal that the traffic in ardent spirits, to be used as a
beverage, is morally wrong, and ought to be universally
abandoned.   In 1834, Congress, in a law passed "For the
Protection of the Indian Tribes," prohibited the sale of all
strong liquors to the red men, and enforced its prohibition
by instructing the Indian agents to seize and destroy all
such liquors introduced for sale into the Indian territory.

The steps which led to the first attempts at prohibitory
legislation by State authority were taken by men whose
moral convictions were outraged, not simply by the liquor
traffic, but quite as violently by the theories of some professed Temperance advocates. In 1832, a State Temperance
Society was formed in Maine, on the then common basis of
Moderation.   As the cause progressed elsewhere and higher
ground than this was taken, the leaders in Maine grew
more and more conservative, and finally compromised more
fatally with wine drinkers.   As a protest against such vir-

* Centennial Temperance Volume, p. 450.

tual abandonment of the work, a new Society was organized on the basis of Total Abstinence, and one of the first acts of its leaders was to attempt to secure prohibitory legislation. They made their first appearance in the Legislature of that State in 1837, when they presented a Memorial, drawn up by Gen. James Appleton, of Portland. In this document they demanded, not only an abrogation of all license laws, " as the support of the traffic," but also " an entire prohibition of all sale, except for medicine and the arts, for the same reason that the State makes laws to " prevent the sale of unwholesome meats, or for the removal of anything which endangers the health and life of the citizen, or which threatens to subvert our civil rights or overthrow the government." This appeal failed to create a law, but it produced a discussion which paved the way for future success.

In 1844, a petition printed and circulated at the personal expense of Hon. Neal Dow, praying for a stringent law, and " that the traffic in intoxicating drinks might be held and adjudged as an infamous crime," was presented to the Legislature, and the committee to whom it was referred, " reported a bill favorable to Mr. Dow's views, which passed the House, but was unsuccessful in the Senate." The following year similar petitions met a like fate; and then it was determined to take an appeal directly to the people. A vigorous canvass followed, a Temperance Legislature was elected in 1846, 40,000 citizens petitioned for prohibition, " and a bill abolishing the license system, and leaving all sale forbidden, was passed by a vote of 81 to 42 in the House, and 23 to 5 in the Senate."

This success, and the advanced sentiment on Temperance in other localities, roused the determined opposition of the liquor dealers in several States. Three suits at law were commenced: Thurlow *vs.* the State, in Massachusetts; Fletcher *vs.* the State in Rhode Island; and Pierce *vs.* the State in New Hampshire. The lower Courts sustained the constitutionality of the laws in these respective States,

Alcohol in History.

whereupon appeal was taken to the Supreme Court of the United States. Six of the nine Judges were on the bench, and their decisions fully sustained the right of the State to regulate to any extent the sale of intoxicants. Said Chief-Justice Taney :

"Every State may regulate its own internal traffic, according to its own judgment, and upon its own views of the interest and well-being of its citizens. I am not aware that these principles have ever been questioned. If any State deems the retail and internal traffic in ardent spirits injurious to its citizens, and calculated to produce idleness, vice, or debauchery, I see nothing in the Constitution of the United States to prevent it from regulating and restraining the traffic, or from prohibiting it altogether, if it thinks proper."

"The law of New Hampshire is a valid law ; for although the gin sold was an import from another State, Congress already has the power to regulate such importations ; yet, as Congress has made no regulations on the subject, the traffic in the article may be lawfully regulated by the State as soon as it is landed in its territory, and a tax imposed upon it, or a license required, *or the sale prohibited*, according to the policy which the State may suppose to be its interest or its duty to pursue."

The opinions of the Associate Judges were in harmony with this. Said Judge Catron :

"If the State has the power to restrain by licenses to any extent, she has the discretionary power to judge of its limits, and may go to the extent of prohibiting altogether."

"It is not necessary," said Judge Grier, "to array the appalling statistics of misery, pauperism and crime, which have their origin in the use or abuse of ardent spirits. The police power, which is exclusively in the States, is alone competent to the correction of these great evils ; and all measures of restraint or prohibition necessary to effect the purpose, are within the scope of that authority. If a loss of revenue should accrue to the United States from a diminished consumption of ardent spirits, she will be the gainer a thousand fold in the health, wealth, and happiness of the people."

Judge Daniel, in reply to the argument that, because the importer had paid his duties, he had a right to sell which the State could not take from him, decided :

"No such right as the one supposed, is purchased by the importer, and no injury in any accurate sense is inflicted on him, by denying to him the power demanded. He has not purchased and cannot purchase from the government that which it could not insure to him, *a sale independently of the laws and policy of the State.*" Of all imports, he said: "They are like all other property of the citizen, and should be equally the subjects of domestic regulation and taxation, whether owned by an importer or his vendor, or may have been purchased by cargo, package, bale, piece, or yard, or by hogshead, cask, or bottles."

Judge M'Lean, on the right of the State to seize and destroy intoxicants, said:

"The acknowledged power of a State extends often to the destruction of property. A nuisance may be abated. It is the settled construction of every regulation of commerce, that no person can introduce into a community malignant diseases, or anything which contaminates its morals, or endangers its safety. Individuals in the enjoyment of their own rights must be careful not to injure the rights of others."

To the same effect, Judge Woodbury said:

"The laws seize the infected cargo and cast it overboard, not from any power which the State assumes to regulate commerce, or interfere with the regulations of Congress, but because police laws for the prevention of crime, and protection of the public welfare, must of necessity have free and full operation, according to the exigency that requires their interference."

The moral power of these decisions was manifest in the general advance of Temperance sentiment throughout the country.

In 1847 the Legislature of Delaware passed a Prohibitory Law, referring it to the people. Subsequently, on account of this reference to the people, the Law was set aside by the Court, as unconstitutional. In 1855, the Legislature created a new prohibitory enactment, which the Courts sustained. It was repealed and a license law substituted, in 1857.

In 1848, the New Hampshire Legislature submitted to the people to vote on the expediency of a Prohibitory Law. The vote throughout the State was light, but three-fourths of the votes cast were in favor of the proposed law. The following year, the Legislature enacted a Prohibitory Law. A still more stringent law was passed in 1855.

In 1849, Wisconsin enacted a law permitting no person to vend or retail spirituous liquors until he shall have given bonds to pay all damages the community or individuals may sustain by the traffic—

"To support all paupers, widows, and orphans, and pay the expenses of all civil and criminal prosecution growing out of or justly attributable to such traffic. A married woman may sue for damages done to her husband. and no suit shall be maintained for liquor bills."

The next year, an additional provision made it the duty of Supervisors to prosecute rumsellers in cases of pauperism and crime. In 1855, the Legislature passed a Prohibitory Law, but the Governor vetoed it. In 1872 stringent laws were enacted, which, in 1873, were so modified as to break their force.

In 1850, the people of Michigan put into their new Constitution, the following provision : "The Legislature shall not pass any act authorizing the grant of license for the sale of ardent spirits or other intoxicating liquors." In 1853, a Prohibitory Law was enacted, which was declared unconstitutional by the Supreme Court. In 1855 a new law was passed, which was repealed in 1875.

The Constitution of the State of Ohio, ratified by the people in 1851, contained the following provision: "No license to traffic in intoxicating liquors shall hereafter be granted in this State, but the General Assembly may, by law, provide against evils resulting therefrom." Based on this the so-called Adair Law was enacted in 1854, a law making owners or lessees of buildings rented for the sale of liquors, as also the sellers themselves, responsible for damages resulting from such sales. Subsequently this was so amended as to require that before such persons can be held responsible, they must first receive notice from the persons liable to be injured, not to sell or give liquor to the person liable to commit the injury when intoxicated. A law has also been passed which allows the sale of native wines and cider, and takes from corporations the power to prohibit ale, beer, and porter-houses.

The same year, 1851, a more thorough and perfect law was adopted in Maine. It was repealed in 1856, but re-enacted in 1857, and has from time to time been strengthened by amendments, as occasion has required.

In March, 1852, Minnesota, while yet a Territory, passed a Prohibitory Law, with a proviso for its ratification by the people, which was accomplished the same year; whereupon the Supreme Court decided that the submission of the Act to the vote of the people was unconstitutional.

In 1871, a law was passed prohibiting the sale of intoxicants near the line of the Northern Pacific Railroad during the construction thereof.

In May, 1852, the Legislature of Rhode Island passed a Prohibitory Law, which it made more stringent the following month, and still further perfected in 1853. It was, however, declared unconstitutional by the Supreme Court. In 1874 another Prohibitory Law was adopted, but was repealed in 1875.

In May, 1852, the Legislature of Massachusetts made Prohibition the Law in that Commonwealth. Some of the provisions of the law having been declared unconstitutional by the Supreme Judicial Court, in 1854, the law was thoroughly revised in 1855, and has since that time triumphantly stood the test of the sharpest judicial contests and criticism. In 1868 it was repealed, and a license law was substituted, but in 1869 the License law was repealed and the Prohibitory law was re-enacted, with malt liquors exempted. Changes in the modifications in favor of malt liquors, were made in 1870, but the law was repealed in 1873. In 1875 a license law supplanted all other legislation, with a proviso that cities and towns might refuse to grant licenses.

Vermont also passed a Prohibitory Law in 1852, which

in 1853, was ratified by the direct vote of the people. In 1880 the law has been made more stringent by the enactment of a Nuisance Act, the most significant sections of which are the following :

"*Sec.* 1. Every saloon, restaurant, grocery, cellar, shop, billiard-room, bar-room and every drinking-place or room used as a place of resort, where intoxicating liquor is unlawfully sold, furnished, or given away, or kept for selling, furnishing, or giving away unlawfully, and every place or room used or resorted to for gambling, shall be held to be a common nuisance, kept in violation of law.

"*Sec.* 2. When, upon trial, it is proved that intoxicating liquor is kept for unlawful sale, furnishing, or giving away, or is unlawfully sold, furnished, or given away in a place named in the preceding section, or that gambling is done in such place, the court shall adjudge such place to be a common nuisance, and the same shall be shut up and abated by the order of the court; and the person keeping the same shall be adjudged by the court guilty of keeping and maintaining a common nuisance, and shall be fined not less than twenty dollars, nor more than two hundred dollars, or he shall be liable to a fine not exceeding twenty dollars, and imprisonment not less than one month, nor more than three months, in the discretion of the court.

"*Sec.* 6. The State's attorney, when such a bond is forfeited, shall prosecute and recover the amount so forfeited on behalf of the State, and when such duty is neglected by the State's attorney for six months after being notified of such forfeiture, any other person may institute proceedings for such recovery in an action of debt in the name of the State, and such person, upon recovery and the payment of such amount into the State treasury, shall be allowed one-half the amount thereof.

"*Sec.* 8. A person who knowingly lets a building, tenement, place, or room, owned by him or under his control, for any of the purposes named in the first section of this act, or knowingly permits the same, or a part thereof, to be so used, shall be fined not less than twenty dollars nor more than two hundred dollars, or he shall be liable to a fine of twenty dollars, and imprisonment not less than one month and not more than three months."

In 1853, the Legislature of Connecticut passed a Prohibitory Law, which was vetoed by Governor Seymour; but the following year it enacted another, which was repealed in 1872.

In 1853, the Indiana Legislature passed a Prohibitory Law, with a provision that it be submitted to the people, a clause which the Supreme Court pronounced unconstitutional. In 1855 the Legislature passed another Law, but the conflicting opinions and weaknesses of its courts have rendered it inoperative. A license law took its place in 1874. In 1854, a Prohibitory Law was passed by the Legislature of New York, and was vetoed by Governor Horatio Seymour. The next year the law was re-enacted, but some of its provisions being pronounced unconstitutional by the Court of Appeals, the Legislature substituted a license law, in 1857.

In 1855, Illinois enacted a law, which, so far as it was applicable to dram-drinking was prohibitory. It allowed, however, the free manufacture of cider and wine, and their sale in quantities not less than five gallons. On its being submitted to the approval of the people, it was rejected.

Iowa also passed a Prohibitory Law, the same year which was ratified by the vote of the people. In 1858 fermented liquors were excluded from its prohibitions, and the law is thereby badly crippled.

In the Territory of Nebraska a Prohibitory Law was enacted in 1855.

As experience has shown that all statutes are liable to modifications and to repeal; that partizan zeal accepts, if it does not solicit, the influence of numbers, regardless of the price paid, and is so bent on immediate success as to be willing to make any compromise in order to secure it; and that Constitutional Amendments which merely deny to the Legislature the power to license the evil of the liquor traffic, are often powerless to prevent the free sale of the liquors which produce drunkenness;—efforts are now being made to place in the Constitutions of the several States, a provision absolutely prohibiting the manufacture of, and the traffic in intoxicants as a beverage.

The first of these attempts as yet crowned with success, was in Kansas, where in the November election of 1880, the following amendment to the Constitution, was adopted by the people · " The manufacture and sale of intoxicating liquors shall be forever prohibited in this State, except for medical, scientific, and mechanical purposes."

In 1881 the people of Iowa adopted the following amendment :

" No person shall manufacture for sale, sell, or keep for sale as a beverage any intoxicating liquors whatever, including ale, wine, and beer. The General Assembly shall, by law, prescribe regulations for the enforcement of the prohibitions herein contained, and shall thereby provide suitable penalties for violations of the provisions thereof."

In 1884 the State of Maine incorporated into its Constitution the following :

"The manufacture of intoxicating liquors, not including cider, and the sale and keeping for sale of intoxicating liquors, are, and shall be forever prohibited; except, however, that the sale and keeping for sale of such liquors for medical and mechanical purposes and the arts, and the sale and keeping for sale of cider may be permitted under such regulations as the Legislature may provide."

In 1885, the following became a part of the Constitution of Rhode Island :

" The manufacture and sale of intoxicating liquors to be used as a beverage shall be prohibited. The General Assembly shall provide by law for carrying this Article into effect."

Attempts in this direction are also being made in other States.

X. LOCAL OPTION LAWS.—In several States, where it has been found impossible to obtain prohibitory legislation, laws have been enacted, allowing the people of the several towns and cities, to determine by popular vote whether the sale of intoxicants as a beverage, shall be allowed or forbidden. In some instances this privilege of Local Option covers all parts of the State, while in others, particular localities are exempt from the operation of the law. The following is believed to be an accurate statement of the order and extent of such legislation.

The honor of inaugurating this form of relief is due to the State of Kansas, whose Legislature, passed a law in 1867, for the regulation and control of the liquor-traffic, in which was the provision :

"That no license should be granted to any individual to sell intoxicating liquors within the State until the party applying for the license, should present to the proper authorities a petition for the same, signed by   majority of the adult citizens, both male and female, of his district, or, if in a city, the ward in which he proposed to engage in the business."

In 1871, on the failure to obtain a general law from the Legislature of New Jersey, several townships had their petitions granted for special legislation giving this privilege to their respective localities. An additional number of towns obtained the same right in 1873.

In 1872, the Legislature of Pennsylvania passed a Local Option law, requiring the vote to be taken by cities and counties, and not by wards and townships. Special acts were also passed allowing the Thirteenth, Fourteenth, Fifteenth and Twenty-ninth Wards of Philadelphia to determine the matter for themselves, irrespective of the aggregate vote of the entire city. The following year, the liquor-dealers made violent efforts for the repeal of the law, but without effect; and on the bringing of a test case to the notice of the Supreme Court, the law was pronounced constitutional. Subsequently the law was repealed by the Legislature. .

The same year (1872), the New York Legislature passed a similar law, applicable to towns only, but it was vetoed by the Governor.

In 1873, action was had in the following States : The Maryland Legislature refused to grant the petition of the people for a general law, but enacted one for five counties, and a number of districts in others. A year or two later additional counties were included in the privilege of Local Option.

In Kentucky, a law was passed requiring an election to

be held in every district, town, or city, upon the application of twenty legal voters; and if a majority of votes be against the sale, then the traffic shall be unlawful. The first opportunity for the application of the law, resulted in the ordering of elections in 259 towns, 207 of which voted against the sale of intoxicants.

In North Carolina, a Local Option law was passed, which in 1874, was amended by a provision that, where prohibition has been carried by a vote of the people, it shall stand good until the liquor interest overturns it by calling an election and voting it down.

Mississippi enacted a statute containing this clause:

"That no license shall be granted or renewed unless signed by a majority of the male citizens over twenty-one years of age, and a majority of female citizens over eighteen years of age resident in the supervisor's district, incorporated city or town."

The Alabama Legislature refused to grant a law covering the whole State, but gave special laws to many of the towns. Subsequently the law was made applicable to several counties.

The Legislature of Tennessee passed a strong law, by large majorities in both Houses; but it was killed by the Governor's veto.

Indiana enacted a similar law, which was repealed a year or two later.

In 1874, the following States took action:

In Georgia, the Legislature passed a bill covering forty counties, in which the sale of liquors was prohibited unless two-thirds of the property holders agreed thereto in writing. At the same session Local Option was extended to thirteen counties, and to twenty-five smaller localities.

In Oregon a bill was passed prohibiting the sale of liquors unless a majority of the legal voters shall petition therefor.

In 1875, the Legislature of Massachusetts provided that the local authorities of towns and cities might give or withhold licenses. It also passed a Civil Damage law.

In the Dominion of Canada, after several unsuccessful efforts in former years, success was attained in passing a Local Option law, applicable to the two large provinces of Ontario and Quebec, in 1864. The law provides that:

" On the petition of thirty rate-payers the municipal council of any city, town or township is obliged to submit a by-law to the electors, asking them to vote either for or against a prohibitory liquor law for such city, town, or township, as the case may be. A by-law passed can only be repealed in the same way, and must remain in force at least one year."

A large number of townships, and several entire counties in both provinces, have established local prohibition under this law. In 1878, this law was amended by providing, among other things, that when the by-law is once adopted, " it cannot be repealed for a period of three years." The legality of the law having been questioned, the Supreme Court of Appeal have decided on its constitutionality.

In Great Britain a struggle has been going on since 1863, to induce Parliament to pass a Permissive Prohibitory Liquor Bill. The idea originated with the General Council of the United Kingdom Alliance, and being laid before Sir Wilfrid Lawson, was put into the form of a law, with the following preamble :

" *Whereas*, The sale of intoxicating liquors is a fruitful source of crime, immorality, pauperism, disease, insanity, and premature death, whereby not only the individuals who give way to drinking habits are plunged into misery, but grievous wrong is done to the persons and property of her Majesty's subjects at large, and the public rates and taxes are greatly augmented ; and *whereas*, it is right and expedient to confer on the rate-payers of cities, boroughs, parishes and townships the power to prohibit such common sale as aforesaid; be it therefore enacted," etc.

The bill then provides that, on application of any district, the vote of the ratepayers shall be taken as to the expediency of adopting the provisions of the act, but that two-thirds of the votes taken shall be necessary in order to make an affirmative decision. After many defeats, the bill was passed to its second reading in the summer of 1880, and strong hopes are entertained that it may soon become a law.

CHAPTER V.

The Right and Duty of the State to Prohibit the Liquor Traffic.
—Prohibition a Success.—Obstacles and Objections to Prohibition considered.—Conclusion.

THE power of the State to deal with the liquor traffic
by laws seeking to restrict and restrain it, has been
confessed,—as see the preceding sketch of the history of
License laws,—for more than four hundred years, by the
English-speaking people. More than this, the duty of
using that power has been acknowledged, and as the hundreds of License laws which have been enacted show, has
been performed according to the light and wisdom which
various forms of human government have possessed. But,
for the reasons announced by Sheldon Amos, in treating on Licensed Prostitution, all license of a confessed
evil has wrought mischief and has failed to secure the ends
which it was hoped that it would reach. Says Sheldon
Amos:

"The merits of legislation cannot be judged by the honest motive of its originators, but wholly by reference to the well-
known operation upon man's nature of causes with which all
are familiar. The whole system of regulating vice, by ascer-
taining the conditions it may alone be indulged in without in-
fringing police rules, gives a transparent legality or 'righteousness' to it, when so pursued, which no counter explanation nor
apologies can ever dissipate. It always seems to be forgotten
by those who advocate these systems that there are sufficiently
strong incentives to vice already existing, which it is the
hardest effort of civilization to counteract." "The licensing
system, in all its possible forms, gives public expression to
(382)

the fact that there are forms of licentiousness which are
in strict accordance with law. Because law is too impotent
to punish, there can be no reason why law makers should
go to the other extreme and protect and encourage." . .
. . . . . "In all other cases it is admitted that where
law cannot keep pace with the promptings of morality, it must
at the least, help, substantiate, and never contradict, common
moral maxims." . . . . "It is only in countries where a
system of licensing and regulating, prevails, that, while the
decrees of morality are held to be absolute in favor of virtue, the
decrees of law are equally decisive only when vice is practised outside certain arbitrary limits assigned. Within these limits a great State machinery, constructed at enormous cost, exists
for determining the persons for whom, and the places, the times,
and the conditions within which profligacy may be freely indulged in without risk of interference with law. Vice is anticipated, provided for, paid for, and hedged round with peculiar
securities by the State." . . . "Be it remembered, in the
licensing system there is no one feature which might gradually
work in favor of its own termination, and of the abolition of
immorality, and which must finally secure them. On the contrary, every feature tends to aggravate immorality, whether
with or without its attendant diseases, and to consolidate it
forever,—*i. e.*, as long as the nation can last." . . . "In the
place of the absolute immorality—comes the notion of its relative immorality only in certain places, at certain times, and
under certain conditions." . . . "It is as bad from many
points of view that a law should seem to ninety-nine persons
out of a hundred to be designed to favor immorality, as that it
should in fact favor it . . . . "You would regulate vice,
but it is of the essence of vice to refuse to be regulated. Vice
violates moral law, and you may expect it will transgress human
rules. It is like a mighty river that has overflown its banks. It
is a torrent whose fury you cannot arrest. You cannot say,
'Thus far shalt thou go and no farther.' It mocks at all your
regulations."*

These statements, written, as has been said, as the basis
of an argument against legalized prostitution, are equally
true, as experience sadly attests, against the legalized sale

---

* "Comparative Survey of Laws in Force for the Prohibition,
Regulation, and Licensing of Vice in England and other countries." Pp. 13, 14, 15, 39, 40, 100, 227, 242.

of intoxicants as a beverage ; for, beyond all dispute, the whole history of such traffic, is a history of waste, of shame, and of sin, with no redeeming feature whatever, to relieve this darkness and horror. Possibly, and very probably, alcohol may have a useful and remedial place as administered by skillful physicians in some cases of disease, but this is very different from its place and work as a beverage. The license system which provides for the sale of intoxicants at hotels, saloons, or groggeries of whatever names, is not a system devised or perpetuated for the benefit of the sick ; nor with any reference to them whatever; but it has its place and power wholly with reference to the depraved and perverted appetites of those who seek intoxicants as a beverage only. And in view of the inevitable consequences of such indulgences,—the poverty, crime, general demoralization and wreck of manhood—the state which legalizes such traffic, is, no matter what it may suppose itself to be doing, a guilty provider for and a direct participant in the consequences which flow from its licenses.

"But free sales," it is said, "ought not to be allowed; we should be overrun with drunkenness if that were the case ; besides, a traffic which causes pauperism and crime ought to be made to pay the expenses of its mischief, and hence the license fee is a fund to this end." The reply to this is the simple statement of the fact that, never, in the whole history of the traffic, has licensing diminished drunkenness ; and also, that the license fee never has furnished more than a drop in the bucket of expense which the licensed evil has filled. As compared with what the sober and industrious portion of community pay for the consequences of licensed liquor drinking, and what the licensed dealer pays for the privilege of selling, the difference is 33 per cent., against one per cent., an appalling result in a humanitarian point of view, a stupid blunder in the light of political economy.

It must also be said, that free rum, by which is meant

the toleration of the sale of intoxicants by law, is not a
positive act on the part of our law-makers, but simply a
neglect, a negative act; while toleration is overstepped the
moment the law is made to interfere by placing the traffic
under regulations, with the avowed intent of legalizing the
sales and protecting the seller from interference.   In such
a case, as Sheldon Amos well says:

"If the law is knowingly allowed to incline, on the whole, in
favor of an immoral sentiment, or if such an interpretation of
it is allowed to be so much as possible, it is.one of the most
heinous of moral offences of which a State or its rulers can be
guilty. The general moral sentiments of a people are dependent on a vast variety of subtle and incalculable influences;
their religion, their traditional customs and institutions, their
social habits, their historical antecedents, the amount and
character of their intercourse with foreigners, the dominant
speculative theories, and the prevalent educational enterprises,
all combine to create and enforce the moral sentiments of the
hour; while these sentiments themselves react powerfully upon
all those influences. But, no one of those influences is so om-
nipresent, so enduring, so persuasive, so directly authoritative,
as the voice of the State uttered either in its laws or its admin-
istrative acts. These laws and acts speak with a deliberateness
of purpose and a magniloquence of style which, while they
compel the attention of all, powerfully impress the imagination
in a way no other private or public utterance can." *

Said Rev. Albert Barnes, in his sermon on 'The Throne
of Iniquity:'

"An evil always becomes *worse* by being sustained by the law
of the land. It is much to have the sanction of law and the
moral force of law in favor of any course of human conduct.
In the estimation of many persons, to make a thing *legal* is to
make it morally *right*, and an employment which is legal is pursued by them with few rebukes of conscience, and with little
disturbance from any reference to a higher than human authority. Moreover, this fact does much to deter others from opposing the evil, and from endeavoring to turn the public indignation against it. It is an unwelcome thing for a good man ever
to set himself against the laws of the land, and to denounce
that as *wrong* which they affirm to be right."

---

* Ibid, p. 223.

25

" Soon after the enactment of the present license law in Massachusetts," says Judge Pitman, " I was holding a term of court, when a deputy sheriff said to me one morning : ' I have just seen a sad sight—a fellow persuading a reluctant comrade to enter a grog-shop.' ' Come along,' said he, ' this is now as respectable a place as any ; the Commonwealth of Massachusetts says so.' "*

In the light of all moral reasoning, the license of the sale of intoxicants as a beverage is an unmitigated curse, and the question whether the State has the right to prohibit such sale, can admit of but one answer : It is clearly the province of law to interfere with and forbid whatever jeopardizes the liberty, prosperity, and property of its subjects. Vice of every kind is in antagonism with these, and is the proper subject of legislation ; how much more, then, shall a traffic confessedly the most prolific of all things in promoting vice, be declared a crime. " Virtue," says Prof. Newman, " must come from *within;* to this problem religion and morality must direct themselves. But vice may come *from without;* to hinder this is the care of the statesman."

The following are accepted principles in the Science of Law.

" The true mission of Government is to regulate the equitable relations of men. This it does by protecting the weak against the strong, and by securing to each member of the community the undisturbed possession of his natural and civil rights. The only limitation to the rights of the individual is when he engages in any calling that interferes with the rights of others, or endangers their life or property."

Says Bentham : " The sole object of government ought to be the greatest happiness of the greatest number of the community. This end is promoted by encouraging every industry and institution calculated to confer benefit ; and discouraging, and even sternly repressing, those of a pernicious, immoral, and dangerous character; in a word, by such wise legislation as shall tend to promote the physical health, the social comfort, and the intellectual enjoyments of the people." †

---

* Alcohol and the State, p. 391.

† Our Nation's Peril, by Professor John Moffat, p. 86.

Lord Chesterfield, in the Debates in Parliament in 1743, on the Bill to reduce the Duties on Spirits, said :

" The specious pretence on which this bill is founded, and indeed the only pretence that deserves to be called specious, is the propriety of taxing vice ; but this maxim of government has on this occasion been either mistaken or perverted. Vice, my Lords, is not properly to be taxed, but suppressed; and heavy taxes are sometimes the only means by which that suppression can be attained. Luxury, my Lords, or the excess of that which is pernicious only by its excess, may very properly be taxed, that such excess, though not strictly unlawful, may be made more difficult. But the use of these things which are simply hurtful, hurtful in their own nature and in every degree, is to be prohibited. None, my Lords, ever heard in any nation of a tax on theft or adultery, because a tax implies a license granted for the use of that which is taxed, to all who shall be willing to pay for it. Drunkenness, my Lords, is universally and in all circumstances an evil, and therefore ought not to be taxed, but punished; and the means of it not to be made easy by a slight impost which none can feel, but to be removed out of the reach of the people, and secured by the heaviest taxes levied, with the utmost rigor. I hope those to whose care the religion of the nation is particularly consigned, will unanimously join with me in maintaining the necessity not of taxing vice, but suppressing it, and unite for the rejection of a bill by which the future as well as present happiness of thousands must be destroyed." *

In the same debate, the Bishop of Oxford said : " To leave the nation in its present state, which is allowed on all hands to be a state of corruption, seems to be the utmost ambition of one of the noble lords who have pleaded with the greatest warmth for this bill; for he concluded with an air of triumph by asking, how we can be censured for only suffering the nation to continue in its former state ? We may be, in my opinion, my Lords, censured as traitors to our trust and enemies of our country, if we permit any vice to prevail, when it is in our power to suppress it. We may be cursed, with justice, by posterity as the abettors of that debauchery by which poverty and disease shall be entailed upon them ; contemned in the present as the flatterers of those appetites which we ought to regulate, and insulted by that populace which we dare not oppose." †

* Gentleman's Magazine, Dec. 1743, p. 628.
† Ibid. Jan. 1744, p. 3.

This principle of the right and duty of law-makers to legislate against such a traffic, is also declared by Bishop, a high authority on criminal law :

" The State, in the enactment of its laws, must exercise its judgment concerning what acts tend to corrupt the public morals, impoverish the community, disturb the public repose, injure the public interest, or even impair the comfort of individual members over whom its protecting watch and care are required. And the power to judge of this question is necessarily reposed alone in the legislature, from whose decision no appeal can be taken, directly or indirectly, to any other department of the Government. When, therefore, the Legislature, with this exclusive authority, has exercised its right of judging concerning this legislative question, by the enactment of prohibitions like those discussed in this chapter, all other departments of the Government are bound by the decision, which no court has a jurisdiction to review." *

It was in perfect agreement with, and in defence of this principle, that the Judges of the Supreme Court gave their decision on the right of the State to prohibit the sale of intoxicants, as quoted in the previous chapter.

A still later decision, rendered by Chief-Justice Harrington, of Delaware, in the case of " The State *vs.* Allmond," not only affirms this principle, but also declares that it has never been judicially denied :

" We have seen no adjudged case which denies the power of a State, in the exercise of its sovereignty, to regulate the traffic in liquor for restraint as well as for revenue; and as a police measure, to restrict or *prohibit* the sale of liquor as injurious to public morals or dangerous to public peace. The subjection of private property, in the mode of its enjoyment, to the public good, and its subordination to general rights liable to be injured by its unrestricted use, is a principle lying at the foundations of government. It is a condition of the social state ; the price of its enjoyment ; entering into the very structure of organized society ; existing by necessity for its preservation, and recognized by the Constitution in the terms of its reservation as the right of acquiring and protecting reputation and property, and

---

* Statutory Crimes, sec. 995. Cited in Alcohol and the State. p. 103.

of attaining objects suitable to their condition without injury one to another.'" *

On this principle all Prohibitory Legislation has been
based, and the common-sense arguments in its favor are
irrefutable. Even Herbert Spencer, radically defective as
his definition of a State is, and shocking to all moral sense
and experience as is his declaration that " Government is
essentially immoral," declares that those who are " voluntarily associated " in a State, are so associated " for mutual
protection ; " a declaration, which,—imperfect as it is as a
full definition of the purpose of State government,—is sufficiently broad and explicit to refute his position that sani-
tary regulations are " a violation of rights ; " and that " the
State has no right to educate ; " † and also, in its suggestion
of the inquiry, " Protection " from what ? necessitates the
conclusion that it is the province of the State to prohibit a
traffic which puts in jeopardy the life, liberty and property
of every citizen.

And John Stuart Mill, in his Essay on " Liberty," giving
substantially the same definition of the purpose of Government as that given by Spencer, and objecting to all laws
" where the object of the interference is to make it impossible or difficult to obtain a particular commodity," (p. 185,)
and carrying his ideas of personal liberty to such an extreme
as to declare that, " Fornication, for example, must be tolerated, and so must gambling," (p. 191 ;) yet in his chapter
on the " Limits to the Authority of Society over the Individual," says :

"Law is to prevent infringement upon personal rights, and
to make each person bear his share of the labor and sacrifice
incurred for defending the society or its members from injury
or molestation. . . As soon as any part of a person's conduct
affects prejudicially the interests of others, society has jurisdiction over it. The question is then open whether the general
welfare will, or will not, be promoted by interfering with it.

---

* Cited in Alcohol and the State, p. 106.
† Social Statics, pp. 230, 303, 361, 406.

. . Whenever in short there is a definite damage, or risk of damage, to an individual or to the public, the case is taken out of the domain of liberty and placed in that of morality or law."

To be sure, he claims that there is no such " definite damage, or risk of damage," in the liquor traffic ; but in this he is contradicted by all the history of crime, and of pauperism. If the liquor traffic produces no " definite damage " to both the " individual " and the " public," it is absurd to attribute " damage " to any source ; as to ordinary observation it is obvious that " definite damage " from this source, to say nothing of indefinite, *i. e.*, incalculable, " not limited," " damage," is greater and more horrible than from any other source known to man. And the fact, that, for centuries this traffic has been placed in the " domain of morality and law," is conclusive proof that the judgment of mankind in regard to the damaging character of the traffic, is wholly opposed to Mr. Mill's unsupported assertion in regard to it.

In the preceding chapter a brief sketch is given of the history of Prohibitory Legislation, indicating not only the order, in point of time, of such laws in different localities, but also the fact that such legislation has often been repealed in toto, or so modified by amendments as to break its force. Such an experience leads many to conclude that the Prohibitory principle thus applied, is a failure ; and therefore, whatever may seem plausible, or even just and desirable in theory, cannot succeed when applied.

We now proceed to show that such a conclusion is not warranted by well attested facts.

I. Alexander Balfour, in a letter addressed to Mr. Gladstone, says of the operation of the Permissive Prohibitory Act operative in parts of Sweden :

" So vigorously have the people outside of towns used their permission to limit and prohibit, that among three and a half millions of people there are only 450 places for the sale of spirits. . . This it is which has so helped Sweden to emerge from moral and material prostration, and which explains the ex-

istence of such general indications in that country of comfort
and independence amongst all classes." (pp. 36-7.)

Of the still earlier attempts at Prohibition in Sweden, the
Chief of the Statistical Office in the Department of Justice,
wrote to the Massachusetts Board of Health, as appears in
their Second Report:

> " A vigorously maintained prohibition against spirits in 1753–
> 1756, and again in 1772-1775, proved the enormous benefits
> effected in moral, economical, and other effects, by abstinence
> from spirits."

In Great Britain, large land-owners have power to prohibit the traffic in intoxicants on their premises. Many use
this power, and the results are most definite and satisfactory. In the Province of Canterbury, having a population
of over 14,000,000, the Committee of the Lower House of
Convocation, reported, in 1869 :

> " Few, it may be believed, are cognizant of the fact—which
> has been elicited by the present inquiry—that there are at
> this time, within the Province of Canterbury, upwards of one
> thousand parishes in which there is neither public-house nor
> beer-shop, and where, in consequence of the absence of these
> inducements to crime and pauperism, according to the evidence
> now before the committee, the intelligence, morality, and com-
> fort of the people are such as the friends of temperance would
> have anticipated."

Of other sections of the Kingdom a writer in the Edinburgh *Review* for January, 1873, says:

> " We have seen a list of eighty-nine estates in England and
> Scotland where the drink-traffic has been altogether suppressed,
> with the very happiest social results. The late Lord Palmerston
> suppressed the beer shops in Romsey as the leases fell in. We
> know an estate which stretches for miles along the romantic
> shore of Loch Fyne, where no whiskey is allowed to be sold. The
> peasants and fishermen are flourishing. They all have their
> money in the bank, and they obtain higher wages than their
> neighbors when they go to sea."

At Low Moor, a settlement established by a large Cotton
Manufacturing firm, prohibition is rigidly enforced. One

of the members of the firm writes under date of February 27, 1871:

"We send some account of the community at Low Moor, which we are happy to say still remains without a beer-shop or a public-house. . . . It has neither stocks nor gaol nor lockup. We have a population of about 1,100. Our people can sleep with their doors open, and we have the finest fruit in the district, in season, in our mill windows, (which are never fastened) without any ever being stolen. Our death-rate is perhaps the lowest in the kingdom; taking the average of the last twelve years, it is under sixteen in the thousand."

A similar experience is known at Saltaire, a town belonging to Sir Titus Salt, Bart., and having about 5,000 inhabitants. All the workmen of the town are in the employ of the landholder, who, from a desire to promote the physical comfort and the moral well-being of the people, banished the liquor-traffic from the town. The best possible results followed, viz., an entire absence of drunkenness, crime and pauperism, and the positive blessings of health, comfort and cleanliness. After a time, some workmen brought from a distance, with a view of carrying out certain improvements in the town, raised such an outcry against being deprived of their customary beverage, that Mr. Salt authorized five grocers to sell small beer, not to be drunk on the premises. Such were the effects, insubordination among the men, and drunkenness among the women, that licenses were not renewed after the expiration of the first year. Of the present condition of the town, Mr. Hoyle, in his "Homes of the Working-classes," says:

"One thing there is which is not to be found in Saltaire, and Mr. Salt deserves as much praise for its absence as he does for anything which he has provided. Not a public-house or beerhouse is there. And what are the results ? Briefly these. There are scarcely ever any arrears of rent. Infant mortality is very low as compared with that of Bradford, from which place the majority of the hands have come. Illegitimate births are rare. The tone and self-respect of the work-people are much greater than that of factory-hands generally. Their wages are not high, but they enable them to secure more of the comforts and decencies of life than they could elsewhere, owing to the facilities placed within their reach, and the absence of drinking-houses."

Tyrone County, Ireland, is also under strict Prohibitory Law. Of it, Lord Claude Hamilton said, at a public meeting, in 1870 :

" I am here as representing the county, to assure you that the facts stated regarding the success of prohibition there are perfectly accurate. There is a district in that county of sixty-one square miles, inhabited by nearly ten thousand people, having three great roads communicating with market towns, in which there are no public houses, entirely owing to the self-action of the inhabitants. The result has been that whereas those highroads were in former times constant scenes of strife and drunkenness, necessitating the presence of a very considerable number of police to be located in the district, at present there is not a single policeman in that district, the poor-rates are half what they were before, and all the police and magistrates testify to the great absence of crime."

So also of Bessbrook, Ireland, a town of about 4,000 inhabitants. John Grubb Richardson, a member of the Society of Friends, is sole proprietor of the town. "The distinguishing feature of the town, is the absence of drink-shops, and consequently the absence of crime, pauperism, pawn-shops, and policemen."

In the United States the results have been none the less decisive and satisfactory.

MAINE has had the longest experience under Prohibition, and the success of the law is beyond all question. In his address to the Legislature, in 1874, Governor Dingley said :

" This system has had a trial of only twenty-two years; yet its success in this brief period has, on the whole, been so much greater than that of any other plan yet devised, that prohibition may be said to be accepted by a large majority of the people of this State as the proper policy towards drinking-houses and tippling-shops.

" Where our prohibitory laws have been well enforced, few will deny that they have accomplished great good. In more than three-fourths of the State, especially in the rural portions, public sentiment has secured such an enforcement of these laws that there are now in these districts few open bars; and even secret sales are so much reduced that drunkenness in the rural towns is comparatively rare."

And again, to the Legislature, in 1875, he said:

"The Attorney-General embodies in his report communications from the several county attorneys, furnishing important official statements and statistics relating to the enforcement of the laws prohibiting drinking-houses and tippling-shops. The statistics show that during the past year, in the Supreme Court alone, there have been 276 convictions, 41 commitments to jail, and $30,898 collected in fines under these laws—more of each than in any other year, and four times as many convictions and ten times as much in fines as in 1866, when the general enforcement of these laws was resumed after the close of the war, which had engrossed the public attention and energies. It is significant, also, that during these nine or ten years of gradually increasing efficiency in the enforcement of the laws against dram-shops, the number of convicts in the State Prison has fallen off more than one-fourth.

"The report of the Attorney-General and the statistics accompanying conclusively show that the laws prohibiting drinking-houses and tippling-shops have for the most part been enforced during the past year more generally and effectively than ever before, and with corresponding satisfactory results in the diminution of dram-shops and intemperance. These results are due, to a considerable extent, to the increased efficiency given to these laws by the sheriff-enforcement act, but more especially to the improved temperance sentiment which has been created by the moral efforts put forth in this State within a few years. It is gratifying to know that this sentiment has become so predominant as to secure the very general suppression of known dram-shops, and the consequent marked mitigation of the evils of intemperance in four-fifths of the State."

Subsequently he published the following:

"The recent amendments of the Maine law, prohibiting dram-shops, so as to increase the efficiency of its enforcement, are calling forth a shower of assaults on prohibition. So interested in our prohibition policy has become the whole country, that not only many of the Boston papers, but even the New York *Times* and *Tribune*, have joined in the cry that 'Prohibition is a failure in Maine.' In response to inquiries from all parts of the country, as to the truth of these allegations, we put on record the following significant facts:

" 1. The fact that our prohibitory system has stood in our statutes since 1851—with the exception of two years (1856 and 1857), when license was tried in its place—and has steadily increased in popularity until no party dares to go before the people on the issue of its repeal, is conclusive evidence that the great body of the people of Maine, who have had the best opportunity to judge of its practical workings, believe that it is the most efficient legal policy ever devised as a supplement of moral agencies in dealing with the evils arising from the use and sale of intoxicating liquors. No one expects that it can take the place of moral agencies, but that it is simply an adjunct of them—just as law, prohibiting houses of ill-fame and gambling resorts, are adjuncts to moral means in promoting virtue. No one claims that it can entirely extirpate the dram-shop evil, any more than the laws prohibiting and punishing theft or murder can uproot these crimes. All these laws aid in removing temptation and in creating a healthier public sentiment, and make it easier to do right and harder to do wrong.

" 2. Our prohibitory laws have unquestionably aided materially in creating a better public sentiment in the matter of the use of intoxicating liquors than exists in any State which has a license law. Whatever is prohibited by law, either directly or indirectly, is thereby deprived of a certain appearance of respectability which attaches to everything that is under legal protection. Mr. Raper, the distinguished Englishman, who spent some time in Maine and other parts of this country a few years ago, stated in a public speech that he did not believe there was in the whole civilized world a State of like population so free as ours from the evils of intemperance, or one possessed of so healthy a public sentiment in the matter of the use of liquors as a beverage.

" 3. Prohibition has stopped effectually the manufacture of distilled and fermented liquors in Maine. In 1830, when our population was less than two-thirds of what it is to-day, there were thirteen distilleries in this State, which manufactured over two gallons of rum to each inhabitant, nearly all of which was consumed in the State. To-day there is not a single distillery or brewery in Maine.

" 4. Prohibition has well-nigh stopped the traffic in intoxicating liquors in the rural districts of Maine. Forty-five years ago all the country taverns had open bars, and all the country stores sold intoxicating liquors as freely as molasses or calico. For example, the town of Durham, with less than 1,500 inhabitants, had in 1832 seven licensed grog-shops. To-day there is not a drop of liquor sold in town. Readfield had in 1832 seven open bars, at which were sold 2,300 gallons of spirits annually. Now none is sold to be used as a beverage. Minot (then including Auburn), with a population of 2,903 in 1833, had thirteen grog-shops. Now these towns, with a population of 10,000, have not a single place where liquor is known to be sold as a beverage.

" 5. Fifty years ago, even in our rural districts, nearly every male drank liquor. Liquors were kept in most of the houses to treat callers. Nobody thought of having company, or a raising, without a supply of ardent spirits. At musters and other public gatherings, drunkenness and drunken affrays were common. Now, three-fourths of the males in the rural districts are total abstinents, and the practice of keeping liquors in houses to treat callers has practically ceased. It would be considered an unpardonable offence to furnish spirits at a public meeting. At large public gatherings, cases of intoxication are surprisingly few, and drunken altercations rare. This improvement is strikingly shown by statistics. In 1833, Secretary Pond, of the Maine Temperance Association, reported that in the town of Alfred there were fifty-five men and three women accustomed to get beastly drunk; in Kennebunk, ninety-five notorious drunkards; in Topsham (population 1,564), forty drunkards; in New Gloucester, forty; Farmington, eighty; Wayne, thirty. Recent reports from these towns show that the present number of notorious drunkards in these and other towns is not one-eighth, and many towns say not one-tenth, of what it was forty years age. The reports also show a marked improvement in the condition of the people.

" 6. In the cities and larger villages, representing less than one-fourth of the population of Maine, the improvement is less marked than in the rural districts, although undeniably real, even there. There are three reasons for this less marked improvement : the greater facilities that vice has to hide itself in crowded populations; the concentration there of a large foreign population, which has come into this State within thirty years; and the resort to the city of the drinking men, still left in the rural regions, for supplies of liquor. If the cities had simply held their own under these circumstances, it would be a great gain. But they have done more than this. As a rule, there are no open dram-shops even here. Occasionally, through the failure to elect both city officials and county sheriffs friendly to prohibition, the law is neglected and open dram-shops appear. That has been the case during the past year in two or three cities whose condition is being quoted to the exclusion of the large part of the State where the law is well enforced. But generally speaking, even in the cities intoxicating liquors are sold only surreptitiously, and are to be found only by those who know the signs and pass-words of the liquor fraternity—and then, mainly, in places kept by foreigners. There are few open dram-shops to tempt. In the cities of Lewiston and Auburn, with a population of nearly 20,000, there is not a single open dram-shop, and no hotel has even a secret bar. In the larger cities there are many cases of drunkenness, but a majority of them are of foreigners, who resort to the most desperate expedients to obtain a supply of liquor. As confirmed inebriates in the rural districts are obliged to resort to the cities to obtain their potations, it frequently happens that the police reports of a city like Portland show nearly all the cases of drunkenness for a populous county.

"7. The charge is frequently made that, so far as the cities are concerned, the traffic has been simply driven out of sight. Even if nothing more had been gained, it is something to banish the temptations of the dram-shop where only those seeking them will find them. It is also occasionally alleged that club-rooms, more dangerous than dram-shops, have taken the place of the latter. After careful inquiry, we cannot learn that club-rooms exist outside of a few cities in Maine, and even there, not so extensively as in many cities of similar size in license States. The new amendments to the prohibitory law will reach this attempt to evade its provisions, and soon serve to make drinking clubs scarce. Setting aside the large foreign population in our cities, it is conceded that the improvement in the drinking habits of the remainder is marked. This is especially so with the bone and muscle of the native population.

"8. It is difficult to obtain reliable statistics of the extent to which the surreptitious sale of liquors is still carried on in Maine. Some of the enemies of prohibition claim that a million and a quarter dollars' worth are sold here annually. But allowing even this, and we have $2 per inhabitant now, against $25 per inhabitant forty years ago, and $16 per inhabitant as the average for the Union to-day. This shows that not more than one-tenth as much liquor, proportionally, is consumed in Maine as there was forty years ago, and not more than one-eighth as much as in the country at large to-day.

"On this point the revenue collected by the United States, from the manufacture and sale of intoxicating liquors in Maine, in comparison with that collected in license States, sheds some light. Prohibitory Maine has about the same population as license New Jersey; yet the liquor tax in the former State is only three cents per inhabitant, while in the latter State it is $2.40, and in the country at large $1.83. In reply to the assertion that tobacco and opium-eating are taking the place of liquor-drinking in Maine, we may mention that the tobacco tax paid by Maine is only seventeen cents per inhabitant, while the average for the country is $1 per inhabitant; and that opium-eating is far less prevalent here than in other Eastern States.

" 9. While it is undeniable that great temperance progress has been made in Maine with the help of our prohibitory policy, yet no one claims that either the sale or use of intoxicating liquors has been banished from our borders. They have been greatly limited, and the great body of the people recognize our prohibitory laws as essential aids in this good work. But much remains to be done. Experience is showing weak spots in our laws, and from time to time these are being strengthened. The recent amendments will increase the efficiency of the laws, and secure better results. In spite of jeers, in spite of opposition, in spite of declarations that the temperance cause is retrograding instead of advancing, the good work will go on in Maine, and year by year will show new triumphs in the great battle against King Alcohol."

And still more recently, he has published the following :

" In 1830, thirteen distilleries in the State manufactured *one million* gallons of rum (two gallons to each inhabitant,) together with 300,000 gallons imported—not including cider and other fermented liquors. Now there is not a distillery or brewery in the State. In 1833 there were 500 taverns, all but 40 of them having open bars. Now there is not a tavern in the State with an open bar, and not one in ten of them sells liquor secretly. In 1830 every store sold liquor as freely as molasses; now, not one.

" In 1832, with a population of only 450,000, there were 2,000 places where intoxicating liquors were sold—one grog-shop to every 225 of the population. Their sales amounted to $10,000,000 annually, or $20 for each inhabitant. Last year the aggregate sales of 100 town agencies was $100,000, or fifteen cents per inhabitant. Including clandestine sales, even the enemies of temperance do not claim that the aggregate sales in the State exceed $1,000,000, less that $2 per inhabitant. This is but *one-tenth* what the sales were forty years ago, and but *one-eighth* what they are on the average in the remainder of the Union, which is $16 per inhabitant. Liquor-selling is almost wholly confined to the five or six cities of the State, so that hard drinkers are compelled to journey thither for their drams. Hence most of the drunkenness of the State is concentrated in those cities where the police arrest all persons under the influence of strong drink, making the number of arrests for drunkenness seem large in comparison with places where few arrests are made for this offence.

" In 1855 there were 10,000 persons (one out of every forty-five of the population) accustomed to get beastly drunk ; there were 200 deaths from delirium tremens annually (equivalent to 300 now ;) there were 1,500 paupers (equivalent to 2,200 now) made thus by drink ; there were 300 convicts in State prison and jails (equivalent to 450 now;) and intemperance was destroying a large proportion of the homes throughout the State. Now not one in 300 of the population is a drunkard—not one-sixth as many ; the deaths from delirium tremens annually are not fifty ; and criminals and paupers (not including rumsellers) are largely reduced, notwithstanding the great influx of foreigners and tramps."

The Hon. W. P. Frye, member of Congress from the Lewiston district, and ex-Attorney-General of Maine, also (1872) writes: " I can and do, from my own personal observation, unhesitatingly affirm that the consumption of intoxicating liquors in Maine is not to-day one-fourth so great as it was twenty years ago ; that, in the country portions of the State, the sale and use have almost entirely ceased; that the law itself, under a vigorous enforcement of its provisions, has created a temperance sentiment which is marvellous, and to which opposition is powerless. In my opinion, our remarkable temperance reform of to-day is the legitimate child of the law."

The Hon. Lot M. Morrill, United States Senator from Maine, writes: "I have the honor unhesitatingly to concur in the opinions expressed in the foregoing by my colleague, Hon. Mr. Frye."

The Hon. J. G. Blaine, Speaker of the House of Representatives, writes: " I concur in the foregoing statements ; and on the point of the relative amount of liquor sold in Maine and in those States where a system of license prevails, I am very sure, from personal knowledge and observation, that the sales are immeasurably less in Maine."

The Hon. Hannibal Hamlin, United States Senator and ex-Vice-President of the United States, writes: "I concur in the statements made by Mr Frye. In the great good produced by the Prohibitory Liquor Law of Maine, no man can doubt who has seen its result. It has been of immense value."

The Hon. John A. Peters, the Hon. John Lynch, and the Hon. Eugene Hall, members of Congress from Maine, substantiate the foregoing testimony.

Mr. Frye further testifies :

"The 'Maine Law' has not been a failure in that, 1st, It has made rumselling a crime, so that only the lowest and most debased will now engage in it. 2d, The rum-buyer is a participator in a crime, and the large majority of moderate respectable drinkers have become abstainers. 3d, It has gradually created a public sentiment against both selling and drinking. 4th, In all of the country portions of the State, where, twenty years ago, there was a grocery or tavern at every four corners, and within a circuit of two miles unpainted houses, broken windows, neglected farms, poor school-houses, broken hearts and homes, it has banished almost every such grocery and tavern, and introduced peace, plenty, happiness, and prosperity. These two things, making the rum-traffic disgraceful both to seller and buyer, the renovating and reforming of all the country portion of the State, are the worthy and well-earned trophies of our Maine Liquor Law, and commend it to the prayers and good wishes of all good citizens. . . . Of this law I have been prosecuting attorney for ten years, and cheerfully bear witness to its efficiency, whenever and wherever faithfully administered. It has done more good than any law on our statute-book, and is still at work. With its provisions you can effectually close every liquor-shop outside your cities, and in them make the selling of ardent spirits a very dangerous and risky business. There cannot be found a man in Maine, who is not prejudiced by reason of being a seller, or drinker to excess, or by party passion, who will not concur with me in saying that its blessings have been incalculable, nor a respectable woman who does not pray for its continuance. Thus briefly I have given my testimony, and I know whereof I affirm."

Hon. Woodbury Davis, Judge of the Supreme Court of the State of Maine, thus replies to the charge that "the Maine Law is a failure : "

" So its opponents have often alleged. 'The wish is father to the thought.' So its friends sometimes have almost conceded. They have been too easily discouraged. They have hoped for results too large, and too soon ; and they have been disappointed. The law has not been a failure. It has already accomplished great results, though it has but just passed the ordeal of political agitation and judicial construction, in its struggle for permanent life. Every new system, though it may ride prosperously in its first success, is subject to the law of reaction. It must enter the lists, and conquer the place it would hold. The

Maine Law has been no exception. Even in Maine, as we shall see, its friends have been, and still are, compelled to spend much of their strength in wringing from its enemies amendments needed for its success, instead of giving their time for its enforcement. Much has been done in this respect since the law was originally enacted; but some things remain yet to be done. The period of growth is not the time for fruit, especially when the whole country has been swept by the storm of civil strife. That as much has been accomplished as ought to have been expected, an examination of the circumstances will show."

"The Maine Law, in its prohibitory form, but without the search and seizure clauses, was first enacted in this State in 1846. This first law was extensively enforced; and it prepared the way for that of 1851. Before that time, the old Temperance reform, and the Washingtonian movement, had each successively reached its climax. And, notwithstanding all the good that was done in reforming the habits of the people, there were still large numbers accustomed to use intoxicating liquors; and there was really no legal restraint upon the sale. It was permitted in almost every town; nearly every tavern, in country and in city, had its "bar," at almost every village and "corner" was a grog-shop; and, in most places of that kind, more than one, where old men and young, spent their earnings in dissipation; men helplessly drunk in the streets, and by the wayside, were a common sight; and at elections, at military trainings and musters, and at other public gatherings, there were scenes of debauchery and riot enough to make one ashamed of his race.

"What has become of this mass of corruption and disgusting vice? It seems so much like some horrid dream of the past that we can hardly realize that it was real and visible until twenty years ago. The Maine Law has swept it away forever. In some of our cities something of the kind may still be seen. But in three-fourths of the towns in this State such scenes would now no more be tolerated than would the revolting orgies of savages. A stranger may pass through, stop at a hotel in each city, walk the streets in some of them, and go away with the belief that our law is a failure. But no observing man who has lived in the State for twenty years, and has had an opportunity to know the facts, can doubt that the Maine Law has produced a hundred times more visible improvement in the character, condition, and prosperity of our people than any other law that was ever enacted."

Hon. Neal Dow, in a speech made in July, 1875, said:
"They say the Maine Law has failed, even in Maine. Now,
26

Mr. President, ladies and gentlemen, there is not a word of truth in that; it is all false from beginning to end. The Maine Law has not failed, directly or indirectly. Is there not any liquor sold in Maine or in any of the other Maine-Law States? Yes, there is; but you do not infer, therefore, that it is a failure. If you can show that there is as much liquor sold in proportion to the population with the same effect as there was before the Maine Law, that would show the law to be a failure. But in the State of Maine there is not one-tenth part as much of liquor sold as there was before the Maine Law. The whole character of the population is changed, as the result of that law. There is liquor sold in Maine, but only secretly. I live in the largest town in Maine, and you see no sign of liquor-selling anywhere at all. If one went into a hotel and asked for a glass of liquor, I do not know but that a person who knew the ropes might get it. They declare, however, that they honestly keep the law, and apparently they do. Wherever liquor is suspected of being kept with intent to sell in violation of law, the officers search for it and seize it. Every two or three days we have some seizure, but usually in very small quantities—a quart, a gallon, and sometimes only the bottle from the pocket of a man who intends to sell that way.

"I remember the time when there were seven distilleries in Portland, running night and day; at the same time vast quantities of liquor were imported, especially in the ship *Margaret*, one of the most famous ships in New England, whose cargo of St. Croix rum was spread out upon the wharves. How is it now? We have not a distillery running in all the State of Maine, nor is there a puncheon of rum imported. I should be warranted in saying that there is not one-fiftieth part of the quantity of liquor sold now as was sold previous to the passage of the prohibitory law, but I will say one-tenth. Senators and representatives in Congress, judges of courts, ministers and merchants, have signed certificates which were sent to England, in which they say the quantity of liquor sold is not one-tenth so great as was sold before."

So, writing to the "*Advance*," in the fall of 1880, Mr. Dow says:

"The evasions of the law are confined almost entirely to the cities and larger towns, which contain a considerable foreign population. The secret rum-shops that exist more or less in these places, are kept almost exclusively by those people. These shops continue in their illicit trade for lack of a few additions to the law, which we shall obtain by-and-by. The distilleries, breweries, and wine factories are all suppressed; there is not one remaining in the State. There were many of them formerly. In the smaller towns, villages, and rural districts of the State, the liquor traffic is quite unknown. Before the law it existed all over the State, on a large scale, wholesale and retail. The penalties of the law, as they now are, suffice to suppress entirely the rum-shops in small towns. One fine of a hundred dollars will use up a country rumseller; but in the larger cities it will require longer terms of jail to do it, in addition to the fines. The benefits resulting from the law are so great and manifest, that there is no organized or respectable opposition to it, in any quarter, or by any party."

MASSACHUSETTS, has reached similar results. In a report made to the Senate of that State, in 1865, one of the ablest lawyers in the Commonwealth said:

" This prohibitory statute, known in its earliest form as the Maine Law, is the fruit of much experience, avoids the practical difficulties discovered by hostile lawyers in the earlier statutes, is minute, thorough, and comprehensive, and is believed to be the only criminal law where the Legislature has provided forms of proceedings: in short, as those who have administered it have testified, it is as perfect as a criminal statute well can be."

Said the Constable of the Commonwealth: " Up to the 6th of November (1867), there was not an open bar known in the entire State, and the open retail liquor traffic had almost entirely ceased. The traffic, as such, had generally secluded itself to such an extent that it was no longer a public, open offence, and no longer an inviting temptation to the passer-by."

In a Circular issued by that official, in Oct. 1867, he said: " To us who are daily observers of the effects of these prosecutions, the fact is not to be winked at or argued out of sight, that very many of the liquor-dealers are utterly discouraged, and were it not for the hope that the approaching elections may afford them some relief, they would at once abandon the traffic."

This hope was based on the organization of all interested in the liquor traffic, and the lavish use of money in the effort to elect an anti-prohibitory Legislature. The effort succeeded, but prior to the election, an address to the People was put forth by Revs. Gilbert Haven, A. A.

Miner, E. P. Marvin, and Judge Pitman, containing facts in regard to the Prohibitory Law and its operation, which it was impossible to controvert. The Address said:

"The Prohibitory Law went into effect in all the towns of the Commonwealth. It was executed in every city except Charlestown and Boston. It received the approval of every legislature. It was carried up to our Supreme Court, and received the indorsement of Chief-Justice Shaw and his associates. It was attacked in Congress and in the Supreme Court of the United States, and both the national legislature and national court recognized its legality. While thus assailed by interested enemies, it was carrying blessings through all the Commonwealth. Three-fourths of our towns, including nearly every small village and most of our large towns, were without any public bars. Almost a generation has grown up in these places without beholding the open sale of intoxicating spirits.

"As a consequence of this law, pauperism and crime had greatly decreased in all localities where it was observed. In not a few of even our largest towns the alms-house had become an obsolete institution.

"The State constabulary was established, and has suppressed the sale of spirits in many of our cities, and greatly reduced it in the city of Boston. During the last two years it has closed hundreds of dram-shops. It has abolished more than twenty-five hundred open bars in this city. It has paid into the treasury of the State within the nine months ending Oct. 1st, 1867, in fines, $199,421 64 ; in value of liquors seized and sold, not less than $40,000—a sum amounting to one hundred and fifty thousand dollars more than all the expenses of the police."

The testimony of Major Jones, formerly chief of State Police, was:

"The law is as well enforced generally through the State as any other law ; but in Boston the liquor-sellers and dealers spend money freely and are well organized. There are about three hundred and sixty towns, and in three hundred of them the law is well enforced, and it exercises an influence upon the others."

General B. F. Butler said: "This law was enforced in all the cities and towns, with the exception of a few of the larger cities, as much and as generally as the laws against larceny."

Rev. Wm. M. Thayer, Secretary of the State Alliance, says:

"Previous to the November election of 1867, the State constables enforced the Prohibitory Law so thoroughly in Boston, that the tax on liquors at the internal revenue district No. 3, including most of the rumselling portion of the city, was reduced from $22,000 per month to $6,000. The month immediately succeeding the election, the receipts at the same office on liquors advanced again to nearly $22,000, showing that a great incubus was lifted from the traffic by the license triumph.

"The positive testimony of over 250 towns in the State visited since last November by myself or some of the agents of the Alliance, is to the very marked increase of intemperance. Even the smaller and more retired rural districts have not escaped the direful consequences of 'free rum.' In one small town, situated five miles from the nearest railroad station, on a Saturday night just previous to our visit, eleven intoxicated men were counted upon the principal street. The oldest inhabitant remembers no such scene of intoxication as that.

"In the county of Suffolk, on the 1st of September, 1867, there were less than 900 places where liquor was sold, and most of these clandestinely. On the first day of September, 1868, nearly 2,500 liquor shops were opened on the same territory, a fact which proves how utterly false was the plea of the license advocates, that their object was to diminish the traffic."

Oliver Ames and Son, one of the largest business firms in Massachusetts, said:

"We have over 400 men in our works here. We find that the present License Law has a very bad effect among our employees. We find on comparing our production in May and June of this year (1868), with that of the corresponding months of last year (1867), that in 1867, with 375, we produced eight (8) per cent. more goods than we did in the same months in 1868 with 400 men. We attribute this large falling off entirely to the repeal of the Prohibitory Law and the large increase in the use of intoxicating drinks among our men in consequence."

Governor Claflin, in his message to the Legislature, January, 1869, said:

"The increase of drunkenness and crime during the last six months, as compared with the same period of 1867, is very marked and decisive as to the operation of the law. The State prisons, jails, and houses of correction are being rapidly filled, and will soon require enlarged accommodation if the commitments continue to increase as they have since the present law went into force."

The Chaplain of the State Prison, in his Annual Report for 1868, says:

"The prison never has been so full as at the present time. If the rapidly increasing tide of intemperance, so greatly swollen by the present wretched license law, is suffered to rush on unchecked, there will be a fearful increase of crime, and the State must soon extend the limits of the prison, or erect another."

The Chief Constable of the Commonwealth, in his Annual Report for 1869, said:

"This law has opened and legalized in the various cities and towns about two thousand five hundred open bars; and over one thousand other places where liquors are presumed not to be sold by the glass."

The Legislature re-enacted the Prohibitory Law; but in 1870 it made a fatal exception in favor of the sale of beer; but returned to the policy of entire prohibition in 1873. In his Report, January, 1874, the Chief of State Police said:

"The law is only partially enforced, but in one-half the towns it has entirely suppressed the sale. There are five hundred less places in Boston for the sale of liquor than there were two years ago."

The District Attorney for Suffolk County bore witness that:

"The law is enforced generally throughout the State in the country towns, and with good effect. The shutting up of the open bar is certainly productive of a great reduction in drinking."

Judge Pitman, writing on the practical working of the law in New Bedford, and showing from official reports, a decrease of 37 per cent. in cases of drunkenness under prohibition, and an increase of 140 per cent. in cases of drunkenness when license prevailed, deduced the following conclusions:

"*First.* It has been fully demonstrated that the prohibitory law can be enforced to the same extent as other criminal laws.

"*Second.* That such enforcement would be productive of the

diminution of crime in general, and the promotion of peace and good order in our communities.

"*Third.* That this can be effected by *electing men to do it*, and in no other way.

"*Fourth.* That to allow the sale of malt liquors is a complete surrender of the battle, and opens the door to all the evils of a free liquor traffic."

Of the working of the Prohibitory Law in VERMONT, Governor Peck, also Judge of the Supreme Court of that State, says:

"In some parts of the State there has been a laxity in enforcing it, but in other parts of the State it has been thoroughly enforced, and there it has driven the traffic out. I think the influence of the law has been salutary in diminishing drunkenness and disorders arising therefrom, and also crimes generally. You cannot change the habits of a people momentarily. The law has had an effect upon our customs, and has done away with that of treating and promiscuous drinking. The law has been *aided by moral means*, but moral means have also been wonderfully strengthened by the law.

"I think the law is educating the people, and that a much larger number now support it than when it was adopted; in fact, the opposition is dying out. All the changes in the law have been in the direction of greater stringency. In attending court for ten years, I do not remember to have seen a drunken man."

Governor Convers said:

"The prohibitory law has been in force about twenty-two years. The enforcement has been uniform in the State since its enactment, and I consider it a very desirable law. I think the law itself educates and advances public sentiment in favor of temperance. There is no question about the decrease in the consumption of liquor. I speak from personal knowledge, having always lived in the State. I live in Woodstock, sixty miles from here, and there no man having the least regard for himself would admit selling rum, even though no penalty attached to it."

In RHODE ISLAND, Governor Howard, addressing a Temperance Convention, said:

"I am here to-night especially for the purpose of saying, not from the standpoint of a temperance man; but as a public man, with a full sense of the responsibility which attaches to me from my representative position, that to-day the prohibitory laws of this State, if not a complete success, are a success beyond the fondest anticipation of any friend of temperance, in my opinion.

"Ladies and gentlemen, prohibitory legislation in Rhode Island is a success to a marvellous extent. I have desired, I have felt it incumbent upon me to make that declaration, and I desire that it shall go abroad as my solemn assertion."

The Providence *Journal* said, just after the law went into effect:

"Whatever may be the ultimate results of the prohibitory and constabulary acts, it cannot be denied that up to this time their working has been rather salutary. There may be as much liquor drunk in private club-rooms and other out-of-the-way places as formerly, but if it is so the dealers are clearly taking pains to keep their workmanship out of sight. There has not been for years such an exemption from the indecencies of intoxication in our streets and the highways of our villages, as we have enjoyed for the last two months."

Similar results were also noticeable in CONNECTICUT. Governor Dutton said of the law, after it had been in operation a few months:

"The law has been thoroughly executed, with much less difficulty and opposition than was expected. In no instance has a seizure produced any general excitement. Resistance to the law would be unpopular, and it has been found in vain to set it at defiance."

In 1855, in his annual message to the General Assembly, Governor Dutton said:

"There is scarcely an open grog-shop in the State, the jails are fast becoming tenantless, and a delightful air of security is everywhere enjoyed."

Governor Miller, in 1856, said:

"From my own knowledge, and from information from all parts of the State, I have reason to believe that the law has been enforced, and the daily traffic in liquors has been broken up and abandoned."

Rev. Dr. Bacon of New Haven, after the law had been in operation one year, said:

"The operation of the Prohibitory Law for one year is a matter of observation to all the inhabitants. Its effect in promoting peace, order, quiet, and general prosperity, no man can deny. *Never for twenty years has our city been so quiet as under its action.* It is no longer simply a question of temperance, but a governmental question—one of legislative foresight and morality."

Edwards County, Illinois, decided twenty-five years ago, that no liquor should be sold in their territory. Recently the Circuit Court says :

"There has not been a licensed saloon in this county for over twenty-five years. During that time our jail has not averaged one occupant. This county never sent but one person to the penitentiary, and that man was sent up for killing his wife while drunk on whiskey obtained from a licensed saloon in another county. We have very few paupers in our poor-house—sometimes only three or four. Our taxes are about 35 per cent. lower than they are in adjoining counties where saloons are licensed. Our people are prosperous, peaceable and sober, there being very little drinking except near Grayville, a license town of White county, near our border. The different terms of our circuit court occupy three or four days each year, and the dockets are cleared. Our people are so well satisfied with the present state of things that a very large majority would bitterly oppose an effort made toward license under any circumstances."

In Tennessee the laws prohibit the establishment of the liquor trade within " four miles of chartered educational institutions which are not located in incorporated towns." Desiring to rid themselves of the curse of the traffic, the town of Tazewell had their charter abolished, that they might come under the beneficent operation of this law. A correspondent of the Morristown *Gazette*, after mentioning among the immediate results, " a noticeable diminution of drunkenness, rowdyism, quarrelling, and profanity," adds :

"Men are improving their property; people—good citizens—are moving into town; citizens who have long endured this curse are encouraged to plan for the future, and invest accordingly. The people of the county are likewise encouraged; have hope that the moral debasement occasioned by this iniquitous traffic will soon be reformed, and that law and order will] again prevail."

Quite recently, Thomas Hughes, Esq., of England, find-ing the best interests of the colony which he is establishing in Tennessee, interfered with by the persistence of liquor-sellers in plying their business in the vicinity of his settle-ment, has availed himself of this law, established a char-tered school, and so delivered his people from the nuisance. The latest information in regard to Tennessee Laws is from the pen of Mr. A. A. Hubbard, of that State, and published in " The National Temperance Advocate," for January, 1881 :

" Some twelve years since the Legislature of Tennessee en-acted that license should not be granted for the sale of intoxi-cating drinks within six miles of any blast furnace in this State, which law has proved of great advantage to the numer-ous blast-furnaces within our borders. Again, five years ago, our Legislature further enacted that the sale of intoxicating liquors should not be licensed within four miles of any chartered academy in this State, and also that any of our common or district schools might charter as such by making due applica-tion to the Secretary of State, which application must be signed by at least five persons, who propose to become trustees of the said chartered academy; and the result is that very many of our counties are so honey-combed by chartered schools that there is no room left in which to set up a groggery ; and still the chartering goes on at a rapid rate, and if academies are a sure index of intelligence, we bid fair to soon outstrip the New England States. A very interesting feature of this business is that the greatest proportion of schools are being chartered in locali-ties where moonshining has been very common, and where even now very few men can be found who would sign a temperance pledge ; and when asked why they apply for the chartering of their schools, reply that it is for the protection of their children.

" Our Supreme Court has decided that whenever a school is chartered all licenses for the sale of liquors within four miles of the said school-houses at once become void; and as the penalty for a violation is one hundred dollars and three months in a work-house or jail, many a vender of the vile stuff at once removes to a portion of the county where there is room to sell for the re-mainder of the time for which he was licensed ; but very often the people of that locality will at once proceed to charter their school, and so, like Noah's dove, he finds no resting-place.

" Another feature of this law is that it is inoperative within the bounds of a chartered city or village, and in consequence of this, very many—I think between fifty and seventy—of our chartered villages, during the last session of the Legislature, surrendered their charters, and, as in many of them there were already chartered high-schools, the rumsellers beat a hasty retreat."

Potter County, Pennsylvania, has been under a Prohibitory law many years. Hon. John S. Mann, bears this testimony to the results of its operation:

" There it stands, a shield to all the youth of the county against the temptation to form drinking habits. Under its benign influence the number of tipplers is steadily decreasing, and fewer young men begin to drink than when licensed houses gave respectability to the habit. There are but few people who keep liquor in their houses for private use, and there is no indication that the number of them is increased since the traffic was prohibited. The law is as readily enforced as are the laws against gambling, licentiousness, and others of similar character.

"Its effect as regards crime is marked and conspicuous. *Our jail is without inmates, except the sheriff,* for more than half the time. When liquors were legally sold, there were always more or less prisoners in the jail."

The same is true of Caroline County, Maryland. Mr. Emerson, of Denton, says of its operation:

" There is not a drop of alcoholic stimulants sold in this county, and the contrast between the past and present is a wonder to those accustomed to behold the scenes of but a few years ago and now. Instead of wranglings, black eyes and bloody noses, enmity and strife, drunken brawls and midnight debauchery, we have a peaceful and quiet community here and throughout the entire county.

" At the late sitting of the grand jury for this county there was not a single case of assault and battery before them, nor a single complaint of a violation of the public peace. Our jail is without a tenant, and has been for the past six months. At the recent session of our circuit court, had it not been for the old business which had accumulated under the whiskey reign, the term would not have lasted three days. The operation of the law has wrought a complete revolution here, and it is the greatest boon ever conferred upon our people by legislative enactment. It is a rare sight now to see any one under the influence of strong drink. Before the operation of the law, it was almost an hourly occurrence to come in contact with some one in this bestial condition."

At Vineland, New Jersey, a colony was commenced by Mr. Landis, in 1861. From the first he determined that the territory should not be cursed by the liquor traffic. He says, in an article prepared for and published in *Fraser's Magazine* for January, 1875, that he is not a total abstinence man, nor did he impose this condition of refraining from the traffic, on the part of those who purchased his land, from any philanthropic principle, but wholly as a business operation, and " solely as it would affect the industrial success of his settlement." He had observed that the tavern was the consumer of the industry of its patrons, and the enemy of their homes. And as his own success " depended directly on the success of each individual who should buy a farm " from him, so whatever militated against that success, must be forbidden. He " had long perceived that there was no such thing as reaching the result by moral influence brought to bear on single individuals; that to benefit an entire community the law or regulation would have to extend to the entire community." " In the first place, I decided to theorize and reason with nobody. . . . I would make the fixed principles of my plans of improvement the subject of contract, to be signed and sealed."

This settlement numbers a population of 10,500, who have come to it from all parts of America, from Germany, France, England, Ireland, and Scotland. Mark the significant result of Prohibition among such a people. The Constable and Overseer of the Poor, in his Report for 1874, says :

" Though we have a population of ten thousand people, for the period of six months no settler or citizen of Vineland has received relief at my hands as overseer of the poor. Within seventy days there has been only one case, among what we call the floating population, at the expense of $4.

" During the entire year there has only been one indictment, and that a trifling case of battery among our colored population.

" So few are the fires in Vineland that we have no need of a fire department. There has been only one house burned in a year, and two slight fires, which were soon put out.

" We practically have no debt, and our taxes are only one per cent. on the valuation.

"The police expenses of Vineland amount to $75 a year, the sum paid to me; and our poor expenses a mere trifle.

"I ascribe this remarkable state of things, so nearly approaching the golden age, to the industry of our people and the absence of King Alcohol."

A more recent colony, founded upon temperance principles, with a perpetual proviso against liquor traffic, is Greeley, Colorado. Like Vineland, it has a miscellaneous population, about 3,000, and is rapidly increasing in numbers. Efforts have from time to time been made to introduce the sale of alcoholic beverages, but with little success. Not long after the colony was founded, a fair was held, and the proceeds ($91) put into a fund for the poor. Two years and a half afterwards there still remained of this fund unappropriated and with no calls therefor, $84. Meanwhile, several churches, Presbyterian, Baptist, Methodist, and Episcopal, three schools, two banks, several extensive stores, two weekly journals and one monthly, and two literary societies, have been established, and are in a flourishing condition. N. C. Meeker, Esq., of the Greeley *Tribune*, projector of the colony, writes, Sept., 1873 :

" No liquor is sold in the town nor on the colony domain. A rum-shop was started the first year, and it was burned down in broad daylight. A few months ago one was opened five miles from town, and one night all the liquor was destroyed."

Similar testimony in regard to the success of Prohibition in these and other localities, might be almost indefinitely multiplied. It has never failed to diminish crime and pauperism, and has always secured and increased prosperity and comfort.*

---

* See The Prohibitionist's Text Book : "Alcohol and the State," and the "Argument of A. A. Miner, D. D., before the Committee of the Massachusetts Legislature, April 2, 1867."

II. The efficiency and success of Prohibitory Laws is further evident in the opposition of the various branches of the liquor traffic to them. "The United States Brewers' Association" was organized in 1862, "To foster and protect the trade from many threatening dangers." Their subsequent proceedings show that those dangers are chiefly two; viz.: Prohibitory Legislation, and heavy taxes levied by the General Government. The introduction to their Constitution, contains the following significant declarations:

"That the owners of breweries, separately, are unable to exercise a proper influence in the interest of the craft in the Legislature and public administration.

"That it appears especially necessary for the brewing trade that its interests be vigorously and energetically prosecuted before the legislative and executive departments, as this branch of business is of considerable political and financial importance, touching national interests generally, and exerting a direct as well as an indirect influence on political and social relations.

"Finally, that the truth, based upon the experience of all civilized nations generally, should be vindicated—that the use of fermented beverages prevents intemperance and promotes real temperance, and thus the manœuvres of the temperance party, which aims at the suppression of freedom of conscience and of trade, be defeated."

At the opening of the seventh session, in 1867, the presiding officer said:

"The Association is opposed by a dangerous foe, who, with a display of means and power, not only endeavors to hinder the development of our trade, but threatens utterly to destroy it. It will, therefore, be necessary to immediately come to some conclusion which will give evidence of the strength, energy and perseverance with which we follow our purpose, and thereby exert such a pressure upon our Legislatures, that an alteration or revocation of the obnoxious law may be expected."

The following resolution was also adopted:

"*Whereas,* The action and influence of the temperance party is in direct opposition to the principles of individual freedom and political equality, upon which our American Union is founded, therefore,

" *Resolved,* That we will use all means to stay the progress of this fanatical party, and to secure our individual rights as citizens, and that we will sustain no candidate, of whatever party, in any election, who is in any way disposed toward the total abstinence cause."

At the eighth session, 1868, they

"*Resolved,* That we will continue in the future, as we have in the past, to battle for the promotion of the cause of civil and religious liberty throughout the United States, that we will use all honorable means to deprive the political and puritanical temperance men of the power they have so long exercised in the councils of the political parties in this country, and that for that purpose we will support no candidate for any office who is identified with this illiberal and narrow-minded element.

"*Resolved,* That an effective organization of brewers and of their business friends should be maintained in every State and county, and that the same should act in concert with every other society and organization whose object is to uphold and promote the cause of civil and religious liberty, and that a committee of five be appointed for each State, with full power to organize local societies and call a State Convention whenever necessary.

" *Resolved,* That we will patronize and sustain all papers advocating the same views entertained by us, and that we will use our best exertions to bring to the notice of our enlightened American public the great advantage this country would derive from a settled governmental policy adopted in accordance with our views."

In 1869, they " reiterated and affirmed " the above, " as our standing creed and unchangeable purpose."

In 1871, they also " reiterate : "

" *Whereas,* Fanatics and religious hypocrites continue their open and avowed agitation for restrictive and sumptuary laws against the use of malt and fermented liquors as a beverage ;

" *Resolved,* That we reiterate," (etc., etc., as by the 8th Session, 1868.)

" *Resolved,* That in order to carry out the views and the objects expressed in the said preamble and resolutions, the Committee on Agitation is hereby authorized and directed to select in each Congressional District, three brewers, residing therein, as a Local and Provisional Organizing Committee for such districts, whose duty it shall be, upon accepting such appointment, to organize, by means best suited to the locality, as they may determine in their discretion, all the 'defenders of the rights of man, of the liberty of conscience and the inviolability of the guaranteed rights of person and property,' in order to defeat at all elections any candidate for office, whose success might give encouragement to temperance fanatics and religious hypocrites to carry out their proposed proscriptive, injurious and dangerous plans. These local-district organizations are farther requested to agitate the question of a liberal change in the laws of their locality for a proper licensing system, for adequate police regulations, for the better protection of our trade and all those in any way engaged in it, and for the establishment of the principle that the sale of beer as a beverage is as legitimate a trade as the sale of any other useful commodity of general consumption, and ought not to be subjected to any other restrictions than trade and commerce in general.

" *Resolved*, That all candidates for public office, of whatever political party, who accept these views as expressed and reiterated in these resolutions, and pledge themselves to adopt them for their rule of official action, whenever and wherever applicable, are hereby recommended to our members for their earnest support, and such candidates may firmly rely upon it."

In 1872, the Executive Committee reported :

" Many dangers threatening from the proposed enactment of laws to regulate the sale of intoxicating drinks, which have been attempted in nearly all the States of the Union, under the pretense of providing a safeguard for public morals.

" You should be well prepared to meet all attempts on the part of these temperance fanatics ; they strike at our trade and to undermine our manly dignity and influence.

" No trade, in view of its political power, is better calculated to exercise a marked influence on the elections than yours ; and it is your duty and a matter of self-defence to take a direct and active part in the political revolution and transformation of parties, so that in this direction, too, the desired reforms may be achieved."

In 1874, the President, in his address, said :

" Repeal your present laws—they are useless ; encourage and foster malt liquors and light wines, for they are the true medium of temperance.

" Urge upon your legislatures to abolish all prohibitory laws, and instead pass healthy license laws. Instead of condemning and prosecuting the saloon-keeper, punish the drunkards, refuse to recognize them as gentlemen, debar them from all society, disfranchise them at the polls. Condemn them to sweep the streets of your city with chain and ball fastened to their feet. Make drunkards criminals, but not the honest producers and purveyors of a necessity of life."

In 1875, although the President of the session declared that, " prohibition has failed, and will ever fail ; " and the chairman of the Agitation Committee, expressed himself certain that " the evils of intemperance could not be cured by prohibitory laws," the official reports showed a reduction in the number of breweries during the year of nearly thirty per cent. In 1873 there were 3,554 breweries, and in 1874, only 2,524. There was also a decrease of 30,194 barrels manufactured during the year. The cause of this reduction was confessed by Mr. Schade, in his address, as being the existence and operation of those laws which the President had said, have " failed, and will ever fail." Mr. Schade said :

" Very severe is the injury which the brewers have received in the so-called temperance States. The local-option law of Pennsylvania reduced the number of breweries in that State from 500 in 1873, to 346 in 1874, thus destroying 154 breweries in one year. In Michigan it is even worse ; for of 202 breweries in 1873, only 68 remained in 1874. In Ohio the crusaders destroyed 68 out of 296. In Indiana the Baxter law stopped 66 out of 158. In Maryland the breweries were reduced from 74 to 15, some few of those stopped lying in those counties in which they have a local-option law. We sincerely hope that the Maryland Democracy, which had yielded too much to the women crusaders, will take an early opportunity to eradicate that unjust law which permits the people of a portion of the State to be put under the tyranny and despotism of those fanatics."

"There is no doubt that the temperance agitation and prohibitory laws are the chief causes of the decrease compared to the preceding year. Had our friends in Massachusetts been free to carry on their business, and had not the State authorities constantly interfered with the latter, there is no doubt that instead of showing a decrease of 116,583 barrels in one year, they would have increased at the same rate as they did the preceding year."

The Agitating Committee also reported the repeal of the

27

Local Option Law of Pennsylvania, as a part of their successful work. The Association also:

"*Resolved*, That where restrictive and prohibitory enactments exist, every possible measure be taken to oppose, resist and repeal them."

So in 1877, the Executive Committee reported:

"Your Executive have closely watched and determinately tracked the efforts that are being made by the so-called temperance party in nearly every State in the Union; they have taken stringent measures to thwart many of the schemes they have adopted to hamper our trade, and in most cases have been successful; but we forbear entering into details which are duly recorded."

At the same session, Mr. McAvoy, delegate from Chicago, informed the Association: "The brewers of Illinois have expended $10,000 to beat the temperance party at the elections." Mr. Pabst, of Milwaukee, remarked:

"The brewers of Milwaukee have also expended a large amount of money to oppose the temperance party." And the President said: "Almost every local association has expended large amounts for this purpose."

So in 1878, Mr. Schade said:

"In my last year's report to you I urged the necessity of the government protecting its principal tax-payers against prohibitory State legislation. I showed that if section 3243, Revised Statutes, would be so amended as to deny the States the privileges therein contained to interfere with the constitutional right of Congress to 'lay and collect taxes, duties, imposts, and excises to pay the debts and provide for the common defence and welfare of the United States,' all your present troubles, caused by the fanatics, would cease at once."

Vice-President Lauer, a pioneer brewer, made an address, in which, though he proclaimed the "failure" of prohibitory legislation, urged—as a most singular comment on such pretended failure—the "greatest vigilance on the part of the National Brewers' Association to secure the necessary protection against the encroachments of fanaticism and misdirected temperance zeal."

And Mr. Lewis Schade, of Washington, thus referred to the efforts to secure a National Commission of Inquiry:

" For the last four years the temperance fanatics have, at the beginning of every session of Congress, introduced immense numbers of petitions from all parts of the country, every one of them asking for the appointment of a Commission of Five to investigate the liquor traffic. At the first glance one might suppose that such a commission could do no harm. But would the fanatics renew their efforts for such a commission every year, if they meant no harm? Is it not apparent that that is to be the stepping-stone to bring this question into the National Congress? Suspecting everything coming from that quarter, I have, through my paper, the Washington *Sentinel*, and also in person, strenuously opposed the adoption of such a bill, and though the latter has passed the Senate two or three times, it has always failed in the House. In the present session a similar effort has been made by the Senate, but fortunately there is less hope for a passage of the bill by the House than ever before."

At the session in 1879, a table showing the quantity of beer made in each State and Territory, mentions that Maine, which formerly manufactured annually from 7,000 to 10,000 barrels of beer, produced the preceding year " *seven barrels.*"

After passing resolutions condemning the " Advocates of Prohibition," who " continue to wage indiscriminate war against the manufacture, sale, and consumption of alcoholic beverages, both distilled and fermented;" the Association also adopted the following:

" *Whereas*, The near future may bring issues gravely affecting the welfare of the brewing business of this country, and requiring united action and strenuous exertion ; therefore,

" *Resolved*, That it is the duty of every member of this Association, by persistent personal effort, to extend its membership and influence as far as possible, and that it is a matter of duty and self-interest for every one directly or indirectly connected with the brewing trade to join its ranks and to labor for the advancement of its aims and objects."

Those anticipated " issues " in " the near future," were doubtless, in a large measure, the Constitutional amendments then being considered by the Legislatures of Iowa and Kansas. At all events, the Beer Brewers of Iowa held a Convention in 1879, at which they denounced the proposed Constitutional Amendment, and raised a fund with which to try and defeat it. They said :

"Never was it more necessary for us to defend our business. Never was our business in greater danger than at present. Now it is for us to decide whether we will stay idle and let our business (heretofore acknowledged as a legitimate and legal one) be ruined by unjust and hypocritical legislation and chicanery, or whether we will, as men and fathers, protect our trade, and so our wives and children, and maintain our liberty and our rights."

These declared purposes, reiterated resolutions of intention to carry them out, confessions of the necessity for protection against Prohibitory State Legislation, reports of damage already received, and of troubles anticipated, are unmistakable confessions that Prohibitory Laws, wherever executed, are a success.

But not the Brewers alone are active to prevent Prohibitory Legislation. The " Wine and Spirit Traders' Society of the United States," is composed of men of great ability in the commercial world, and of immense wealth ; having its President, several Vice-Presidents, a Council of twenty members, many standing Committees, eminent legal counsel, and *a Committee on Legislation.* Honorary membership is granted to persons and business firms doing business outside the United States, by the payment of one hundred dollars, a privilege which many of the largest manufacturers of intoxicants in various parts of Europe, have availed themselves of. There are also many State " Liquor Leagues," " Saloon-Keepers' Associations," " Protective Unions," etc., representing the various branches of the trade in Ardent Spirits, all of which are active in attempts to manipulate Legislatures, and to obstruct and defeat the operations of the laws against the manufacture and sale of intoxicants.

Early in 1879, what is called the " Merchants' Protective

Union," composed of the whiskey dealers of Topeka, Kansas, issued the following circular:

"DEAR SIR: We have organized a society in this city known as the Merchants' Protective Union, for the purpose of defeating the prohibition amendment to the Constitution of the State of Kansas, which is to be voted on in 1880. We propose organizing similar societies in each city and town throughout the State, and for the purpose of organizing the different societies in unison with each other, we think it necessary to call a mass convention, to be held in this city at an early date, so that all parties interested may have an opportunity of advising each other as to the best plan of carrying on our campaign. We therefore urgently request that you confer with all parties interested in your vicinity, giving us your views as early as convenient.

"Yours very respectfully,
"C. R. JONES,
"*Corresponding Secretary.*"

In accordance with this call, a Convention was held, attended by one hundred and thirty-eight delegates, who organized a Society called: "The People's Grand Protective Union." They unanimously adopted the following:

"*Resolved*, That the Prohibition Amendment to the Constitution of the State of Kansas, if adopted, would be a law, in its practical application, far beyond the public sentiment of the people, and would be inoperative; that its adoption would take the whole subject of temperance out of the power of the Legislature, leaving the people without a remedy. Laws so stringent that they cannot be enforced are destructive of all good, because it teaches men not to respect the restraining power of law. The laws now upon the statutes of the State are as stringent as can be enforced, and may be amended or repealed as public interest and public sentiment shall demand. The amendment, if adopted, would do what no Constitution in any State of this Union does; it would legalize the manufacture and sale of liquor, unrestrained by law, and the liquor once purchased and in the hands of the purchaser, its use cannot be controlled, thereby offering a premium to falsehood, perjury, and intemperance."

The Ohio State Liquor Dealers' Association adopted the following resolution:

"*Resolved*, That we, the liquor dealers of Ohio, in convention assembled in Akron, do hereby pledge and affirm that in the future we will not support any but the most outspoken, honest, and just acting and thinking men in behalf of liberal legislation on the liquor traffic."

In Chicago they have organized what is called "The Spirit and Wine Manufacturers' and Dealers' Society," with the following object:

"To encourage societies and co-operation among the trade in other places; to advocate a national organization; to remove unjust, obstructive, and needlessly complicated laws; to devise appropriate legislation, local and national; to oppose intolerance and fanaticism; to see that the laws are respected and enforced; to support the broadest liberties consistent with good government and social tranquillity."

There is also, "The Saloon-Keepers' and Liquor Dealers' Association of Illinois," which, at a Convention held in Chicago, September, 1880, adopted the following:

"*Resolved*, That this association will watch, with the greatest vigilance, the action of our representatives in the halls of the legislature, holding them to a strict account for every vote, or neglect to vote upon all laws respecting our liberty and just rights; and that we will use our united power, and that of our friends, as well as that of all our resources, to prevent the election of men either too cowardly to resist the allurements of temperance women, or too stupid to comprehend the vicious effects of sumptuary legislation."

The following was issued as its date indicates:

"NASHVILLE, TENN., August, 1880.

"DEAR SIR—We beg to call your attention to the importance of co-operation among liquor dealers throughout the State, to secure, if possible, such representation in the next Legislature as will not be opposed to our interests. The indications are that determined efforts will be made to secure the passage of Local Option, or Prohibitory Laws, and to repeat in the State the folly and failure of sumptuary legislation, which has in the past been productive of no substantial or moral good, but has only harassed the masses, and destroyed their material interests without benefit to those for whose reformation the laws were passed. If you will give your attention to it, you can elect men to the Legislature who are not in favor of fanatical experiments to interfere with a traffic in which there is invested so much capital, and which is the chief source of revenue to our State. We suggest that you consult with liquor dealers, and friends in the country, and endeavor to send fair men to the next General Assembly who are practical and reasonable, and not disposed to set on foot a code of laws to make criminals of yourselves and others engaged in the liquor traffic, and destroy the business, in the absurd and ever failing attempt to eradicate the evil of drunkenness by law.

　　" Respectfully,　　　　LIQUOR DEALERS OF NASHVILLE.

" This communication is confidential, and for your own advisement, and is not intended for publication."

After the adoption of the Constitutional Amendment in Kansas, prohibiting the Manufacture and Sale of Intoxicants in that State, " The Western Brewer," published several articles in its issue of November 15, 1880, in which it bemoaned the fate that awaited its business there.   In one article it said :

"Kansas has become the Maine of the West by the decree of a majority sufficiently large to appal the friends of personal freedom and progression throughout the world.   The new act becomes a part of the organic law of the State—a part of the constitution—shrewdly voted upon in this shape by the fanatics, in order that it cannot be repealed or blown away, even after their oppressive majority leaves them, and people get sick of their bargain, except by a two-thirds vote.   It is, therefore, safe to infer, that however great a revolution may occur in the sentiments of the citizens of Kansas, the present generation will not see the law repealed.   What action the thirty-five brewers of Kansas will now take, remains to be seen.   There is but one possible way for them to do.   They must close their doors, put out their fires, and seek other lands for an opportunity to earn bread for their children by practicing the only trade they have learned, and by means of which alone they can make a living.　. . .   These men are forced to ruin by a power unknown in any despotic country, any effete monarchy ; the power of a *majority*, which the American people are slowly coming to understand is a personal despot, of less conscience and greater means of oppression than any ruler, be he Roman, or Bourbon or Guelph. . . .   Iowa comes next.   The precedent established, it is no longer a matter of reasonable doubt how that State will **vote on a similar law. Iowa rolls up a fanatical majority** always and every time, to glorify the saints and advance the growth of long hair. . . . It was largely accomplished by money contributed by temperance fanatics. One hundred thousand dollars alone having been contributed by the short-haired, goggle-eyed and rich temperance women of Boston."

And in another article :

" The thirty-five breweries in Kansas will now put out their fires and lock up their doors. Moth and rust will take possess-ion. The proprietors, many of whom have spent a lifetime in building up their business, will have to look elsewhere for bread for their children. So say the people of Kansas at the polls. There is no despotism like the despotism of the majority."

And again : " When and where will fanaticism in this coun-try be checked ? Apparently not at the polls. Ballots have settled the brewers in Kansas."

And once more : " No wonder prohibition won in Kansas. At Winfield the polls were taken charge of by the women, who appeared in full force, and remained all day with tickets in their hands, soliciting votes for the amendment with tears in their eyes. They waltzed round the ballot boxes, ogled the mas-culines, and lifted their dresses just high enough to keep them out of the tobacco juice. At home their husbands cared for the yelling infants, and washed up the dishes. And this is one reason why Kansas has become Maine. Not even the ballot-box can withstand petticoats."

How all this venom indicates that the business which makes home a hell for woman, dreads Prohibitory Law !

III. The testimony of men who have had great experi-ence in various departments of Temperance work, as to the necessity and value of Prohibition, is no slight proof of its success.

Of Dr. Lyman Beecher, one of the most able and active pioneers in the Temperance work, Dr. Charles Jewett says, in his " Forty Years Fight with the Drink Demon," record-ing an incident which occurred in 1859 :

" At one of the prayer meetings held at the Old South Church, he gave a terrible shock to the usual decorum which charac-terized those meetings, by a burst of enthusiasm over the Maine Law. He had pictured, as he only could, the conflict which had been going on in the Universe for centuries, between the powers of light and darkness, of good and evil, and the anxiety and dismay which he, as well as millions of others, had felt at times, notbwithstanding their trust in God and the promises of His word, in view of the fierceness of the struggle and the seeming advantage sometimes gained by the powers of evil. 'But, brethren,' said he, 'let us rejoice and be glad, for the powers of hell are just now in dismay. That glorious Maine Law was a square and grand blow right between the very horns of the Devil, and from the moment of its reception I seem to see him falling back—stubborn and terrible, but falling back!—and the consecrated host of God's elect pressing close upon him!' While thus giving vent to emotions too strong for words alone to express, the grand old man was advancing on the floor, swinging his big cane with a powerful energy, which showed very clearly the spirit in which he would fight the biggest devil in existence had he been there. He wound up magnificently. 'So shall it be, brethren—I believe it—I *see* it—they will crowd him back, and crowd him back—(still advancing and swinging his cane)—until they shall push him over the battlements, and send him back to the hell from which he came forth! And then shall come up from a redeemed earth the shout, 'Glory to God in the highest, and on earth peace and good will to men.'" Pp. 363-4.

In the same work, Dr. Jewett, one of the most efficient champions of the Temperance cause, thus utters his own convictions: "God be thanked for the Maine Law! and the grand inspiration, energy, and honest devotion to the public weal by which it was created! May no backward step ever be taken in that noble State, which now bears the flag of prohibition, in the advance of our temperance host." P. 324.

Said Father Mathew:

"The question of prohibiting the sale of ardent spirits, and the many other intoxicating drinks which are to be found in our country, is not new to me. The principle of prohibition seems to be the only safe and certain remedy for the evils of intemperance. This opinion has been strengthened and confirmed by the hard labor of more than twenty years in the temperance cause. I rejoice in the welcome intelligence of the formation of a Maine Law Alliance, which I trust will be the means under God of destroying this fruitful source of crime and pauperism."

John B. Gough, on the occasion of his last visit to Edinburgh, Scotland, said:

"I wish to put myself right on prohibition. I am a thorough prohibitionist, for we must not only abstain, but educate public opinion to vote right at the ballot-box on this question. We often hear it said that the Maine law is a failure; that is false, for it is a grand success. True, in Massachusetts it has been reversed; but why? Simply because temperance people sat at ease after the law was passed and neglected to educate the rising generation, so that when the vote came we were in a minority. Let us sink all differences, let bygones be bygones, and by education of personal abstinence for the individual, and prohibition for the nation, success is certain."

Said Rev. Henry Ward Beecher, on the occasion of the passage of the Prohibitory Law, some years ago, in the State of New York:

"We might be baffled and balked a great while before we could make all the teeth of this law meet with a good subject between them; we might have to deal with men who would come, and disappear, as spirits do; but there was one thing they could not reverse; after years of discussion, the people in this Empire State had declared, that the making and selling of intoxicating drinks, for such purposes, was a crime. The principle was born; and there was nothing born on the face of this earth that carried such joy as the birth of a moral principle."

Hon. Henry Wilson, said, in 1867:

"Do you think that Christian men can pray for the license law? Does any man dare take that law into his closet, and read his Bible, and on bended knee ask God to bless it? I would like to see the man who would do it. I tell you, gentlemen, that what the people of Massachusetts, the great masses, cannot pray God for, cannot go on the statute book of this State, and stay there."

Said Dr. Henry A. Reynolds:

"I always voted for Prohibition, and I always intend to. I hope if God ever sees me start with a ballot for license or free rum, He will take my life before He permits it in the ballot-box."

Francis Murphy, in an Address given at the International Temperance Conference, in Philadelphia, July, 1876, which he commenced by eulogizing the Prohibitory Law of Maine, concluded thus :

"Every sacred feeling in us stands up arrayed against the liquor-traffic. The Constitution of the country stands against it. In the courts of the nation, where the honor of the nation has been vindicated, rum has been brought to the bar of justice, and I thank God to-night that the legal profession, the men who have been draped with the ermine of the Judge, who have sat upon the seat of justice, have been equal to their duties, and stand to-day before this nation the pride and honor of it in the administration of justice for right against this cursed traffic."

Says Rev. Joseph Cook:

" While we embrace every opportunity to call out the efforts of the church in personal visitation of the poor, and in the founding of self-supporting religious institutions, let us not forget the responsibility of the civil arm for the shutting up of the dens of temptation."

Miss Frances E. Willard, as President of the " Illinois Woman's Christian Temperance Union," thus, in an article in the *Advance*, urging the closing of the liquor shops by law, gives expression to the general sentiment of the women who are actively at work in the cause of Temperance.

"The saloons that were closed are open once more—and why ? Because the laws of a Christian republic make liquor selling reputable. The tens of thousands who yearly, since 1874, have been reformed by efforts of the Women's Christian Temperance Unions, have many of them gone back to their cups, and why ? Largely because, on every street corner, the wide-open door of the saloon, sheltered under the ægis of law, invites them back to the old indulgence and the coveted comradeship. Boys and young men in the slippery path of inexperience and the special danger of imagined strength are forming habits of drinking in far greater numbers than moderate and immoderate drinkers are forming habits of sobriety, and why ? Because, again, so long as the traffic in any commodity is respectable—and in some States nothing short of ' a good, moral character ' entitles a man—by the law—to be a saloon-keeper ! so long will it be respectable to buy and use that commodity. From all these hard facts of experience the temperance women of Illinois have deduced a very rational conclusion, namely: that they will use the great influence they are acquiring over public sentiment to hasten the time when the liquor-traffic shall be put under ban by the laws of Illinois. They have often seen their whole year's work on the moral suasion line endangered or upset by the falseness to principle of the average voter when election day came around. They have been taught 'by the argument of defeat' that you must fight fire with fire ; meet the enemy in his stronghold of power by overwhelming numbers ; and offset bad voters by good ones, if ever the liquor-traffic is to go down. Therefore, they are rising up in the might of moral power and Christian womanhood, and saying: "On this question of license we want the ballot. We ask it *only on this one issue;* as temperance women we have nothing to do with other phases of the mooted question of women's rights, and no matter what our private opinions on the general question may be, we do not bring them into this discussion. The ballot for 'Home Protection,' so far as regards license or no license of the grog-shop, merits the approval of all good men. They tell us that they like the phrase and are glad that the petition for a vote on the question bears the name of 'Home Protection Petition.'"

And Judge Pitman, in the closing chapter of his masterly work on "Alcohol and the State," after passing in review the various methods which have become historic in the treatment of this great problem of Intemperance, and showing that the extinction of the evil " requires the intervention of law, and that moral suasion, educational and religious instrumentalities, are all inadequate without the aid of legislation," thus hopefully concludes :

" Before the aroused conscience of the people, wielding the indomitable will of a State, the ministers to vice, the tempters of innocence, the destroyers of soul and body, shall go down for ever." (p. 405.)

He had previously said : " May I not rightly sum up the duty of those who believe the liquor traffic to be a curse, as this: Wherever license prevails, contest every inch of territory you can for prohibition ; where prohibition prevails, never surrender an inch to license, except from dire necessity." (p. 213.)

In England, ex-Bailie Lewis, after speaking of the energy and zeal of the Church, the moral efforts of Temperance reformers, the multiplication of organizations in

Edinburgh, and the increase of intemperance, in spite of all these, says :

"The conclusion of the whole matter is this—until there is sufficient patriotism among the leaders of the people to demand the statutory prohibition of this license enormity, society must make up its mind to bear all the accumulated horrors of the drink curse." *

In a memorial not long since presented to the Archbishop of Canterbury and the other Bishops, members of the House of Lords, asking for a law for the stop of the sale of intoxicants, and signed by over thirteen thousand clergymen of the Church of England, occurs this significant sentence :

"We are convinced, most of us, from an intimate acquaintance with the people, extending over many years, that their condition can never be greatly improved, whether intellectually, physically, or religiously, so long as intemperance extensively prevails amongst them, and that intemperance will prevail so long as temptations to it abound on every side." †

Dr. Temple, Bishop of Exter, puts it thus : "Men who are hard at work, whose frames are exhausted by their toil, who feel within them the natural weariness and lassitude that labor produces, and who are then shown something that will give them temporary relief; who know, that for at any rate a short time, they may have something like real pleasure, though it be but of vicious kind—men who are worn and weary, and taken as it were at their weakest moment—is it just to thrust in their faces this temptation, which in their own consciences they know they ought not to approach."

Said Archbishop (now Cardinal) Manning :

"I agree most heartily and cordially, that the great curse which withers our people, that the pestilence which is devouring them, is drunkenness. I feel that to labor to put it down is our duty, and I am convinced that to put it down, *legislation is absolutely necessary.*" ‡

And Canon Farrar, in his brave utterances on the subject, has said :

---

* Cited in Alcohol and the State, p. 148.   † Ibid, p. 146.
‡ Cited in Bacchus Dethroned, p. 253.

"I say unhesitatingly, that the grounds on which Parliament does not interfere with the sale of drink are theoretically untenable as well as practically disastrous." . . . "That bill," (Sir Wilfrid Lawson's,) " is simply intended to enable the people to protect themselves from that which they have found by long and bitter experience to be an overwhelming peril. Hitherto Parliament has utterly refused to help us. They cannot and will not refuse if the demand comes to them in a nation's voice, and if that voice speak in the accents of men who are resolutely and indignantly determined to use every means in their power to save a new generation from a sin which has been, to an extent so utterly deplorable, the ruin and curse of this generation in which our lives are cast." *

" If the Permissive Bill be so 'bad' as statesmen have told us it is, why, in heaven's name, does not some statesman come forward and give us a better ? I am perfectly sure Sir Wilfrid Lawson—who ought to have the sympathy of all good men, because he has the abuse of all bad men—I am quite sure he would be the very first to welcome such a bill. And I am very sure the statesman who should pass such a bill would wear through the rest of England's history a greener laurel than was worn even by Chatham's self. Oh! for one 'still strong man in a blatant land,' who is not afraid of prejudice, of abuse, or to fight the battle of the people in the fight with their besetting sin." †

But, notwithstanding all these concessions on the part of Temperance leaders, and all the fears of those interested in the liquor traffic, it is obvious that many Temperance men put obstacles in the way of the enactment and enforcement of Prohibitory Laws, and that the liquor interest is a unit in both arguing and working against them. Why is this? What are the grounds of this opposition ? and what are their validity?

I. It is not unjust to say—since it is so fully avowed in the quotations already made from their numerous declarations—that self-interest is at the bottom of the opposition of the liquor maker and the liquor seller; but is

---

* Talks on Temperance, American edition, pp. 119, 133.
† The Duty of the Church, American edition, pp. 21, 22.

not this also the basis of opposition to all law on the part of the wrong-doer, of every hue and grade? The publisher of obscene literature, the keeper of the brothel, the proprietor of the gambling-house, the counterfeiter, the burglar, the highwayman, and each and every man and woman who, like the shrine-makers of Ephesus, " get great gain" from their craft, put in the same plea against interference with their business.

> " No rogue e'er felt the halter draw,
> With good opinion of the law."

But, by common consent, it is the business of the Legislature to suffer no man's self-interest to war against the general good; and it must be obvious to all who are candid in the examination of the appalling facts which show that pauperism, crime, burdening taxes, and innumerable evils, legitimately flow from the liquor traffic, that less than all others can those who engage in that traffic be justly shielded by the plea of self-interest. The old legend of the man who was tempted by the devil to do one of three things, either to kill his neighbor, commit adultery, or become intoxicated, and who, thinking that he was chosing the lesser evil, elected to get drunk, and so was led both to adultery and murder, is no exaggeration as illustrative of the fact that drunkenness leads to all other crimes. And the only legitimate and radical ground for law-makers to take is that such a traffic should be crushed out, no matter what poverty comes to the few by so doing; and that, if more than a few are interested in it, by just so much as these increase facilities for drunkenness, should the laws be more thorough and unsparing. The State has no right to encourage any monopoly, and least of all to protect a few in warring against everything that is of interest and worth to the people at large.

"It appears to me," said Lord Chesterfield, in debating this question in the British Parliament, "that since the spirits which the distillers produce are allowed to enfeeble the limbs, and vitiate the blood, to pervert the heart and obscure the intellect, that the number of distillers should be no argument in their favor; for I never heard that a law against theft was repealed or delayed because thieves were numerous. It appears to me, my lords, that if so formidable a body are confederated against the virtue or the lives of their fellow-citizens, it is time to put an end to the havoc, and to interpose, while it is yet in our power to stop the destruction.

" As little, my lords, am I affected with the merit of the wonderful skill which the distillers are said to have attained, that it is, in my opinion, no faculty of great use to mankind, to prepare palatable poison; nor shall I ever contribute my interest for the reprieve of a murderer, because he has, by long practice, obtained great dexterity in his trade. If their liquors are so delicious that the people are tempted to their own destruction, let us at length, my lords, secure them from the fatal draughts, by bursting the vials that contain them; let us crush, at once, these artists in slaughter, who have reconciled their countrymen to sickness and to ruin, and spread over the pitfalls of debauchery such baits as cannot be resisted." *

And Lord Lonsdale said, in the same debate:

" When it is once granted that spirits corrupt the mind, weaken the limbs, impair virtue, and shorten life, any arguments in favor of those who manufacture them come too late, since no advantage can be equivalent to the loss of honesty and life. When the noble lord has urged that the distillery employs great numbers of hands, and therefore ought to be encouraged, may it not upon his own concession be replied, that those numbers are employed in murder, and that their trade ought, like that of other murderers, to be stopped ? When he urges that much of our grain is consumed in the still, may we not answer, and answer irresistibly, that it is consumed by being turned into poison, instead of bread ? And can a stronger argument be imagined for the suppression of this detestable business, than that it employs multitudes, and that it is gainful and extensive ?" †

These considerations lose nothing by their age, for they voice the moral sense of all the ages, and a failure to hear and heed them is sure to demoralize and render worthless

---

* Gentleman's Magazine, January, 1744, p. 9.
† Ibid, February, 1744, p. 63.

all that our civilization makes possible to any form and administration of human government. The great Commentator on American Law, lays down and defends the same principle :

"The Government may, by general regulations, interdict such uses of property as would create nuisances, and become dangerous to the lives, or health, or peace, or comfort of the citizens. Unwholesome trades, slaughter-houses, operations offensive to the senses, the deposit of powder, the building with combustible materials, and the burial of the dead, may be interdicted by law, in the midst of dense masses of population, on the general and rational principle that every person ought so to use his property as not to injure his neighbors, and that private interest must be made subservient to the general interest of the community." †

II. Another ground of opposition to Prohibitory Laws is, that they interfere with personal liberty. By this is meant, if anything, that the fact that a man has engaged in any traffic, establishes, of itself, his right to continue in that traffic. But this is too glaringly absurd to deceive any one. It cannot possibly be a universal rule, else every man's house might be exposed to a nuisance, every rascality known among men would have a valid plea for non-interference, protection of life and property would be an impossibility, law could take no cognizance whatever of such a thing as the public good. This is so self-evident that there is no man who does not demand that his family, his neighborhood, and himself shall be protected by laws which shall interfere with this liberty of others to do just as they please. Hence, on this principle of self-protection, we have laws prohibiting the sale of immature or tainted meats, prohibiting lotteries, gambling and gambling-houses, brothels, the sale of poisonous or adulterated drugs, the traffic in unwholesome and light-weighted bread, and many other things.

In short, all law proceeds on the fact that there is some

---

† Kent, II., p. 240.

wrong or some danger of wrong which the public must be guarded against. "The law," says an inspired authority, " is made for the lawless and disobedient, for the ungodly and for sinners, for unholy and profane, for murderers of fathers, and murderers of mothers, for manslayers, for whoremongers, for them that defile themselves with mankind, for men-stealers, for liars, for perjured persons." (1 Timothy, i. 9.) And every application or enforcement of the law interferes with the offender's freedom to do the things which the law is aimed against ; and because of this we uphold, vindicate, and rejoice in the law as protecting ourselves and all others, and making true liberty a possibility.

We say, then, that the plea of personal liberty avails no man anything, if in the exercise of that liberty he is jeopardizing the rights, the security and the happiness of others. The safety of the people must always, in any equitable government, be the supreme law ; and under this ancient but perpetually significant rule, the liquor traffic must be outlawed, put beyond the pale of toleration, crushed out, as being, more than all things else that can be mentioned—since it is the incitement to all conceivable evil—the curse of the land and the destroyer of its people. Better, far better, that we allow this plea of Personal Liberty as a bar against interfering with any other known vice, than that we listen to and heed it as urged by the liquor-seller in defence of his vile traffic. He is, of all others, the greatest foe to the liberties of the land, and to the prosperity and protection of the homes and the lives of the people ; and therefore, his Personal Liberty to do as he pleases in establishing and perpetuating his traffic, ought to be interfered with and forbidden.

Mr. Gladstone is reported as having said, and justly, that the sphere of government in regard to man is " to make it easy for him to do what is right, and difficult for him to do wrong ; " and by all means, this " sum of all wrong," should be hedged about with such difficulties as to make its continuance an impossibility. This, Prohibition aims to do, and no less radical dealing with the vice can, or intends to, accomplish this.

III. Somewhat akin to this last considered objection, is the plea that Prohibitory Laws are of the nature of a Sumptuary Law, seeking to interfere with and determine what a man shall eat or drink. This, if not wholly a mistake, is equally valid against all laws for the punishment of drunkenness, and especially against such as the Brewers' Association (as see previous quotations, from their proceedings in 1874,) desire to have enforced against drunkards.

But in point of fact, no law for the Prohibition of the traffic in intoxicating liquors, has ever said, or ever intended to say that the liberty of any man to drink Intoxicants shall be interfered with, any more than the law against the sale of immature meats, poisonous or adulterated drugs, unwholesome or light-weight bread, shall be an interference with a man's eating or using all that he wants to of such unwholesome and cheating things. But, inasmuch as the traffic in these exposes the people to disease, or is the practice of a fraud upon them, the law, which is intended to shield the people from harm and imposition, prohibits the trade in such noxious and fraudulent articles. To cure the people of any disposition to indulge in the consumption of such things, they need, it is true, to be enlightened with regard to the personal injury which they sustain, and, especially in the case of the use of drugs for other than medicinal purposes, their moral sense must also be appealed to. But, a failure to so enlighten and convince, or a perversity, arising either from love of indulgence in the hurtful thing, or of weakness of will to resist the temptation to indulgence, does not alter the fact that either a free or licensed traffic in these things, is fraught with danger and mischief to the community, and should therefore be forbidden.

So is it, on precisely the same principle, with the intent

and operation of Prohibitory Liquor Laws. They are in no sense a dictation to any man as to what he shall not drink. They are the expression of the decision of the law-makers that the traffic, because it creates disease, pauperism, crime, general insecurity of life, constant jeopardizing of property, increased taxation, and an unsettling and disturbance of the very foundations of society, is incompatible with the welfare and safety of the State, and therefore should not be allowed. If men will use such liquors, let them do so, and let them obtain them as they may; but the State, satisfied of the ruinous consequences of such use, as just described, has a right to say, and is guilty of injustice to its citizens, if it does not say, "the sale of such a source of mischief shall be prohibited." If men are tempted to drink and persist in drinking, either from love of the oblivious condition into which drunkenness places them, or from inability, on account or self-enfeebled will, to resist, or from inherited tendencies and weakness, or from any other cause, whatever, they are to be directly reached by enlightenment of mind and conscience, by moral suasion, by medical treatment and care, or by any other personal appeal, and attention; and the fact that the State wisely does its duty " in making it difficult for them to do wrong," is a powerful help in the use of these agencies; but if even with all these instrumentalities and this facility for their use, reform is not effected, the duty of the State remains the same, the right of the citizens to protection against the evils of the traffic is unchanged, and its prohibition is none the less a duty and a necessity.

IV. It is frequently urged that Prohibitory Laws are at war with the financial interests of the country, inasmuch as the General Government derives a large revenue from the duties and tax paid on imported and home manufactured intoxicants, and each State, town or city, is enriched from the fees paid for licenses. If it is conceded that the above is true, and that there are no offsets for the consequences of drinking, to be paid out of these receipts, the mere fact of revenue and tax would not be a wise argument for accepting or continuing it, unless it could be shown that the liquor traffic was productive of more good than harm to the country at large, or to any particular section of it. But since this is impossible, since, as has been abundantly shown in preceding pages, the history of the traffic is unrelieved by a single instance of good result, but all is shameful and ruinous, the fact of revenue derived therefrom is no reason for its continuance; or if it is urged as a reason, it has no more valid plea than has any other crime and outrage to justify itself because of its willingness and desire to put money in the public treasury. There is no conceivable immorality or crime which would not gladly pay for the privilege of an unmolested career, even a larger sum than the liquor traffic pays.

But, so fearful are the offsets of personal loss, general demoralization, and actual expense paid in dollars and cents for the detection and punishment of crime, and the support of pauperism, occasioned by the drink traffic, that no public treasury is in any sense enriched by the duties, taxes and license fees, which that traffic produces. Pitt justly characterized the attempts of the British Ministry to tax America, as miserable financiering, " a boast of fetching a peppercorn into the exchequer at the loss of millions to the nation!" Our plea of Revenue from the Liquor Traffic is still more short-sighted. Our liquor dealers pour out a beverage for the citizens, the use of which incapacitates them for labor, reduces their families to beggary, makes the drinker a nuisance and a criminal, and from the money thus received, for which no useful equivalent has been given, turns over the worse than wasted capital of the country into the public treasury, to become less than a " peppercorn," in the " millions " which must be paid for the direct and indirect results of the traffic. The mere loss on labor alone, from the use of liquor in the United States, based on actual census returns of the value of the labor of those engaged in the business, time lost during drunkenness, the insanity, idiocy, sickness and death, caused by intemperance, is annually $1,244,395,000, as a return for the privilege of causing which, the liquor manufacturers pay the General Government, $61,225,995.53, and the liquor seller pays the States for licenses, $50,000,000, leaving an excess of loss for the privilege of receiving this insignificant revenue, of $1,133,169,004.47. Add to this the direct expenses of pauperism and crime, and we have a showing that should satisfy anybody of the blunder and crime of any attempted revenue from the liquor traffic.

We may approximate an idea of the expenses throughout the country in supporting the pauperism, and prosecuting the crimes caused by intemperance, by one instance. Albert Barnes says, in a note to his discourse on "The Throne of Iniquity:" "The exact sum received in the city and county of Philadelphia for tavern licenses in the year 1851, was $66,302; the whole sum in the State was about $108,000. The expenses for prosecuting the crime, and for the support of pauperism, consequent on intemperance, in the city and county, was, for the same year, as accurately as it can be computed, $365,000." That is, for every $181.65 paid into the city and county treasury as a contribution to the revenue by the liquor traffic, the treasury paid out $1,000 for poverty and crime occasioned by that traffic. A wonderful enrichment!

And what is true in the United States, is equally true in other countries. In Great Britain the revenue from the tax on Intoxicating Liquors, "amounted, in 1868–9, to £25,603,-160. Now what does it cost the nation to get at this sum? Probably £259,000,000, equivalent to paying 1,000 per cent., for collecting the tax. The following are the particulars:

1. The retail value of the liquor sold.......... £103,000,000
2. For the detection and punishment of crime caused by intemperance............ 3,000,000
3. In poor-rates and police-rates, extra on account of drunkenness, and drink-made paupers....................................... 10,000,000

4. Losses incurred through intemperance to shipping, commerce, and the productive industry of the nation.......................... 112,670,000
5. Cost of disease, physical and mental, both in public hospitals and in private practice..... 6,000,000
6. Voluntary taxes, in support of ragged schools, local charities, etc............................6,000,000
7. Extra expenses incurred through intemperance in the army and navy......................... 2,422,000
3. Cost of corn imported to replace that destroyed in distillation, etc. 16,000,000

Total *..................£259,092,000

No wonder, that, in view of this enormous waste, the London *Times* should have said, in 1853 :

"Neither supplying the *natural wants of man*, nor offering an adequate substitute for them—a system of voluntary and daily poisoning—no way so rapid to increase the wealth of nations and the morality of society could be devised, as the utter annihilation of the manufacture of ardent spirits, constituting, as they do, an infinite waste and an unmixed evil."

And the *Daily Telegraph* confessed, in 1862, that : "Our revenue may derive some unholy benefit from the sale of alcohol, but the entire trade is, nevertheless, a covenant with sin and death."

Similar to this is the opinion of the Courts in the United States :

"The whole course of legislation on this subject prevents any presumption being indulged that this traffic, like other employments, *adds to the wealth of the nation*, or to the convenience of the public. The presumption is thus declared in almost express terms, to be that the *traffic is injurious to the public interests*, and hence the rule protecting other employments does not apply to this one, and therefore it cannot be said to be within the rule."—Supreme Court of Indiana, Harrison *et al. v.* Lockhart.

And Justice Grier, as before cited : "If a loss of revenue should accrue to the United States from a diminished consumption of ardent spirits, she will be a gainer a thousandfold in the health, wealth, and happiness of the people."

And Gladstone is reported to have said to a deputation of Brewers, when they triumphantly asked him what he would do without the liquor revenue: "Give me a nation of sober Englishmen, and I will take care of the revenue."

Powell quotes the late Canon Stowell, as saying, in a lecture:

"If the Government can control drunkenness, it ought to do so. If it does not, it is afraid of its revenue. What will be lost will come back tenfold, in consequence of the promotion of honest industry." And he adds: "This opinion received ample confirmation some years' ago in Ireland, where, through the labors of Father Mathew and other great and good men, the consumption of liquor decreased amazingly, and yet the revenue improved. In the year ending January 5th, 1839, shortly before which period the reformation commenced, the produce from licenses was £128,494. Year by year this amount was reduced, till the year ending January 5th, 1842, the produce was only £95,980, being a total reduction upon the three years of £32,514. In the year ending January 5th, 1839, the amount received from the tax on malt was, £289,869: in the year ending January 5th, 1842, it stood at £165,153, making a total decrease in the three years, of £124,716. With regard to spirits, the revenue for the year ending January 5th, 1839, was £1,510,092; in the year ending January 5th, 1842, the amount was reduced to £964,711, being a decrease in the three years of £545,381. The whole decrease of the revenue from spirit licenses, malt, and spirits, during the five years ending January 5th, 1842, amounted to £682,611. Yet notwithstanding this very heavy reduction, arising from the success of the temperance movement, there was a large increase of revenue from the increased produce of other excisable articles; the revenue for 1841 was, £4,107,866, which increased in 1842, to £4,198,689, showing a total increase of £90,823. The revenue on tea alone for the year ending January 5th, 1842, had increased by £80,639." *

There is probably, therefore, no more senseless plea, than that the revenues of a country will fall off if the liquor traffic is destroyed; as there is certainly no more false pretence than that duties, taxes, and license fees for intoxicants, enrich any public treasury.

---

* Ibid, p. 260.

V. A very common objection, and one sometimes urged quite as persistently by men who claim to be opposed to the liquor traffic, as by those engaged in that traffic, is, that a Prohibitory Law is in advance of, and opposed by, public opinion. If it be conceded, for the purpose of a full and fair consideration of this objection, that public opinion is opposed to Prohibitory Liquor Laws, we shall also be confronted by this fact: that there is nothing peculiar in such laws to distinguish them from legislation levelled against other vices antagonistic to private and public good, and that the fact that ethical laws in general, arouse opposition, is never urged by right-minded men against their enactment and enforcement. Says the Rev. James Smith, M. A.:

" Among the ancient Jews and the early Christians, ethical principles had to be applied in an age and in circumstances very different from ours ; but these principles themselves do not partake of the narrowness by which every individual, nationality, or period, must be more or less characterized ; they are catholic in their character and adaptation, and are at all times in advance of the highest attained morality. They possess, moreover, an educating and elevating power, so that the honest application of them, in the most untoward circumstances, and from the very lowest starting point, will tend to ameliorate the condition and to elevate the character of those who apply them." *

Take the laws set forth in the Ten Commandments, as illustrative of the truth of this. These laws, given primarily to the Israelites, were promulgated at the time of their most manifest lack of harmony with an endorsement of them ; and they were held to a rigid observance of their requirements, even when with almost entire unanimity they attempted to violate them, and often foolishly thought themselves successful in their revolt. Centuries of slavery had degraded and demoralized the Hebrews, and the opinions formed in such a state of servitude, were for a

* The Temperance Reformation and its claims upon the Christian Church. Pp. 256-7.

long time in opposition to the high demands of the laws
of Jehovah. But by degrees these laws became their
educators, and led them to better thoughts and to purer
deeds. Who shall say that Almighty Wisdom was at
fault in framing laws so far in advance of popular opinion?

We have already seen that those who have been most
active in the Temperance cause, and especially those who
have faithfully advocated Moral Suasion, and have sought
by such means to educate the popular mind, confess a conviction that law is needed to supplement their efforts even
in this direction. Their judgment is certainly entitled to
consideration. The practical operation of the Prohibitory
Law confirms the wisdom of that judgment. Senator Frye,
State Attorney of Maine, for ten years, says of the law in
that State:

"It has gradually created a public sentiment against both
selling and drinking, so that the large majority of moderate,
respectable drinkers, have become abstainers." And he adds:
"No law will enforce itself, but if enforced, its tendency is to
create public sentiment."

And Judge Davis, of the same State, says of the Prohibitory Laws:

As teachers of the public conscience, the standard of which is
seldom higher than human law, their value is above all price.
Many a man refrains from buying intoxicating liquors when he
wants them simply because he must buy of a violater of the
law; and this is often the secret of his opposition to the law.
He does not like to give his conscience a chance to appeal to.
such a law. It tends to make both buying and selling disreputable. It holds up the standard of right, and puts the brand of
infamy upon the wrong. He is a blind observer of the forces
that govern in human life who does not see the moral power of
penal law, even when extensively violated, in teaching virtue
and restraining vice."

The quotations previously made from the writings of
Sheldon Amos, show conclusively that all license laws
demoralize public sentiment, by giving to vice the sanction
of legalized protective legislation. A review of the history
of the American people for the last thirty years corroborates

the fact that law is a most effective educator. When the politicians of 1850 secured the enactment of the infamous Fugitive Slave Bill, how it debauched and demoralized the public opinion of the country. Everywhere it made eminent men and women leaders in society, bitter pro-slavery apologists. But since that law has been abolished, and laws in defence of universal liberty have taken its place, what a change has come over public opinion; for who is not now a defender of such liberty, and also quite solicitous to be regarded as having always maintained that position!

Sheldon Amos has some remarks on this subject, which are worth remembering:

"This subtle agency of public opinion does not owe its power to the width, the representative character, nor still less to the inherent worth of the opinion itself. It owes its power rather to the nearness of the public concerned, and to the concentration of its movements. Thus to a school-boy, his school-fellows, at least as much as his masters, and far more than his parents, supply the guage of right thinking and right acting. To a workman, his fellow-operatives; to a soldier, his comrades; to a lawyer or doctor, the members of their several professions; to a whole people, their public writers, their magistrates and their legislators, determine the standard not only of right or wrong conduct in the greater crises of action, but of just or unjust, benevolent or harsh, becoming or unbecoming, thinking, feeling, and acting at every moment of life. The bearing of this on the present subject, (Legalized Prostitution,) is obvious. Vice grows quite as much by moral as by material opportunities. A clear and broadly diffused public sentiment in favor of purity is one of the strongest fences against impurity; while on the other hand, public callousness, not to say laxity, on the subject, is a vehement stimulus to vice. It need then hardly be pointed out that, whether in the language of law or of literature, all formal recognition of social vice as anything but an evil determinedly to be combatted at every point, as a gross, temporary and unnatural excrescence on civilized society, buoys up the interested public opinion already pledged to countenance it, and affords to vice itself the most direct and unremitting stimulus."--Pp. 12, 13.

The application of this to public opinion on the subject of Temperance, especially its bearing on Prohibitory

Legislation, is as significant as it possibly can be to any other theme. Neglect of the use of every educational facility, and especially neglect or refusal to put the legislative stamp of crime on the liquor traffic, vitiates and destroys all effort at intellectual enlightenment or moral suasion; while determination to deal with it by law, as an appalling vice and a foe to the interests of society, is the best and wisest guide to a just public opinion in regard to its enormity.

VI. Another common objection to the Prohibitory Law, is, that it is not, and cannot be, enforced. This is frequently stated by the various Liquor Associations, as see preceding extracts from their proceedings. Why, then, the creation of such associations on purpose to prevent the enactment of such a law? Why the confessed expenditure of tens of thousands of dollars to defeat attempted legislation of this kind? Are not such facts far more significant of what is true in regard to the efficiency of the law, than the blustering denials of its power can be? Men are not likely to band together and lavishly spend money in opposition to that which is inoperative. "The children of this world" are far too wise to be guilty of such folly; and we may be sure, that, whatever their pretensions to the contrary, they are not thus fighting what they believe to be a shadow.

But, if it could be made to appear that the Prohibitory Law is successfully evaded in some localities, that attempted prosecutions of its violations failed- of conviction in some instances, that combinations, secrecy, the timidity or dishonesty of the officers of the law, or any other cause, operates to produce the non-enforcement of Prohibitory Laws, what then? Is that a fate peculiar to those laws and unknown to other enactments? Or, is it not just what happens to all kinds of criminal legislation? Would you, therefore, repeal all criminal law? And if not, why make an exception in this instance? Judge Davis, of Maine, says truly:

" Those who denounce the Maine Law, because it is not enforced, little know how plainly it can be seen what spirit they are of. The whole secret of their opposition is generally a fear that it will be enforced, or a desire for indulgence, without feeling that they are causing it to be violated. For they are still more dissatisfied when it is enforced. In this State the law was executed vigorously from 1851 to 1856—a rather long time, Dr. Bacon might presume, for the broom to 'sweep clean' because it was 'new.' And it has never been so thoroughly executed as it was in 1855, when the same men were fiercest in their opposition who now oppose it on the ground that it is not enforced.

" And they show the same inconsistency in another way. For there are other laws, of which they never complain, to which this objection might be made with equal force.

" Penal laws are divided into two classes in this respect. Those of one class are enforced without any general effort in the community; while those of the other class are not. The reason is easily stated.

" 1. When the offence injures some one, in person, or property, like larceny, arson, or murder, the friends of the injured party, and the whole community, are interested in bringing the offender to punishment.

" 2. But in the other class of crimes, like gambling, licentiousness, and selling intoxicating liquors, there is no injured party anxious to have the guilty punished. These offences can be committed secretly; and all the parties are interested in concealing them. They are, therefore, detected with difficulty, and are punished only by special effort. And, especially in cities and large towns, the laws against them are but partially enforced.

" The Maine Law is not peculiar in this respect. There is not a large city in the country in which there are not scores of gambling-houses and houses of ill-fame, the existence of which is well known to the inhabitants, and to the authorities; and yet the laws against them are not enforced. Are the laws, therefore, wrong? And ought they to be changed into license laws? The truth is (and Temperance men must not forget it,) this class of laws will always be extensively violated. The Maine Law, even now, is enforced far more thoroughly than the license laws ever were. In proportion to the number of people participating in the evil to be suppressed, it is enforced as well in this State as are the laws to prevent licentiousness." *

---

* The Maine Law Vindicated, pp. 5, 6,

Elsewhere he is reported to have said, and at a later date than the above :

"The Maine Law has produced one hundred times more visible improvement in the character, condition, and prosperity of our people than any other law that was ever enacted."

Recently, (the Fall of 1880,) there have been some crooked experiences in the execution of the law in one locality in Maine. Hon. Neal Dow, thus calls attention to what is said of them, and to reasons explaining them, in the *"Advance:"*

"The following paragraph is cut from the *Congregationalist.* Similar paragraphs have been going the rounds of the religious and secular press, the object being, I suppose, to show that the liquor traffic, the grog-shop, the drunkard-factories, cannot be suppressed by law, nor even diminished in number; and that, therefore, they ought to be licensed by the State, and sanctified by statute. It seems to me not a little singular that many of the papers which eagerly give currency to such paragraphs as this, afford no room for anything upon the temperance side. Now for the slip alluded to :

"'A most trustworthy Western friend, who has been spending some little time in Bangor, Me., and who had considerable interest in satisfying himself how the 'Maine law' is now working in practice in a place of that size, informed us, this week, that he is satisfied that there are from two to three hundred places in that city where intoxicating liquors are sold as openly as in Boston or New York. As there must be nearly or quite 20,000 inhabitants in Bangor, that would give an average of one open rum-shop (say) to about every *eighty-three* inhabitants. He further stated that the usual intoxicants had their usual place upon the printed bill of fare at the hotel where he boarded. It would seem to be in order for some one to rise and explain.'

"The writer of the above paragraph gives it as a sample of the way the 'Maine Law is now working in practice!' I wonder the writer did not see that it was a sample only of the way the grog-shops flourish in places where the Maine law does not work at all, where it is not enforced. Portland has a population of about 35,000; why did not the writer point to it as a demonstration of the failure of the law? Here (Portland) the sheriff and police hunt rumsellers vigorously, wrest from them heavy fines, and inflict upon them long terms of imprisonment. In Bangor the officers and the authorities, high and low, the churches assenting to it, make bargains with the rumsellers, giving them immunity from the law in exchange for votes. Bangor is eminently a city of churches and piety—I mean what goes by that name—and yet the women and children and the dearest interests of thousands of the people there, are deliberately given up to wretchedness and ruin, in exchange for votes. That's the way 'the Maine law operates,' says the writer of the above paragraph, and thousands of others as foolish, as thoughtless, as illogical as he.

" No, that is the way ' license ' operates; the understanding is between the authorities and the rumsellers, that the shops shall be shut up, sharp, at ten o'clock at night, and that no drunken man shall be permitted to come out of them into the street. The city authorities have deliberately resolved that the Maine law shall not ' work ' there at all, directly or indirectly. They deliberately violate their oath of office, which requires them to enforce the law—all laws. They deliberately nullify the law; they deliberately assume to set aside the authority of the Legislature and to trample the constitution under foot. By doing this they deliberately, in the full knowledge of what they do, set an example to every rowdy, blackguard and rascal, of disregard of law and order, of contempt for law, and the rights of the people. It is the right of the people that the laws shall be observed; it is the right of every man to enjoy the protection of the law against the liquor trade, ' the gigantic crime of crimes.' But all this is ignored in Bangor, with the deliberate assent and consent of Bangor piety and morality, without which it could not be done."

This, then, is the secret of alleged failure,—the success of interested liquor dealers in debauching public officers ! Smugglers might, perhaps, have in some instances, thus corrupted customs officers, and with equal justice and fairness boast of unsuccessful revenue laws, but the boast is not to the credit of either the tempters or the tempted, and least of all does it argue defect in the law. A conspiracy to overthrow the Government is no proof that the Constitution of the land is weak and worthless.

Something akin to this determination on the part of those interested in the infamous business to break down the force of the law against their trade, was manifest nearly a hundred and fifty years ago, in Great Britain. Lord

Chesterfield, in the famous Parliamentary debates already referred to, in answer to an argument that in consequence of the non-enforcement of the provisions of the Gin law, another and different enactment was necessary, said:

"In order to discover whether this consequence be necessary, it must first be inquired why the present law is of no force ? For, my lords, it will be found, on reflection, that there are certain degrees of corruption that may hinder the effects of the best laws. The magistrates may be vicious and forbear to enforce that law by which themselves are condemned ; they may be indolent and inclined rather to connive at wickedness, by which they are not injured themselves, than to repress it by a laborious exertion of their authority ; or they may be timorous, and, instead of awing the vicious, may be awed by them. In any of these cases, my lords, the law is not to be condemned for its inefficiency ; since it only fails by the defect of those who are to direct its operations. The best and most important laws will contribute very little to the security or happiness of a people, if no judges of integrity and spirit can be found amongst them. Even the most beneficial and useful bill that ministers can possibly imagine : a bill for laying on our estates a tax of a fifth part of their yearly value, would be wholly without effect, if collectors could not be obtained. I am, therefore, my lords, yet doubtful whether the inefficacy of the law now existing necessarily obliges us to provide another ; for those that declared it to be useless, owned at the same time that no man endeavored to enforce it, so that perhaps its only defect may be that it will not execute itself." *

And Lord Carteret, after citing the attempts of the liquor sellers to weary out the Magistrates by submitting to the fines and imprisonments for violations of the law, and at once putting some other seller in their shops, thus describes another expedient resorted to by them to render the law inoperative :

" At length, my lords, instead of wearying the magistrates, they grew weary themselves, and determined no longer to bear persecution for their enjoyments, but to resist that law which they could not evade, and to which they would not submit. They therefore determined to mark out all those who by their

---

* Gentleman's Magazine, January, 1744, p. 7.

information promoted its execution, as public enemies, as wretches who for the sake of a reward, carried on a trade of perjury and persecution, and who harassed their innocent neighbors only for carrying on a lawful employment for supplying the wants of the poor, relieving the weariness of the laborer, administering solace to the dejected, and cordials to the sick. The word was therefore given that no informer should be spared; and when an offender was summoned by the civil officers, crowds watched at the door of the magistrate to rescue the prisoner, and to discover and seize the witness upon whose testimony he was convicted; and unfortunate was the wretch who, with the imputation of this crime upon him, fell into their hands. It is well remembered by every man who at that time was conversant in this city, with what outcries of vengeance an informer was pursued in the public streets, and in the open day: with what exclamations of triumph he was seized, and with what rage of cruelty he was tormented. One instance of their cruelty I very particularly remember. As a man was passing along the streets, the alarm was given that he was an informer against the retailers of spirituous liquors. The populace were immediately gathered as in a time of common danger, and united in the pursuit as of a beast of prey, which it was criminal not to destroy. The man discovered, either by consciousness or intelligence, his danger, and fled for his life with the utmost precipitation; but no housekeeper durst afford him shelter; the cry increased upon him on all hands, and the populace rolled after him like a torrent not to be resisted, and he was upon the point of being overtaken, and like some others, destroyed, when one of the greatest persons in the nation, hearing the tumult, and inquiring the reason, opened his doors to the distressed fugitive, and sheltered him from a cruel death. Soon afterwards there was a stop put to all information; no man dared afterwards, for the sake of reward, expose himself to the fury of the people, and the use of these destructive liquors was no longer obstructed." *

VII. But perhaps the most formidable obstacle to the success of Prohibitory Law, is found in the fact that the liquor traffic has obtained control of the great political parties of the United States. The action, so often repeated, of the Brewers' Association, in resolving to vote for no man of any party who is in favor of Prohibitory Legislation, has

---

* Ibid, Dec. 1743, pp. 635, 636.

been no idle and unmeaning boast; but has been carried into effect not only by themselves, but very generally by those engaged in various branches of the liquor trade. And the result has been two-fold,—the immediate success of the party which has nominated men known to favor the liquor interest,—and a zealous effort on the part of Democrats and Republicans to bid for and secure the votes of the liquor-makers and the liquor-sellers. Except in the State of Maine, where for several years it has been impossible for any party to carry an election without the aid of the votes of Prohibitionists, the Democracy has for many years uniformly arrayed itself on the liquor side.

The Democratic National Convention, in 1876, declared: " The vital principle of the Republic is . . . . in the liberty of individual conduct from sumptuary laws."

Occasionally, as in Massachusetts, in the nomination of Thomas Talbot for Governor, the Republicans have dared to put forth Prohibitory candidates ; but since the result of the special action just indicated—the desertion of a sufficient number of Republican distillers, brewers and liquor-sellers to the Democratic side, and the consequent defeat of Mr. Talbot,—the experiment is not repeated.

The Republican National Convention of 1872, made the following declaration in its platform :

" The Republican party propose to respect the rights reserved by the people to themselves as carefully as the powers delegated by them to the State and to the Federal Government. It disapproves of a resort to unconstitutional laws for the purpose of removing evils by interference with the rights not surrendered by the people to either State or National Government."

What did this mean ? was a question soon urged with so much persistence in various quarters, as to elicit an answer from its author, in which, although he was confronted by the fact that every Prohibitory law submitted to the highest tribunals in the land,—the Supreme Court of the United States, and the Courts of Final Appeal in the several States, they had declared that those laws were in harmony with the Constitution,—the author of the resolution, Mr. Raster, of Illinois, gave this explanation of its meaning: " It was adopted by the platform committee with the full and explicit understanding that its purpose was the discountenancing of all so-called temperance (prohibitory,) and Sunday laws."

The "Brewers' Association," in session the same month, in New York City, were addressed by their President, in regard to party politics. Alluding to the Democratic party, he said:

"The Presidential election which takes place this fall may change the aspects of that party. At the Cincinnati Convention they have placed at the head of their ticket, a man (Horace Greeley,) whose antecedents will warrant him a pliant tool in the hands of the Temperance Party; and none of you, gentlemen, can support him. It is necessary for you to make an issue at this election throughout the entire country; and although I have belonged to the Democratic party ever since I have had a vote, I would sooner vote for the Republican ticket than cast my ballot for such a candidate."

The Executive Committee also reported concerning the "many dangers which threatened their trade" from adverse legislation, and added:

"No trade, in view of its *political power*, is better calculated to exercise a marked influence on the elections than yours, and it is your duty and a matter self-defence to take a direct and active part in the political revolution and transformation of parties, so that in this direction, too, the desired reforms may be achieved."

The leading public journals of the Republican party endorse this non-prohibitory position of that party. In 1875, *Harpers' Weekly* said:

"The Republican party is not a prohibition party. As the best sentiment of the country agrees that the subject shall be legislatively treated by authorizing a license system, the Republicans would make that system as just and efficient as practicable. Further than this as a national party it will not go, and the attempt to buy the prohibition support by adopting a prohibition platform, could end only in the destruction of the party. This is perfectly well understood by the bulk of Republicans, and they will act accordingly."

And again, referring to those interested in the liquor traffic, it said:

"Unless the Republican party is ready to announce its own death, it cannot consent to legislate adversely to the interests of this class of people."

The New York *Times* said: "None of the probable candidates are likely to be in favor of prohibitory laws. The temperance societies could not possibly get an out-and-out temperance man nominated. They know this as well as we do."

And the Chicago *Tribune:* "More than the third term, more tnan the Credit Mobilier, more than salary grabbing, more than Butlerism, more than all other causes put together, Prohibition has undermined and destroyed the Republican party ; Prohibition must be prohibited by the Republican party."

In view of this, how marvellous it is that whenever an attempt is made to induce independent action on this great question of the liquor traffic, the cry is invariably raised : "Do not thus jeopardize the success of the Republican party, for it alone is the party favorable to temperance views, the only party from which the Prohibitionists can hope for legislation favorable to their views." And how stupid professed temperance men in the Republican ranks are to believe and act on such an assurance.* And what a deception is practised when the assertion is made, as it so often is, that all the Prohibitory Legislation we have ever had has been made for us by the Republicans. The facts in the case are not difficult to ascertain. See how plainly they put the stamp of falsehood on this assertion :

"1st. Not a single Prohibitory law, (we don't mean qualified license) now on a statute book in this country, is the work of the Republican party. It must be recollected that the Republican party is less than twenty-five years old, dating back only to 1856, and every Prohibitory statute we now have was passed before that time. The Democrats controlled Maine, when the law of that State was adopted, as they did the State of Ohio,

---

* See Centennial Temperance Volume, pp. 339–341.

when the provision against license was inserted in its constitution. To the Democratic, Whig, and American parties, and frequently to a combination, are all our present Prohibitory statutes attributable. In Massachusetts, and one or two other States, the Republican party has done some temperance legislating, but when the lager beer pressure was brought to bear upon it, it has made haste to reverse its action.

"2d. While the Republican party has added nothing to our list of Prohibitory laws, it has, on the other hand, materially reduced it. In Michigan, Connecticut, Massachusetts and Rhode Island, it is responsible for the repeal of such laws, thus carrying four States over to Rum's side. In Pennsylvania it had the Governor and Senate, while the Democrats had the lower House, when local Option was stricken down. In a number of the States, including New York, it has made important legislative concession to the Rum interest.

"3rd. But the Republican party is chargeable with more than four rumselling States. During the time it has had the power, the President and both Houses of Congress being of its politics, several new States have been added to the Union, the fundamental laws of which it controlled, but all of which it has permitted to come in under the Rum curse. In one instance, it was careful to insist that the constitution of a new State with less than fifty negroes and mulattoes should have a perpetual prohibition against slavery; but required nothing against Rum, although over one thousand gin mills were in operation in the district legislated on.

"4th. And beside new States, it has organized several Territories, for all of which a Republican Congress has made the laws, and in not one of which is Rumselling prohibited. The great crime of the Democratic party, in Republican eyes, used to be that it was willing to let slavery enter new Territory.

"5th. For nearly twenty years, the Republican party has controlled and legislated for the District of Columbia, and during the whole of that time the Nation has been dishonored by notorious drunkenness under the very shadow of the Capitol.

"6th. For nearly twenty years it has permitted intoxicating liquors to be openly brought into the country, and the quantity annually imported has been constantly increasing.

"7th. It has been the first national party to put a plank in favor of Rum, and against the Sabbath in its national platform. (See the 16th resolution of the Philadelphia Platform of 1872, and its author's explanation, elsewhere given.)

"8th. In many States it has legislated directly in the interest of Rum. In New York it put the Wine and Spirit Traders' act on the statute book.

"9th. In many states it has got the vote of temperance people for its ticket on the express promise that it would, if successful, legislate for Prohibition, (in New York and Ohio its promise to give Local Option is well known,) and then has purposely violated its word, and worked directly for Rum."

This is the record of the party which makes such boasts of interestedness in the Temperance cause. This is the manner in which it has helped along the cause of Prohibition! The following table, compiled by Hon. James Black, of Lancaster, Pennsylvania, shows when, and by what political parties the several Prohibitory Laws have been enacted:

Order	States	Dates of Enactment	Governor	Political Status	Political Char. of Legislature.
	Maine............	....1846..	Anderson....	Democrat..	Democrat
	Delaware......Feb.	1847..	Houston.....	Whig......	Whig
1st..	Maine..........June 2,	1851..	Hubbard....	Democrat..	Democrat
2nd.	Minnesota.....March,	1852..	Ramsey.....	Democrat..	Democrat
3rd.	Rhode Island..Mar 7.	1852..	Allen.......	Democrat..	Democrat
4th..	Massachusetts.May 22,	1852..	Winthrop....	Whig......	Dem. & Free Soil.
5th..	Vermont.... ..Nov. 23,	1852..	Fairbanks...	Whig	Whig
6th..	Michigan......Feb. 12,	1853..	McClelland..	Democrat..	Democrat
7th.	Connecticut.. June 16,	1854..	Dutton......	Whig......	Dem. & An-Neb
8th..	Indiana ......Feb. 8,	1854..	Wright......	Whig.. ...	" "
9th..	Delaware......Feb. 27,	1855..	Causey	American...	American
10th..	Iowa.... ......Feb.	1855..	Grimes......	Whig and..	Whig & Repub.
				Repub	
11th..	Nebraska.....Apr 1,	1855..	Izard........	Democrat..	Democrat
12:h..	New York. ...Apr 9,	1855..	Clark........	Fusion.....	Whig
13th..	New Hamp....July 14,	1855..	Metcalf	American..	Amer. & Repub.
14th..	Illinois ....Feb. 16,	1855..	———.....	American..	Whig & An-Neb.

The laws in these States numbered one to fourteen, were all in operation when the Republican Party became a National Party, in nominating John C. Fremont for President in 1856. Those laws, except in the States of Maine, Iowa, Vermont and New Hampshire, have all been repealed by the Republican Party. Behold its record, and boast of it if you can!

CONCLUSION.—We have thus passed in review the History of this great Scourge, Intemperance; have put ourselves in possession of some idea of its numerous instrumentalities, of its age, extent, and disastrous influences wherever it exists; have noted its effect on religion, morals, education and general welfare ; its responsibility for crime, pauperism, idiocy, and all other forms of disease and degradation ; and gained a glimpse of its enormous drain on the resources and prosperity of the nation, and its damage to the physical, mental and moral vigor of the individual. We have also traced the history and development of the many agencies, which, from most remote antiquity to the present time, have been employed for the suppression of the Drink Traffic and of Drinking Customs and Habits ; and have seen the inadequacy and worthlessness of many of them ; the character and extent of the opposition arrayed against those which are most efficient ; the blindness of political parties to the great evil, and their willingness to sacrifice the best interests of Society at the bidding of the Rum Power.

From all this array of evidence, we shall, if wise, draw some conclusions both with regard to fact, and to duty. The following are believed to be legitimate and necessary :

I. The Liquor Traffic is an unmitigated curse, without one bright spot or redeeming feature in all its history, and ought, therefore, to be regarded and treated as a crime.

II. Total Abstinence from all Alcoholic beverages, is the only wise rule for any man or woman to adopt.

III. The State, if it does its duty to its citizens in relieving them from oppressive taxation, from the peril of insecurity in their possessions, of danger to their persons, of general demoralization and shame, must prohibit the manufacture and sale of all intoxicants as beverages.

IV. Every citizen, knowing that, in a Republic, he is a part of the Government, and that everything pertaining to its laws and their execution, is determined by the citizens' votes, must feel and manifest a personal responsibility for what the State is and does ; and knowing, also, that Intemperance is a Crime more seductive and far-reaching in its influence, and more disastrous in its results than are all other evils that can possibly oppose the best interests of a people, must, in his political theories and acts, be a Pro-hibitionist.

V. Provision should be made in all our educational institutions, for the instruction of the young in the Nature and Effect of all Alcoholic Beverages.

INDEX.

www.ingramcontent.com/pod-product-compliance
Lightning Source LLC
Chambersburg PA
CBHW031814270326
41932CB00008B/419